# Manjaro Linux User Guide

Gain proficiency in Linux through one of its best
user-friendly Arch-based distributions

**Atanas Georgiev Rusev**

BIRMINGHAM—MUMBAI

# Manjaro Linux User Guide

**Group Product Manager**: Pavan Ramchandani
**Publishing Product Manager**: Prachi Sawant
**Book Project Manager**: Neil Dmello
**Senior Editor**: Athikho Sapuni Rishana
**Technical Editor**: Yash Bhanushali
**Copy Editor**: Safis Editing
**Language Support Editor**: Safis Editing
**Proofreader**: Safis Editing
**Indexer**: Pratik Shirodkar
**Production Designer**: Prafulla Nikalje
**Marketing Coordinators**: MaryLou De Mello and Shruthi Shetty

First published: October 2023

Production reference: 1171123

Published by Packt Publishing Ltd.
Grosvenor House
11 St Paul's Square
Birmingham
B3 1RB, UK.

ISBN 978-1-80323-758-9

www.packtpub.com

*Dedication:*

*To my daughter, Isabella; my wife, Mariela; and my family, for all the support for this book's creation.*

*Thank you:*

*To Philip Müller and the Manjaro team, who dedicate their time and resources so all users can learn and benefit.*

*To Linus Torvalds and the thousands of Linux developers who write free and open source software for the benefit of all of humanity.*

*To Richard Stallman, who created and developed the idea of free and open source software, the GNU GPL licenses, and started this all!*

# Contributors

## About the author

**Atanas Georgiev Rusev** is a senior software engineer with a master's degree in embedded electronics and vast experience in multiple device projects over the last 17 years. He has been a Linux user for over 14 years and is the founder of an Embedded Linux start-up company. After trying many distributions, he has been a dedicated Manjaro fan for the last five years. He enjoys writing articles and has professionally written thousands of pages of technical documentation. None of the topics in the book can be found online in one location; instead, they are spread over thousands of articles, manuals, and posts. As he believes Manjaro is great, writing a full reference guide for beginners and advanced users was his personal mission.

*I'd like to express a big thank you to the entire Packt team and my technical editor, Kaloyan Krastev. Their consistent support and dedication were vital in steering this project to completion. Their guidance has been invaluable in upholding the book's quality throughout the entire process.*

# About the reviewer

**Kaloyan Krastev** is a Linux user and software developer with over 20 years of experience in the area. As a Ph.D. physicist with multiple publications related to elementary particle physics, he did all his research based on free and open source software and Linux. His endeavors have led him through many Linux distributions, and his writing experience and deep analytical skills were the main drivers for improving the book's text quality and accuracy.

# Table of Contents

# 3

# Editions and Flavors                                77

# 4

# Help, Online Resources, Forums, and Updates         115

# Part 2: Daily Usage

## 5

## 6

# 7

## All Basic Terminal Commands – Easy and with Examples          173

# Part 3: Intermediate Topics for Daily Usage

# 8

## Package Management, Dependencies, Environment Variables, and Licenses          211

# 9

## Filesystem Basics, Structure, and Types, NTFS, Automount, and RAID                                                                                      231

# 10

## Storage, Mounting, Encryption, and Backups                                                                                      253

# Part 4: Advanced Topics

# 13

## Service Management, System Logs, and User Management    345

# 14

## System Cleanup, Troubleshooting, Defragmentation, and Reinstallation 387

# 15

## Shell Scripts and Automation 407

# 16

# Linux Kernel Basics and Switching     433

# Preface

Hello, citizens of the world! Linux, as many people know, runs on billions of devices. It **powers** PCs, laptops, servers, **Android**, smart TVs and set-top boxes, micro-computers, embedded devices, robots, cloud services, **Amazon**, **Google**, and **Facebook** servers, network equipment, IoT devices, all the top 500 supercomputers, industrial automation, the Tesla Autopilot, and much more. After the **GNU** tools, it is the most widespread open source SW ever. Practically, there is no modern person in the world who doesn't benefit at least indirectly from its thousands of applications. This is all thanks to its adaptability and thousands of developers and companies contributing and sharing with everyone.

More than a decade ago, learning Linux was a challenge. It required hundreds of hours reading hard-to-digest books and accepting that compared to its counterparts, **macOS** and **Windows**, regular user software was not that good. Despite that, it was always known as a rock-solid OS, offering complete user privacy and the ability to build any software.

Thanks to its great community, it has now spread to all known major device types and is being used more and more. Its great community has also developed thousands of free applications, offering the best user experience we can dream of. They also developed a wide variety of software development tools and frameworks, offering a plethora of work and business opportunities. Knowing some **BASH** is often a job requirement nowadays, and having Linux experience is often an advantage.

In addition, in this age of digital surveillance, **privacy** and **security** are hard to achieve. Practically, Linux is the only OS offering complete privacy and security to its users, along with constant improvements, keeping it at the highest possible level. Otherwise, millions of servers would not use it, as their users' data would be vulnerable.

The last great advantage of Linux comes from developing itself into the most effective and scalable OS kernel, running both on microdevices and all top 500 supercomputers in the world.

Considering that the regular users and tools offered for Linux now cover all possible daily needs, including **office** work, digital **art**, **audio**, **video**, **text**, and **gaming**, we only need a great book. Here it is, written to introduce you easily to Linux and based on one of the greatest user-friendly Linux distributions of the last eight years. This is the book with which you will know it all – from installation through daily use to all the advanced topics.

# Who this book is for

*This book is written for the following:*

- **Regular users** *who want to switch from macOS or Windows to one of the best Linux alternatives! For them, I have described in detail all the basic GUI features, and presented the Manjaro installation and the thousands of available software applications installation basics with screenshots. A comparison of and information about the best tools in a few basic categories is also provided. Some of the topics such as* **backup**, **storage** *management,* **encryption**, **firewalls**, *and a great* **Terminal** *primer, all with examples, will turn them into intermediate power users.*

- **Developers** *willing to jump into Linux. Along with what is covered in the previous point, parts 3 and 4 of the book describe many of the basic design principles in an easy* **learn-by-example** *approach list, without going into development. These parts include the basics of the* **internet**, **network** *fundamentals,* **service** *management, system* **logs**, *shell scripts, automation, and the Linux kernel itself.*

- **Students** *in schools and universities who want to know the basics and later continue further with advanced Linux development books. For them, all the topics targeting regular users and developers will give them the basics to later turn into highly paid professionals. Some may argue AI will overtake this, but without people to use, develop, and manage it, it will never exist. In addition, Linux itself is developed by people, and AI can leverage this but never create it by itself. Educate yourselves and stay in the top 5%, who will lead the next generation of discoveries based on the best OS kernel in the world!*

- **Manjaro Linux** *users, who want to know every basic aspect of this distribution, and potentially turn themselves into* **power users**.

- **Linux users** *with less experience, who want to know more and turn into power users.*

- **Anyone** *willing to learn Linux. With simple explanations and examples, everyone can learn it.*

*As for the required background, none in particular is needed. Basic computer knowledge is helpful but not required to benefit from a great user-friendly distribution, such as Manjaro, and with a detailed guide, such as this book.*

# What this book covers

*Chapter 1, Introduction to Manjaro and Linux*, provides basic information on how and why Linux, and Manjaro in particular, was created, covering their basic characteristics and comparing them to all other operating systems. It then presents a key feature description of all major Linux distributions.

*Chapter 2, Editions Overview and Installation*, is the chapter to help you understand the key differences between the official Manjaro flavors and all the details on their installation. It covers all installation cases and provides detailed explanations for beginners and advanced users, including dual boot with Windows.

*Chapter 3, Editions and Flavors*, previews the basic GUI environment controls of each official Manjaro flavor. After reading it, you will know how to control networks, external devices, edit the view in the UI, and know all the options you have at your disposal.

*Chapter 4, Help, Online Resources, Forums, and Updates*, presents the online resources provided by the Manjaro team, covering guides and the main website features, and describing in detail the forum, which is the place to find or get support for any issue you encounter. It finishes by explaining the release model and updates, critical to keeping yourself with the best security and newest features.

*Chapter 5, Officially Supported Software Part 1*, starts by explaining all the basics of installing applications and packages via the GUI, covering the different application containers in the Linux world, and providing information on the Manjaro servers. It further covers office tools with a justified choice for the best suite. The next part is about browsers, with a brief history overview and test results of over 15 browsers. Its last part presents the best free imaging and video software that professionals use worldwide.

*Chapter 6, Officially Supported Software Part 2, 3D Games, and Windows SW*, continues with a short presentation of audio and music applications, listing tens of options. The chapter then covers messaging software, text editors, drivers, tools, and simple games. A dedicated section for advanced 2D and 3D gaming provides a detailed explanation of the extreme development of the area on Linux, supported by multiple big names in the industry. The last two parts are dedicated to converting `.deb` and `.rpm` packages and running Windows software on Linux.

*Chapter 7, All Basic Terminal Commands – Easy and with Examples*, is the first exploration of the Linux roots, presenting over 50 Terminal commands. It explains first the basics of shells and terminals. Then, it continues with a learn-by-example approach, covering the basics of operations with files, permissions, paths, long outputs, streams, autocomplete, command history, and sudo. It then continues with application installation, grep, piping, editors, and find. The last part also covers system and hardware reporting, processes, running applications in the background, how to get detailed command help, and information on the Terminal.

*Chapter 8, Package Management, Dependencies, Environment Variables, and Licenses*, presents the basics of Linux's modularity, achieved via advanced package management. The chapter gives all the details on how to work with pacman and pamac, unveiling their advanced options, including the installation of software from AUR. It then taps into environments and environment variables, which are at the roots of the process hierarchy. The chapter's last part briefly explains open source licenses, which are the root cause of free software existence.

*Chapter 9, Filesystem Basics, Structure and Types, NTFS, Automount, and RAID*, dives into how filesystems work, starting with the concept of "everything on Linux is a file," Unix file types, links, drives, partitions, inodes, and file attributes. It then describes the most widely used filesystems on Linux, providing their history and main feature descriptions. The next stop is the Linux directory structure, which is essential for orienting yourself in it, and in particular on Manjaro. The chapter then covers external storage usage, NTFS, and automount. The last part explains the basics of RAID.

*Chapter 10, Storage, Mounting, Encryption, and Backups*, presents in detail storage management, covering formatting, partition creation, mounting, ownership, resizing, and encryption. It then taps into backups, briefly presenting multiple tools and explaining in detail how to work with the best one, which is also used for Manjaro initial installation.

*Chapter 11, Network Fundamentals, File Sharing, and SSH*, explores network basics, which include IPv4 and IPv6 addressing, DNS and WWW, ping, and routers. It then continues with static, dynamic, local, and pubic addresses with examples, and explains local network scanning and subnets. A big portion of this chapter is dedicated to network sharing via NFS, providing complete examples on the client and server side, including from Windows and macOS clients. It then explains briefly Samba servers, and then moves on to SSH for remote Manjaro access (including from Windows) and SSHFS for file sharing.

*Chapter 12, Internet, Network Security, Firewalls, and VPNs*, starts with how the internet works. It then deep-dives into protocols, port scanning, a good explanation of the different network attacks, and proof of Linux's security features. The next section provides security advice, which, when followed, guarantees your privacy, and then explains all the types of firewalls. Further, the chapter provides a detailed ufw firewall setup guide, providing at the end a full configuration and coverage of its GUI module. The last part is dedicated to VPNs, with use-case descriptions, legal point explanations, quality points, potential providers, and finally, a presentation of the best worldwide VPN provider for the last few years, officially supported on Manjaro.

*Chapter 13, Service Management, System Logs, and User Management*, continues with the basics of processes, daemons, and systemd. It then moves on to service management and explains systemd configurations, units, and targets, and then explains how to analyze the OS startup sequence after boot. The chapter's next stop is the `systemctl` command and a few other essential `systemd` commands. The chapter continues with Linux virtual TTY consoles and dives deep into journalctl and all system logs, including kernel ring buffer messages. The last part of the chapter is dedicated to user management, passwords, groups, ownership, and root account privileges.

*Chapter 14, System Cleanup, Troubleshooting, Defragmentation, and Reinstallation*, begins with operating system and filesystem cleanup, explains when and how to achieve this, and includes moving the home directory to a separate partition. We will then briefly cover when defragmentation of ext4 is necessary and move on to troubleshooting, an essential Linux topic. We will also review `inxi` and `mhwd` and provide two practical examples. The last part of the chapter explains how to reinstall Manjaro and keep your home contents intact.

*Chapter 15, Shell Scripts and Automation*, presents the basics of shell scripting, which is in the essence of automatic task execution on Linux and dates back to the origins of Unix. The chapter explains what a shell script is and how to run it, providing a few simple examples. It then presents the basic BASH script elements – variables, arguments, arithmetic operators, file testing, logical operations, and command substitution. As this topic is deep, the chapter offers sources from where to get more information and examples, as BASH by itself is a topic for a whole book. The second part of the chapter explains in detail how to work with cronie and systemd for calendar and time scheduling of scripts and commands.

*Chapter 16, Linux Kernel Basics and Switching*, ends the book by presenting how an operating system kernel works in a simple way, explaining the task distribution and management, different types of memory, the kernel and user space, Linux architecture, and the task scheduler. It then presents the system calls, kernel modules, and drivers, how to inspect, load, and unload them, and kernel versioning. The last practical part is related to the Manjaro kernel change approach and the RTLinux kernel version.

## To get the most out of this book

The only hardware requirement is a working PC or laptop, with the following minimal requirements – Dual Core Intel Celeron, Intel Core i3, or AMD Ryzen 3 CPU, 2 GB of RAM, and at least 30 GB of free disk space. To have an excellent experience, it is strongly advised to work on a PC with at least a four-core Intel Core i5 or AMD Ryzen 5 CPU, 8 GB of RAM, and at least 50 GB of free disk space. The only other requirement is a working internet connection.

OS requirement: Manjaro Linux.

## Conventions used

There are a number of text conventions used throughout this book.

This is a **keyword** or an application or package **name**.

`Code in text`: Indicates code words in text, database table names, folder names, filenames, file extensions, pathnames, dummy URLs, user input, and Twitter handles. Here is an example: "Open the main menu, type `terminal`, and press Enter",

A block of code is set as follows:

```
$ ls -la
-rw-r--r-- 1 luke luke 18K Jul 11 16:15 img.png
drwxr-xr-x 3 luke luke 4.0K Jul 11 15:39 git
```

When we wish to draw your attention to a particular part of a code block, the relevant lines or items are set in bold:

```
pacman {-h --help}
pacman {-v --version}
pacman {-D --database} <options> <package(s)>
```

Any command-line input or output is written as follows (whole commands with parameters always start with a dollar symbol, but the dollar shall never be typed in your Terminal):

```
$ sudo pamac install neofetch
```

**Bold**: Indicates a new term, an important word, or words that you see on screen. For instance, words in menus or dialog boxes appear in **bold**. This style is also used for port states and network models' coinciding levels. Here is an example: "The **GNU project** wanted to provide a free alternative and combined its tools with the Linux kernel designed by Linus Torvalds"

Important expressions, definitions, or emphasized parts of sentences are in italics: "…keep in mind that *if the package requires dependencies*, pamac *will download them…*"

> **Tips or important notes**
> Appear like this.

# Get in touch

Feedback from our readers is always welcome.

**General feedback**: If you have questions about any aspect of this book, email us at customercare@packtpub.com and mention the book title in the subject of your message.

**Errata**: Although we have taken every care to ensure the accuracy of our content, mistakes do happen. If you have found a mistake in this book, we would be grateful if you would report this to us. Please visit www.packtpub.com/support/errata and fill in the form.

**Piracy**: If you come across any illegal copies of our works in any form on the internet, we would be grateful if you would provide us with the location address or website name. Please contact us at copyright@packt.com with a link to the material.

**If you are interested in becoming an author**: If there is a topic that you have expertise in and you are interested in either writing or contributing to a book, please visit authors.packtpub.com.

## Share your thoughts

Once you've read *Manjaro Linux User Guide*, we'd love to hear your thoughts! Scan the QR code below to go straight to the Amazon review page for this book and share your feedback.

https://packt.link/r/1803237589

Your review is important to us and the tech community and will help us make sure we're delivering excellent quality content.

# Download a free PDF copy of this book

Thanks for purchasing this book!

Do you like to read on the go but are unable to carry your print books everywhere? Is your eBook purchase not compatible with the device of your choice?

Don't worry, now with every Packt book you get a DRM-free PDF version of that book at no cost.

Read anywhere, any place, on any device. Search, copy, and paste code from your favorite technical books directly into your application.

The perks don't stop there, you can get exclusive access to discounts, newsletters, and great free content in your inbox daily

Follow these simple steps to get the benefits:

1.  Scan the QR code or visit the link below:

https://packt.link/free-ebook/9781803237589

2.  Submit your proof of purchase.
3.  That's it! We'll send your free PDF and other benefits to your email directly.

# Part 1:
# Installation, Editions, and Help

Dear reader, let me invite you on an exciting journey into the world of Linux and Manjaro. As you have purchased this book, you probably know at least a bit about Linux. However, a few key points shall be mentioned.

For the last three decades, Linux has been the most successful operating system, used on personal computers, servers, microdevices, supercomputers, robots, and many other systems. As a result, it has spread to hundreds of industries, including the automotive industry, IoT, industrial machinery, home appliances, and so on. I have used Linux for over 14 years, and in 2018, I found Manjaro for the first time, quite quickly admitting to myself that it is one of the greatest user-friendly Linux distributions I have ever used. As a result, I considered writing a book to help any current or future user dive into Linux with Manjaro as my personal mission.

In *Part 1*, we will start our journey in, *Chapter 1*, with a brief history overview, which helps us understand why Linux was created, why it is so successful, and what makes Manjaro so good. *Chapter 2* switches to the editions overview and installation, with both beginner and advanced user reading paths, making the installation easy, regardless of previous experience. *Chapter 3* continues with a detailed review of editions and flavors, where each primary graphical environment feature is presented with screenshots. This way, we can quickly orient in the three official Manjaro flavors. As the introductory part, we finish it with a detailed review of all Manjaro online resources, emphasizing help, news, and announcements. The last section of the final chapter presents Manjaro's update model and how this is of great use to all users.

This part has the following chapters:

- *Chapter 1, Introduction to Manjaro and Linux*
- *Chapter 2, Editions Overview and Installation*
- *Chapter 3, Editions and Flavors*
- *Chapter 4, Help, Online Resources, Forums, and Updates*

# 1

# Introduction to Manjaro and Linux

Why Linux, and why **Manjaro**? This book was written to answer this question perfectly with all the reasons why. Manjaro is one of the best Linux distributions due to its simplicity, speed, security, great amount of available **software** (**SW**), and user-friendliness. Thanks to this, it is one of the best distributions to start learning Linux. In addition, the book covers many topics, generic for most Linux distributions.

In this chapter, we will first look at the book's structure. We will then start our journey with a short introduction to Arch and Manjaro Linux characteristics. We will understand why Linux-based **operating systems** (**OSs**) run on more devices than any other existing OS, looking into alternative OSs. The last part of the chapter reviews Linux's history and key points of its main variants.

If you want, you can skip this chapter. The chapters' topics are as much as possible separated, and most can be read independently by readers with at least basic knowledge. On the other hand, many reference points are clarified earlier or later in the book, so for beginners, I recommend reading the book's chapters consecutively.

The topics we will cover in this chapter are as follows:

- Book structure and contents
- Arch Linux and how Manjaro is related to it
- What about Linux, Windows, macOS, and Unix/FreeBSD?
- A brief Linux history and what a distribution actually is
- Key points for each major distribution

# Book structure and contents

The purpose of this book is to present most of the magnificent features of Manjaro and most modern Linux distributions in an easy and fast way for all users – *beginners* and *intermediate/advanced*. If you are a Windows user, a beginner, or a newbie of any kind, this book is for you. If you are an intermediate or advanced user, you will find most of the topics engaging and with up-to-date information. This book can also serve as a basic Manjaro/Linux reference guide for the presented common topics.

There are **Terminal** fans, **mouse** fans, and people who think one or the other is good/bad/unknown, and so on. Historically, the Terminal with commands was the tool of choice for any features and settings set up for Linux. As Linux keeps pace with developments, we have hundreds of nice graphical tools to make our lives easier. So, for newbies and people looking for graphical menus, all important *basic* settings with a **Graphical User Interface** (**GUI**) will be presented with screenshots in the first two parts of the book.

I provide an excellent Terminal primer in *Chapter 7*. After that, the rest of the chapters are primarily Terminal-based. As such, they will include the necessary commands with relevant parameters. For a beginner, this means you will have an easy *learn-by-example* approach; for the advanced user, this book will be a reference guide.

To justify for beginners the usage of the Terminal in the age of GUI, many advanced or regular features are *faster and easily controlled* via the Terminal; some are even still available only via the Terminal. Such features are easy to handle with hints and guides like this book. Therefore, to become an advanced user, you must learn Terminal commands, which will all be well explained and with examples.

If you are a beginner, rest assured that all the advanced features will be explained in detail, sometimes with a bit of history. This is also helpful for advanced users, as the history often explains why a given function works or is designed in a certain way.

*Part 1* covers installation and GUI characteristics. Some chapters contain separate parts dedicated to beginners or advanced users. It ends with a detailed overview of online resources, forums, and updates.

*Part 2* starts with GUI SW installation. It then covers all general classes of user applications (office tools, browsers, audio, messaging, text, games, etc.), with recommendations for the best and some test results. It ends with the great Terminal primer as a first step toward advanced features.

*Part 3* is a deeper dive into Manjaro and Linux's design, starting with package management, dependencies, and environment variables. It then investigates the Linux filesystem basics, structure and types, and continues the journey with storage, mounting, encryption, and backups. Its last part is a detailed explanation of networking fundamentals, file sharing, SSH, network security and firewalls, and VPNs.

*Part 4* is dedicated to advanced topics, starting with service management with **systemd**, processes, system logs, and user management. It then switches to system maintenance, troubleshooting, and reinstallation. We then explore shell scripts and automation, and end with Linux kernel basics and switching.

## Versions

This book is based on *Manjaro versions from 21.3.0 Ruah to 23.0.1 Uranos, updated until September 2023*. As some of the installation-related and presented SW screenshots are from versions *21* and *22*, versions from *24* on might have slightly different graphical designs without main feature changes.

Generally, a rolling release distribution like Manjaro significantly differs in basic features after at least two years. In newer versions, most of the time, *the old features remain*, bug fixes are added, additional functionalities are sometimes introduced, and the GUI changes slowly. Thus, this book will be *completely relevant at least until* the end of *2025*.

Regarding the parts related to general topics such as filesystems, systemd, and others – in the Linux world, once established, they rarely change. Some of them had their design fixed between *1990* and *2010* and *haven't changed since then*. Others have their roots in **Unix**, based on designs *before 1990*. In addition, *new tools are mostly backward-compatible* – the user can work with them flawlessly. I also try to mention new additions from the last three years when they are available. As a result, *80% of the book will be relevant at least until the end of 2027*. I state this based on my 14 years of Linux experience and extensive usage of four distributions before Manjaro.

## A polite request

Please be merciful to me, as I can't cover all possible facts, commands, options, screenshots, and so on in a book of this size. In addition, structuring all the information is difficult as Linux is an infinite universe.

I will *do my best to cover all the important points* on each topic, but for *particular details*, you might need to do additional research.

Lastly, if you like my approach and find this book valuable, I would appreciate it if you left a review on the platform you purchased it from.

So now, shall we begin?

# Arch Linux and how Manjaro is related to it

**Arch** Linux is one of the fastest and most lightweight major Linux distributions. *Major* means that it serves as a **parent** to multiple **child distributions**.

A **distribution** is a combination of the Linux kernel, a large amount of additional SW, graphical environment SW (the type of windows, menus, and other GUI features), and tools. The combination of all these additional SW and tools is what makes a significant difference. Some are designed mostly for servers, while others are designed for PCs. Some are rich in features (e.g., 3D effects on a desktop by default), while others are lightweight or even for microcomputers (such as Raspberry Pi, ODROID, and Pine). You can learn more about the major Linux distributions in the last part of this chapter.

Arch is famous for its official principles, which are **simplicity**, **modernity**, **pragmatism**, **user centrality**, and **versatility**. In practice, they are implemented with many useful features. These include *a secure OS* with the option of a full OS and user data encryption, *fast* and designed to be *optimal in resource usage* (regarding CPU, RAM, and OS load), configurable in every possible aspect, and with SW and kernel kept up to date with the latest changes. Arch, and also the Manjaro community, name this last feature **bleeding-edge SW**.

Unfortunately, Arch is not convenient for regular users as, despite its powerful usage and configuration, it is complex and by default a Terminal-based distribution without a Graphical Environment. In addition, it is so rich in features that learning how to use it takes many months.

**Manjaro** sits as an **Arch**-based distribution with hundreds of preset configurations, several graphical environments to choose from, and lots of additional tools and SW. This makes it a far more user-friendly and easy distribution for any user, straight out of the box. It also adds many unique additional features and so serves as *a lot more than "just pre-configured Arch"*. **Manjaro** also sticks to the initial principles behind **Arch** a lot, so the user only gains from all the additions.

Here are the most important reasons why Manjaro is a great choice:

- A high-speed/lightweight OS optimized by default. Thus, the user rarely needs to do anything to improve its already *great* **speed** and **efficiency**.

- A great amount of the newest **bleeding-edge** SW packages, following the Arch principles. The Manjaro team regularly configures updates and adds new packages, provided via an easy install and update GUI SW. The update system is robust, with configurable automatic functions.

- Perfect **security**, including a full hard disk system and user data **encryption**, and 100% user privacy (the distribution doesn't collect user data by any means).

- Regular kernel updates provide the newest kernel with all the features and fixes. It also allows you to switch the kernel between the latest versions and older LTS releases, providing additional flexibility when needed.

- Many regular hard Linux configuration tasks go via GUI modules, so the user is not forced to learn many Terminal commands (unlike Arch, which is configured only via the Terminal). The GUI control is excellent for all daily tasks.

- 2D and 3D **Gaming** is flawless. Since *2017*, several big game companies have started developing games for Linux, and there are multiple open source frameworks for game development under Linux.

- Regular computer, PC, and laptop HW is supported automatically, so there is no difference between Manjaro and Windows in this regard.

- Manjaro has a **rolling release model** based on Arch, so OS and SW updates (including security and hotfixes) are always applied ASAP.

- A lot of legacy and some relatively modern Windows SW can be run.

- The Manjaro community provides a good amount of online manuals and an extensive forum database, which provides news and answers to all common issues. When you post in the forum to seek help, thousands of users can assist you if a solution for your issue is not already found.

- Finally, all the preceding features have been supported for over *12 years*. The Manjaro team retained them completely even after creating a private Manjaro company in *2019*.

According to `https://distrowatch.com/`, Manjaro was ranked in the *top 3 distributions* for five years (from 2017 until the end of 2021). Now, it is in the top 5 and was, until 2021, the only Arch-based distribution in the top 5. Although the DistroWatch rank is not an absolute evaluation, it is a serious achievement measure.

For a regular user, Manjaro can offer more than Debian, Slackware, Ubuntu, and other distributions. Essentially, it is Arch Linux with a lot of splendid configurations on top, with a great GUI, stability, tons of additional available SW, and graphics.

## What about Linux, Windows, macOS, and Unix/FreeBSD?

In this section, we will review some common characteristics of these four major types of PC OSs to compare the essential features of each one.

### Linux

Linux-based OSs grew from the idea of being free, effective, versatile, open source, and not funded by any corporation OS. It was based on some principles from **Unix**, famous for its efficiency and great design. Unfortunately, it was also proprietary and paid. The **GNU project** wanted to provide a free alternative and combined its tools with the Linux kernel designed by Linus Torvalds. Here are some crucial facts about Linux:

- It has had regular updates hundreds of times per day for the last 30 years. All vulnerabilities and bugs are found and fixed as soon as possible by the community.

- It has the world's biggest community, constantly developing improvements for the last 30 years.

- The community has developed features for Linux that took years to become available on Windows.

- It is free and open source, so thousands of users check its quality, security, and vulnerability.

- Tens of thousands of companies use it for business. They even pay freelancers and people from the community to develop new features for them, frequently made available and open source for everyone.

- It has so many variants that it can sometimes be difficult for a person to find a suitable one. Conversely, there are flavors for everyone and for every possible task! There are plenty of articles and proposals to make the choice easy. One suggestion is to use Manjaro.

- GUI installers are available for most famous distributions, so installation is easy and usually takes 10-30 minutes. Short **YouTube** tutorials are available for all of them, usually 5–10 minutes long.

- Nowadays, many Windows applications can run on Linux.

- We even have serious game development and gaming support under Linux for over five years.

- Since 2018, according to TOP500 (`https://www.top500.org/statistics/details/osfam/1/`) and several other websites, *all the top 500 supercomputers globally are based on Linux*.

- Linux is scalable and runs on thousands of types of devices (medical, scientific, automotive, IoT, industrial, embedded, robots, Tesla Autopilot, and so on).

- **Android** is based on Linux. This means a multi-trillion-dollar business with mobile phones, smart TVs, tablets, and tens of other smart devices is here thanks to Linux.

- Many major websites, including Google, Amazon, Facebook, Twitter, Wikipedia, and PayPal, run on customized Linux, as stated in official statements by each mentioned company.

- According to `w3tech.com` (`https://w3techs.com/technologies/comparison/os-linux,os-windows`), in June 2022, Linux was used as a backend server OS by 37.4% of all websites whose OS *is known*, compared to 20% for Windows. The rest used Unix, FreeBSD, or others. We can conclude that Linux provides speed, security, and quality for free. It is so good that many companies choose it as the preferred OS for their servers.

- It is really hard to trace how many people, companies, and servers use Linux, as it has hundreds of variants and is entirely free. This means the creators don't care whether 10 or 500 million people download/use Debian (or any other) Linux this year. As the usage is not tracked, people's privacy is maintained.

- There are thousands of articles on the internet regarding which OS is used more – Windows, Linux, Unix, FreeBSD, or others. One thing is sure – **Linux** is extremely widely used and has been the cause of hundreds of thousands of improvements to OSes, SW, and technologies in the last 30 years.

## Windows

This Microsoft OS is the standard choice for thousands of state administrations and companies. It gained popularity initially as it was the host of Microsoft Office tools. Here are some facts about it:

- It is closed SW, so you don't know what happens inside the OS and whether or how user actions are traced and collected.

- It is proprietary and paid.

- It took a decade to catch up with Linux and macOS regarding stability and security.

- It has so many vulnerabilities that, for decades, hacking a Windows system was easy, while doing it with a Linux machine with proper security was much harder and near impossible. This is still the case today; there are thousands of examples, as illustrated in this YouTube video about Windows 11: `https://www.youtube.com/watch?v=_wBPxxI29R0`.

- Microsoft uses your data only to improve your customer experience, as stated officially on their website: `https://www.microsoft.com/en-us/trust-center/privacy/data-management`. However, the fact that they collect any data is already one con compared to Linux.

- Windows was adopted fast thanks to being one of the first commercial GUI-based OSs, packed with the popularity of **MS Office**. Consequently, it secured big contracts with major PC/laptop companies and many government organizations worldwide. Here are some interesting cases from the USA and EU:

  - In 1998, an antitrust complaint against Microsoft was filed at the US Department of Justice: `https://www.justice.gov/atr/complaint-us-v-microsoft-corp`

  - A study by the *American Economic Association* from *2001* on the tight relations between Microsoft and US government authorities: `https://www.aeaweb.org/articles?id=10.1257/jep.15.2.63`

  - Here is an excellent article from `ComputerWeekly.com` on the *European Commission* series of deals with Microsoft over 20 years, starting in 1992: `https://www.computerweekly.com/blog/Public-Sector-IT/How-Europe-did-20-years-of-backroom-deals-with-Microsoft-1993-EC-rubber-stamps-Microsoft-monopoly`

- When Microsoft and MS Office gained massive adoption by government and other authorities, businesses communicating with them were indirectly **forced** to use the same products (MS Office), as there were no format converters at the time. The clients of these administrations were **never allowed** to use any other office SW.

- Even **Internet Explorer** (**IE**) (a Microsoft Windows SW) slowed down the development of **web technologies** (a long story and brutal fact unknown to many). It took over a decade for **Free SW** and the **WWW consortium** to force improvements in *HTML* and *web technologies* despite IE's inability to support them. Eventually, Microsoft stopped using it a few years ago when they introduced their new, modern Microsoft Edge (since *2015* and based on the Google Chromium Framework). Officially, IE was stopped in *2022* (`https://blogs.windows.com/windowsexperience/2022/06/15/internet-explorer-11-has-retired-and-is-officially-out-of-support-what-you-need-to-know/`). In comparison, all other old browsers such as Firefox, Chrome, Opera,  Safari, and so on, continue their existence.

- Due to the monopolization approach of Microsoft, the FOSS (Free and Open-Source SW) and Linux communities *took steps to force Microsoft to open source some of their interfaces and formats*. This allowed **free SW** to be developed for Windows. Ultimately, Linux and FOSS won.

- Since version 7, Windows is more stable, secure, and much better. In addition, it has a Windows Subsystem for Linux – a virtual machine that supports several major Linux distributions by default. Again, I would say Linux won here since Microsoft realized they could not ignore it and must cooperate.

- Yes, I do use Windows quite a bit in my daily activities. Many industries require this for historical reasons, and MS Office is the default document processor SW for most official businesses. However, Linux is now a better option, and it offers multiple free office tool packages, of which two of the most famous are LibreOffice and Apache OpenOffice.

## macOS

**macOS** is the OS of all Apple personal computers. It started as a Unix-based OS and thus has good architecture and high security. Here are some interesting facts about it:

- macOS is stable and almost flawless for daily use (like most good Linux distributions).

- It regularly updates its **security** (like all Linux distributions do).

- It is a Unix-like OS, that is, quite secure. Since **Linux** is also Unix-like, that's quite good as well.

- It has excellent **hardware (HW)** integration, as Apple's HW is custom-designed for their devices, and the SW is tailored precisely for this custom HW.

- It has a good deal of SW, and some of it is built only for macOS. Of course, many SW tools exist exclusively for Linux.

On the other hand, macOS's disadvantages are as follows:

- It's really **expensive**.

- It can't be tweaked.

- The use of SW other than the Apple-allowed SW is officially forbidden.

- Normally, you would use macOS only on their HW. It's officially forbidden to install macOS on any regular PC.

- It is closed and proprietary, and despite being secure, Apple collects usage data, just like Microsoft, officially to improve the user experience.

My verdict is that **Linux** is preferable to Windows in dozens of ways. Compared to macOS, Linux users have the freedom to do whatever they want, don't have to adhere to what the company allows, and can do this for free.

## Unix/FreeBSD

Before we continue, we also have to look at **Unix**. It is important because it is the father of **macOS** and **Linux**. Unix started in *1969* and is practically *the oldest widely used OS*. It was primarily used for servers and big computers, initially only on DEC computers. Over time, it became multi-purpose and multi-user. Later, licenses were given to multiple companies and authorities, the most famous of which is **Berkeley Software Distribution** (**BSD**).

It is still used on servers, and while the commercial variants are slowly fading away, FreeBSD and multiple open source variants remain widely used. It has many command-line tools in common with **Linux** and **macOS**. In the Unix world, one of the most important developments with recent updates is the **Single Unix Specification** (**SUS**) standard, which defines certain features and characteristics of an OS. Its latest version is relatively new, from *2018*. Unix was also the main driving force for creating the **Portable Operating System Interface** (**POSIX**) standard, supported by Linux and Windows. Here are a few essential points about Unix and FreeBSD:

- It has the **highest security**.

- It primarily targets servers, not PCs.

- It does resemble Linux, but it is not Linux. A lot of the open source SW developed for Linux is ported to Unix or FreeBSD, but a significant part is not.

- It is only for advanced users; by default, it only comes with the command line and without a graphical environment. The user would have to install one themselves.

# A brief Linux history and what a distribution actually is

**Linux** by itself is an OS **kernel**; in a car, the equivalent would be the engine. For a complete car, we would also need a chassis, tires, a wheel, doors, seats, a frame, a trunk, and so on. Equally, if we have the knowledge, parts, and tools, we can build a whole OS *with the* Linux kernel.

In *1983*, Richard Stallman created the **GNU Foundation** to popularize **Free and Open Source SW** (**FOSS**). **GNU** stands for **GNU's Not Unix**, and this distinction was because Unix and other kinds of reliable SW were proprietary, closed, and expensive. The GNU Foundation was the first organization to establish a public *free SW license* called the GNU **General Public License** (**GPL**). Nowadays, we have tens of licenses. The five most widely used licenses are *GPL 2 and 3 (with all their variants)*, *Apache, Mozilla, BSD*, and *MIT*. Since *1983*, hundreds of thousands of people have developed FOSS under those and other licenses.

The GNU collection of free tools grew *but lacked a kernel and an OS*. Thus, when **Linus Torvalds** created the free Linux **kernel** in *1991*, a group of developers combined it with the free GNU tools and created the first two Linux distributions – **Slackware** (since *July 1993*) and **Debian** (since *September 1993*). In other words, somebody combined a great engine with tools, seats, a frame, windows, a wheel, and pedals and created a complete car. They also made this available to anyone who wants to join the projects, revealing *100% of the* **knowledge**. The number of enthusiasts grew fast and steadily. As a result, *hundreds of thousands of people* have contributed for free in the last *30 years* and continue to do so.

A **distribution** combines the Linux kernel with all the necessary features, tools, and SW for a standard PC. It can be installed from scratch on a single pass on a computer to provide all that a user might need (just like a brand-new Windows or macOS).

To be precise, *most Linux distributions are officially defined as* **GNU/Linux** due to the combination of the kernel with hundreds and thousands of GNU tools. Due to a lack of knowledge and for convenience, many people skip the words **GNU** and **distribution**.

Creating a distribution from scratch is *hard*, often requiring a collaborative team effort and existing major collections of tools and SW. Many community servers host freely available FOSS collections, some of which have accumulated a significant amount of SW over time. Among the most prominent and longstanding collections are the ones of **Debian**, **Ubuntu**, **Arch**, **Fedora**, and **Slackware** distributions.

Additionally, each distribution adds a lot more – almost always, they *have at least one* official **graphical desktop environment**, require an extensive set of **drivers**, and a default choice of user applications. For desktop environments, additional tools and data collections exist on different servers.

Many distributions have a few different graphical desktop systems, just like Manjaro. While some are suitable for tablets, others are only for PCs or servers. We also have custom-modified distributions for hundreds of **special devices**. For all of them, the *common parts* are the Linux **kernel** and a set of basic SW packages and tools.

Some of the most famous Linux distributions include **Debian**, **Ubuntu**, **Fedora**, **Red Hat**, **Slackware**, **Mint**, **openSUSE**, **Manjaro**, **CentOS**, **Puppy**, and **Kali**.

The libraries with free SW are supported mainly by the teams that created the initial distributions, and we can call only a few distributions **major**. They serve as the master libraries for almost all other child distributions. The source for this information is the reliable and periodically updated page at `https://en.wikipedia.org/wiki/List_of_Linux_distributions`. It shows that in *2023*, we have *over 230* Linux distributions. Over 15% of them aren't based on any major distribution and have pre-built libraries of their own.

# Key points for each major distribution

Here is a list of major distributions (and when they were started): **Slackware** (July 1993), **Debian** (September 1993), **Red Hat** (1995), **Gentoo** (2002), **Arch** (2002), **Fedora** (based on Red Hat – 2003), and **Ubuntu** (based on Debian – 2004). The following are a few important points about each of them:

- **Slackware**:

  - This is the first Linux distribution.

  - Slackware's initial goals were design stability, simplicity, and becoming the most "Unix-like" free OS.

  - The ability to customize is maximized, but no GUI-based or other tools are provided, so it is unsuitable for any regular user.

  - In the last 10 years, it has not received kernel updates more than once a year. This results in very slow updates in general but also provides stability.

  - It is considered a very hard distribution to learn.

- **Debian**:

  - It is famous as one of the *oldest* distributions and for its **stable** branch. This means its SW has been thoroughly tested by thousands of people, normally for a year.

  - It has the biggest number of child distributions. Over **35%** of the *existing* and *discontinued* Linux distributions are *based directly or indirectly on it*. This is also thanks to its "ultra"-**stable** branch.

  - One of the most extensive databases for SW packages can be found on Debian's servers, where you can also find **tools** *to create distributions*. With the available SW and their "stable," "testing," and "unstable" branches, you can create whatever distribution you want.

  - Debian also has a great community bound by its values. We all owe them a great deal for this.

- **Red Hat**:

  - Red Hat was the first commercial and open source Linux distribution.

  - Red Hat discontinued its Linux line in 2003, replacing it with the proprietary **Red Hat Enterprise Linux** (RHEL) for enterprise environments.

  - **Fedora** Linux, developed by the community-supported Fedora Project and sponsored by Red Hat, is the free alternative designed for home users.

- **Gentoo**:

  - It began as a Linux-based OS that can adapt and be customized to any HW, while most Linux distributions only target servers or desktop PCs. Nowadays, there are a lot more Linux distributions targeting multiple HW platforms.

  - It is extremely difficult for non-professionals to use, as it is intended to be custom-built by the users themselves.

- **Arch**:

  The Arch community clearly defines its values: simplicity, modernity, pragmatism, user-centrality, and versatility. This equates to the following characteristics:

  - The users can install and configure any extra features by themselves.

  - It is a playground to test all possible new features and provides one of the richest repositories, the **Arch User Repository (AUR)**.

  - Its **Rolling Release development model** guarantees regular updates. Over the last decade, Arch *has produced at least five releases annually*.

  - The update is automated, smooth, and easy.

  - Arch provides maximum configuration with custom tweaks and setup.

  - A great deal of security features can be customized.

  - Many Arch users recommend this as the best distribution if you want to learn about Linux in depth and with the command line. However, this means that you will be mainly using the terminal.

  - Finally, this distribution is mainly for **advanced users** despite being excellent.

- **Fedora** (based on Red Hat):

  - It is sponsored primarily by **Red Hat** (now owned by IBM) and the upstream source for **Red Hat** Enterprise Linux.

  - It contains SW distributed under various FOSS licenses and aims to be at the forefront of open source technologies.

  - It focuses on innovation, integrates new technologies early, and collaborates with upstream Linux communities. The upstream collaboration means that updates and improvements are available to all possible distributions and users, just like **Arch**, **Debian**, **Ubuntu**, and so on. However, for this distribution, it is officially stated.

  - Since the release of Fedora 30 (in *2019*), besides the Workstation (for PC) version, we now have editions for Server, IoT, CoreOS (for cloud computing), and Silverblue (for an immutable desktop specialized for container-based workflows).

- A new version is released every six months, and for many years, it was famous as Linus Torvalds' distribution of choice.

- Additionally, it aims to build SW with advanced security and support several desktop graphical environments.

The following is a comparison between Fedora, Arch, and Manjaro:

- Fedora has a different **package manager**.

- It refuses to include **non-free SW** in official repositories due to its dedication to free SW (although through third parties, it can be installed). However, Arch and Manjaro are more lenient toward non-free SW, allowing users to choose.

- Both Arch and Fedora are intended for experienced users and developers, and they strongly encourage their users to contribute to project development. Fedora is, by default, integrated with a graphical environment, so it has higher HW requirements. Manjaro has several official graphical environments and better GUI-based settings managers.

- Fedora has a great community and is one of the most significant kernel contributors. At the same time, Arch provides much more flexibility and an improved online documentation base. Arch has one of the greatest SW repositories (AUR) available to anyone who needs it. Manjaro has its additional official SW repository, but it also includes AUR and again has a great online community.

- Fedora is sometimes not recommended for an inexperienced user, as using all the newest experimental features can be unstable and may present some challenges.

- To summarize, Manjaro is excellent for all regular users who want an Arch-based distribution with direct access to AUR, a larger amount of available SW, and the easiest start. Fedora is aimed more at system administrators, developers, and advanced Linux users.

- **Ubuntu** (Debian-based):

  **Canonical Ltd.** developed **Ubuntu**, marketed initially as *Linux for everyone*. The South African entrepreneur and millionaire Mark Shuttleworth funded the project for the greater good of everyone. This was the most famous distribution from 2005 until 2010 (according to DistroWatch `https://distrowatch.com/`). Thanks to Canonical's marketing efforts and significant GUI SW contributions, regular users started to use Linux more and more. Then, its derivative **Linux Mint** gained popularity, and as it is practically a better Ubuntu, it was preferred by many users and was at the top of the DistroWatch rankings from 2010 until 2017.

  Here are a few facts about Ubuntu:

  - Ubuntu is open source, based on the Debian stable, and Canonical generates profits through enterprise services.

  - It provides seven modifications for PCs, as well as for servers, cloud computing, and the so-called **core** (for IoT and robots).

- As a "mainstream," "common," and "famous" OS, it supports a lot of HW by default and is pre-installed on many laptop models.

- Canonical does as much as it can to make a great deal of SW available, but this results in an *overloaded-with-SW* OS, which needs more powerful HW than Arch or Manjaro. I used it for around two years (between 2008 and 2011), but eventually, the number of additional features and services became so unnecessary that I switched to the Ubuntu-based Linux Mint.

- Canonical also developed the Ubuntu-touch variant for **smartphones**. As a result, a few more distributions copied their idea.

- Due to its strict schedule – bleeding-edge SW is not an option.

- Despite Mint being better, without Ubuntu, it would not exist.

- The positive aspects are that it is stable and has thousands of SW modules and additional SW packages. It is well supported, with a large user database, so most problems can be solved easily.

Compared with Ubuntu, both Manjaro and Arch have lower HW requirements, better system flexibility, more recent stable SW, and better security and privacy.

## Summary

Now that we've covered all the basics, you have an overview of how **Linux** has evolved. We have compared it with the other OSs and got a detailed summary of the primary key points of each major Linux distribution.

Reviewers have described **Manjaro** as one of the best Linux distributions of the last five years because of its speed and lightweight design, stability, security, and additional features. As a result of its amount of GUI modules, it offers seamless switching between Windows and Linux. For beginners, it offers a rich GUI in general. All users can access the rolling release development model and stable, testing, and unstable branches. Developers have access to all **Arch** resources.

In the next chapter, we will continue with a brief overview of the different Manjaro editions, also known as **flavors**. We will then cover in detail the basics of Manjaro installation, both for basic and advanced users, on a PC and a virtual machine, and in parallel with Windows.

# 2
# Editions Overview and Installation

The starting point for each **operating system** (**OS**) is its installation. **Manjaro**, as one of the top Linux distributions, has two essential features in this regard: easy installation and several different **graphical user interfaces** (**GUIs**), or *graphical frontends*, named **flavors**. The installation is packed with *automatic* **hardware** (**HW**) *recognition*, and each *GUI flavor has a similar setup*. Even though we'll try to cover all possible scenarios, some custom setups might need additional tweaking. In this case, check the internet as millions of Manjaro users write on the Manjaro forum and elsewhere, covering all problematic cases. Remember that when you're searching, you shouldn't rely on only one or two articles – always check several posts for any particular issue.

In this chapter, we will cover the following topics:

- Editions overview
- Preparing for any installation
- Installing on a USB stick
- BIOS/UEFI setup for installation on a PC
- The installation itself – automatic and manual, on a virtual machine, and dual boot with Windows

The chapter is long *mostly due to the screenshots*, so don't be frightened.

Here are a few tips for beginners:

- Read the *Editions overview* and *Preparing for any installation* sections.
- Regarding the *Installing on a USB stick* section, read the introduction, and then read the *Etcher* subsection under the *For all OSs* section.
- Read the whole *BIOS/UEFI setup for installation on a PC* section.
- In the *The installation itself* section, there is a hint of what to read. This is a total of around 30 pages. In general, skip any section for advanced users or installation cases irrelevant to your situation.

Regarding the tip for advanced users, this chapter is long as we'll cover all possible scenarios. You will know which parts to skip. In any case, going through the explanation on *BIOS/UEFI* and partitioning before starting the installation *is strongly advised*.

## Editions overview

The editions differ mostly in terms of *graphical environments*. However, each environment provides a set of GUI *configuration tools* and a set of *default* **software (SW)** *applications*. They also bring at least some additional CPU, memory, and HW requirements.

For a regular PC or laptop of any kind, Manjaro officially provides the **Plasma**, **Xfce**, and **GNOME** desktop environments.

In addition, for desktops and PCs, the Manjaro community develops **Budgie**, **Cinnamon**, **I3**, **Sway**, **MATE**, and Manjaro **Docker**.

Here, we will take a brief look at the official graphical environments.

These are the official minimum HW requirements for *all desktop environments*:

- 1 GB of *memory*. This is the absolute minimum, but using 2 GB is a more realistic minimum.
- 30 GB of *hard disk space*. Consider having at least 50 GB.
- A 1 GHz *processor*. Practically, using at least a dual-core 1.7 GHz processor is a more realistic minimum requirement.
- A **high-definition (HD)** *graphics card* and monitor.
- A broadband internet connection.

There are no separate HW requirements for each edition, but some recommendations are written in each overview.

It is essential to mention that the default look of each edition is customizable. Each edition has flavors, themes, colors, menu options, and plenty of graphical options. You can also modify the behavior. This is, after all, one of the significant advantages of most Linux distributions, and in particular **Manjaro**.

### Xfce

Let's look at a few basic facts about Xfce:

- It is for *weaker machines* – that is, machines with the following characteristics:
  - Old HW
  - RAM under 4 GB
  - A CPU such as Intel *Celeron* or Core *i3*, or *AMD Ryzen 3*

- It slightly resembles regular Windows 7. It is designed to be *lightweight*, *stable*, and *fast*.

- It has been in development *since 1996* and is based on the **GIMP Toolkit (GTK)** graphical library.

- It's a great environment if you don't want fancy graphical features or rich menus and settings.

The default look of Xfce is shown in *Figure 2.1*:

Figure 2.1 – Xfce flavor default look

## The Plasma desktop

Here are its characteristics:

- It is based on the **K Desktop Environment (KDE)** and **Qt** graphical libraries.
- Qt is a proprietary framework, but it is free for open source usage. This flavor has official Qt 5 and Qt 6 support and configuration menus.
- It offers the richest UI /GUI configurations and options from all environments. This means it has a lot of additional menus and settings, so it is a bit heavy.
- Regarding HW, it is recommended to use this edition if you have at least the following:

    - 4 GB of RAM (2 GB would also work, but 4 GB is strongly recommended)

    - Though 30 GB is the minimum, I recommend at least 50 GB of hard disk space here

    - At least an average dual-core CPU of 1.7 GHz – for example, Intel Core i5 or AMD Ryzen 5 5500

    - It will work on weaker HW, but it could be a bit slow

- Again, the UI resembles a classic Windows layout a bit.
- As it's been in development *since 1996*, there have been five generations of the Plasma desktop.

*Figure 2.2* shows how it looks by default:

Figure 2.2 – KDE Plasma flavor default look

## The GNOME desktop

GNOME has the following basic characteristics:

- It is a desktop environment with a more *modern look*
- It's been in development since 1997 and at the time of writing, there have been four major generations of the project
- One of the reasons it was started was due to the avoidance of the **Qt** graphical library, so GNOME chose **GTK**
- Until 2019, **GNOME** was officially a **GNU foundation** project
- Regarding HW, it's just like Plasma in terms of what is needed:
  - 4 GB of RAM (2 GB would work as well, but 4 GB is strongly recommended)
  - Though 30 GB is the minimum, I recommend at least 50 GB of hard disk space here
  - At least an average dual-core CPU of 1.7 GHz – for example, Intel Core i5 or AMD Ryzen 5 5500

- Its main menu is, by default, centered at the bottom of the screen, while the status bar is at the top

- At the time of writing, it officially supports Qt applications and has official Qt 5 and Qt 6 configuration menus as separate settings applications

- Its application listing is more like what you'd see on a smartphone

*Figure 2.3* presents how it looks with the default applications menu open:

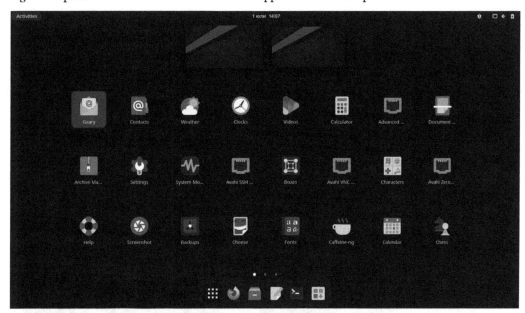

Figure 2.3 – GNOME flavor main menu look

Even the cheapest modern laptops (around 200-300 USD) with a Celeron CPU are dual-core, and in January 2023, I cannot find a new laptop with less than 4 GB of RAM. The only potential issue is that there are machines on the market with a 32 GB SSD – I would only buy one if I had a reason to buy such a low-grade machine.

I always recommend a minimum of 128 GB HDD, 8 GB of RAM, and a CPU – if it's *dual-core*, at least Intel Core i5/AMD Ryzen 5, but *quad-core* would be better.

A good basic machine for me would have at least 256 GB HDD, Intel Core i5 *quad-core* or AMD Ryzen 5 CPU, and 8 GB of RAM. With such machines, Manjaro will run magnificently. This sort of machine, at the beginning of 2023, costs around 500 USD.

## Environments for ARM-based micro-computers

In June 2023, the complete list of supported ARM-based HW is Generic, Generic EFI, Khadas VIM2, Khadas VIM3, ODROID-C4, ODROID-HC4, ODROID-M1, Orange Pi 3 LTS, Orange Pi 4 LTS, Orange Pi 800, Pinebook PRO, Pinephone, PinePhone Pro, Quartz64 Model A, Quartz64 Model B, Radxa Zero, Raspberry Pi 4B/400/3B+/3B/ZERO 2, ROCK 3A, and Ugoos AM6 Plus.

We have the GNOME, Plasma, MATE, and Xfce graphical environments for them.

While all of those have been modified significantly and are more lightweight than the desktop versions, if you consider that Xfce is famous for being the lightest for years, I would recommend it for any weaker microcomputers.

For advanced users using such micro-computers, and explicitly for weaker machines, we have two more special flavors:

- **Minimal**: No graphical environment and completely terminal-based
- **Sway**: A limited desktop environment that provides good mouse and keyboard support, a Wayland limited graphical system, and is intended to be used mainly via the Terminal and with a keyboard

## Preparing for any installation

Before we proceed to the installation, we need to download an image. We'll look at this in this subsection.

### Choosing the right image

To install Manjaro on a PC, you first need to select an image of your choice. So, go to `https://manjaro.org/download/`, select the **X86_64** pop-up option, and for your chosen flavor, click **Download**. The web page design currently looks like what's shown in *Figure 2.4*:

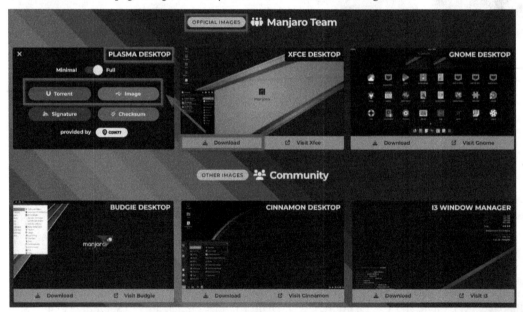

Figure 2.4 – Manjaro downloads web page view

Let's say we choose the Plasma KDE edition. Once we've done this, we have to choose whether we want a full or minimal version. Further, we must click either the **Torrent** button or the **direct image download**, which is depicted with a blue button. The torrent may be faster to download most of the time.

In June 2023, the ISO file size of the *full* version is approximately 3.6 GB for Plasma, while the *minimal* version is 3 GB. The minimal version is smaller as it lacks some applications (for example, Python-based libraries, the `manjaro-printer` package, the Mozilla Thunderbird mail client, a camera application, and more).

For comparison, the Xfce *full* version is approximately 3.5 GB, and the *minimal* version is 2.5 GB. It is a bit smaller than KDE Plasma since Xfce is lighter.

**For beginners**: If you have enough hard drive space on your PC, I *strongly recommend* the *full version*. In this way, you will avoid potentially missing additional SW packages, for example, to make your printer work. Regarding the *signature* and the *checksum*, those are provided for you so that you can verify your download. I have calculated the checksum for all three official editions, all of which are correct. Enough people check them periodically, so, with the quality and security of modern Linux servers, I would not bother with these.

**For advanced users**: The complete list of packages included in each release can be found at the following links:

- **KDE**: `https://gitlab.manjaro.org/profiles-and-settings/iso-profiles/-/blob/master/manjaro/kde/Packages-Desktop`

- **Xfce**: `https://gitlab.manjaro.org/profiles-and-settings/iso-profiles/-/blob/master/manjaro/xfce/Packages-Desktop`

- **GNOME**: `https://gitlab.manjaro.org/profiles-and-settings/iso-profiles/-/blob/master/manjaro/gnome/Packages-Desktop`

You can check what is not included and get the minimal version if it is acceptable for you so that, later, you install additional SW only if you need it.

The **signature** is **GNU Privacy Guard** (**GPG**)-based. In short, it is a way to ensure that the image you download comes from Manjaro servers and has not been hacked.

The **checksum** button will download a **sha1** checksum file on your PC. You can use this to compare whether the downloaded image became corrupted or encountered any errors while it was being downloaded. Manjaro offers a guide for this here: `https://wiki.manjaro.org/index.php/Check_a_Downloaded_ISO_Image_For_Errors`.

## The Calamares installer

Despite the differences between the flavors, how you install them is practically the same. Only the view of the installer (and some options) might have a bit of a different look. The reason for this is that the installation of Manjaro is unified with the **Calamares** installer framework (`https://calamares.io/`) since the basic installation settings are not related to the graphical environment.

To install it on a PC, we must first install an ISO image on a **USB memory stick**. Then, PCs and laptops need a simple **BIOS** *settings* change so that they can *boot* from the USB stick. Booting means starting the PC *from the given stick* and not from any already installed OS from its **hard disk drive** (**HDD**). This is also called starting a **live distribution**. With this feature, most Linux distributions allow you to test them *directly from a USB stick* without needing to make any changes to the HDD.

For dual booting, we have to install Manjaro on a separate part of the HDD so as not to wipe the rest of the OSs and their data. Dual booting means that the PC or laptop will have a menu to choose whether to start Manjaro or any other Linux or Windows OS.

## Automatic HW recognition

Refer to whichever of the following levels applies to you.

### For beginners

**Automatic HW recognition** is a feature of many modern OSs. This means you don't need additional manual setup or installation to run a basic OS view, including improved graphics, sound, network, and other essential services. The manual setup would involve installing a dedicated SW for the exact network controller or sound card model you have on your machine.

Automatic HW recognition is done by **Manjaro Hardware Detection** (**mhwd**), which is unique to Manjaro and is an executable piece of SW. Note that on Linux, executable SW is defined by the file permissions and not by file extensions as it is on Windows (`.exe`). **mhwd** starts automatically during installation. On its Manjaro wiki page `https://wiki.manjaro.org/index.php/ Manjaro_Hardware_Detection`, it is stated that it is still under development, but considering that it was started in 2012, it has been stable for years and thus run by default. As explained on the wiki page, it will recognize any connected PCI and USB HW.

### For advanced users

You can find more information at the following two links:

- `https://linuxhint.com/manjaro_hardware_detection_tool/`
- `https://wiki.manjaro.org/index.php/Manjaro_Hardware_Detection_ Overview`

# Installing on a USB stick

Again, Manjaro's installation is easy and packed in a live USB like most widespread Linux distributions. We must write an image of our choice on a USB stick, test it, and if we like it, start the installer. If not, we can get another one. This way, we can test KDE Plasma, Xfce, GNOME, and other flavors and install the one we like more.

Always prefer a *USB 3.0 stick* and a *USB 3.0 port* – both have blue plastic inside them. This will guarantee fast writing (2-3 minutes) and, later, fast installation. USB 2.0 might take 10-20 minutes to write a 3.6 GB image onto the USB stick.

All the tools in this chapter are **free and open source software (FOSS)**.

For **Linux**, I recommend using **Ventoy** or **Etcher**.

For **Windows**, in my experience, **Rufus** has been one of the best tools for years. However, **Ventoy** and **Etcher** are also excellent and available for Windows.

In the end, if the given installation doesn't work when written on the USB stick with one of these tools, just try another.

For **macOS**, we have **Etcher** (supported since 2014) and **UNetbootin**. In addition, you can use the dd command-line tool from the *macOS Terminal*. There are guides available online on the topic, but I always recommend the GUI tools since they're easier, unless you are an expert or have some specific reason to use dd.

*Table 2.1* presents a list of USB image writer applications with supported OSs and support for two main features:

| | Linux | Windows | macOS | Keeps the Information on the USB Stick | Supports Multiple Images |
|---|---|---|---|---|---|
| Ventoy | **Yes** | **Yes** | No | No | **Yes** |
| Etcher | **Yes** | **Yes** | **Yes**, version 10.10 or newer | No | No |
| Rufus | No | **Yes** | No | No | No |
| YUMI Multiboot USB Creator | No | **Yes** | No | No | **Yes** |
| UNetbootin | **Yes** | **Yes** | **Yes** | **Yes**, but it might have some risks! | No |

Table 2.1 – Comparison of flashing tools for all OSs

Here's some information about the two additional columns in *Table 2.1*:

- USB data is deleted with most tools. Even if the given tool keeps your data, the stick won't contain important data unless you create a backup. After all, writing boot-related information and ISO images changes the basic partition table of the USB stick.

- If you want to be able to test several editions, **Ventoy** or **YUMI** is your tool of choice.

- If you are on macOS, use Etcher, UNetbootin, or dd (from the Terminal).

- The Manjaro images are **ISOhybrid** (this means they can be booted both in **BIOS** or **UEFI** mode – more on this in the next section). You might get a warning with some tools that **DD** mode for ISO image writing is forced. This is expected. **DD** mode is the *direct copy mode*, while ISO mode *first makes partitions* and then copies the files one by one. As a result, *a USB stick written in DD mode will not be readable by Windows after the process* – this is nothing to worry about.

## How to choose the right tool

If you want a single OS and the most simplistic interface, use **Etcher**.

If you want multiple OSs available to install (also with simple interfaces), use **Ventoy**.

If you want more options and you are using Windows, use **Rufus**.

If you are on macOS, you have no option regarding multiple OSs. **Etcher** is recommended in this case.

### For Linux and Windows

**Ventoy** is one of the easiest tools I have ever seen. It creates two partitions on your memory stick. One is for a sophisticated bootloader, and the other is to add as many .iso OS image files as possible. The Ventoy GUI is only for preparing the stick by partitioning and installing the bootloader. Then, you simply copy and paste ISO files from **Windows** or **Linux** to the second partition.

For **Linux**, most distributions might have the Ventoy installer package already available in their upstream repositories. This means you can open an application for adding SW from official repositories and find the package. For Manjaro, we have the **Add/Remove Software** application, and if we look for Ventoy, it will automatically be found, as shown in *Figure 2.5*:

Figure 2.5 – Ventoy installation in the Pamac GUI

If not, or if you experience problems, download it from `https://github.com/ventoy/Ventoy/releases` – particularly the file ending with `******-linux.tar.gz`. Extract the file in your current working directory. It will provide you with a sub-directory containing all the necessary executables. Then, double-click the `VentoyGUI.x86_64` file and enter your administrator password. The GUI will start, as shown in *Figure 2.6*:

Figure 2.6 – Ventoy application GUI

More instructions are available in the **README** file in Ventoy's directory. Apart from that, we have this link: `https://www.ventoy.net/en/doc_linux_gui.html`.

For Windows, you can download the `******-windows.zip` file from `https://github.com/ventoy/Ventoy/releases` and then run `Ventoy2Disk.exe`. It has the same GUI. Select your already connected USB stick and click **Install**.

Once you've done this, you will see **Status – READY** and the version that's been installed on the USB stick, as shown in *Figure 2.7*:

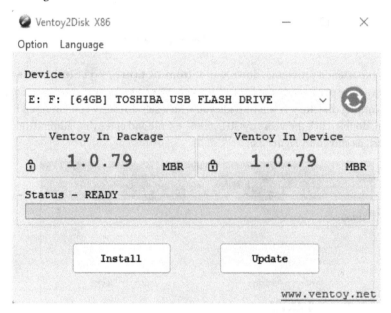

Figure 2.7 – Ventoy – write finished view

Then, copy all your desired ISO images to the Ventoy partition (that is, the `Ventoy` drive in File Explorer, *not* to the `VTOYEFI` drive/partition!). That's all you need to do!

### For all OSs

Both **Etcher** and **UNetbootin** provide simplistic GUI interfaces.

### Etcher

Etcher's GUI is shown in *Figure 2.8*. To flash with it, select **Flash from file**, then click **Select target** to choose your USB stick, and, finally, click **Flash!**:

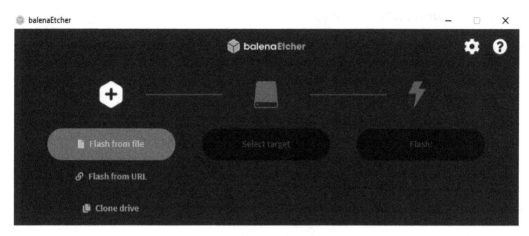

Figure 2.8 – balenaEtcher GUI view

## UNetbootin

As shown in *Figure 2.9*, select the **Diskimage** option and, from the three dots menu, choose the location of your .iso file. Then, choose the correct drive partition and click **OK**:

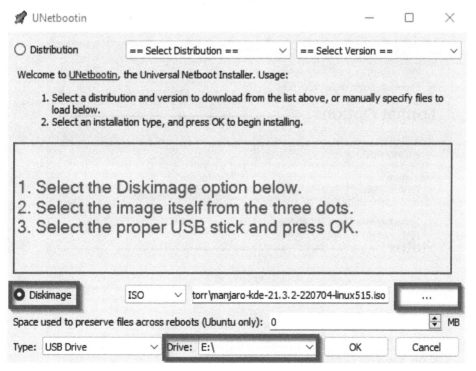

Figure 2.9 – UNetbootin GUI

### For advanced users who use Windows

**Rufus** (which provides *more features than any other tool*) is a FOSS tool that's licensed under GPLv3. I have used it for over 8 years and use it to flash a live USB stick from Windows. The tool is *over 10 years old, fast*, and *intuitive*. At the time of writing, it hasn't been ported to Linux due to the lack of developers' resources. Let's cover what you need to do.

Prepare a USB stick that's *at least 4 GB in size*, considering that all the data on this stick will be lost (*even if it is a lot bigger – for example, 64 GB!*). Download the tool from `https://rufus.ie/en/` and run it. It has only one window, and if you have only one USB stick plugged into your PC, it will detect it automatically. If you have more sticks, you must choose the right one from the first drop-down menu, named **Device**. Then, you have to choose the already downloaded image via the **SELECT** button. Its view is presented in *Figure 2.10*:

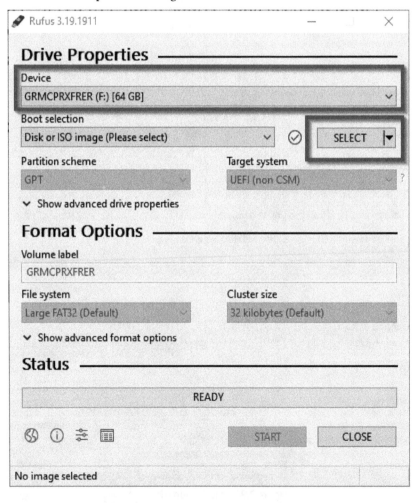

Figure 2.10 – Rufus for the Windows GUI view

Now, select the image of your choice. I have downloaded the Xfce version here as I will flash it on an older *fanless, low-end PC*. Rufus gave me the following *DD image writing mode* information message, as shown in *Figure 2.11*:

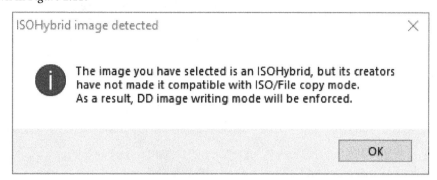

Figure 2.11 – Rufus DD image mode warning

This practically leaves us with no options in the control interface of Rufus. So, you should click **OK**, then **START**, accept the warning that all the data on the given USB stick *will be lost after the operation*, and wait a bit.

### For advanced users – about image type, MBR, GPT, BIOS, and UEFI

As mentioned previously, Manjaro's releases are currently of the **ISOHybrid** type (which supports both BIOS/MRB and UEFI/GPT partitioned drives) and require the *DD writing mode*. In Rufus, this will disable the possibility of choosing **MBR** or **GPT** for the partition scheme and the ability to choose the target system (**UEFI** or **BIOS**). This is not a problem at all. Let's learn a bit more about these two options.

### MBR and GPT

**MBR** stands for **Master Boot Record** and works with only up to 2 TB of space. If your HDD is larger, the rest of the space won't be usable under the MBR scheme.

Under **GPT**, which stands for **GUID Partition Table**, you can work with approximately *9.7 billion TB*. It is the *newer and better* scheme/standard.

You should always prefer **GPT** to **MBR**, though for a regular PC or laptop, MBR still works perfectly (if your hard disk is a maximum of 2 TB).

MBR and GPT are the ways to partition or manage your HDD for a home PC for the last 15 years.

## BIOS and UEFI

**BIOS** and **UEFI** are the basic SW that runs on your PC *before you start your OS*. They are tightly related to the **HW** of your PC, so you cannot choose which to use. This SW checks the connected HW and available *disks*, *ports* (for example, USB), *graphics cards*, *CPU*, *chips*, and more. Then, it performs a *basic HW setup*, chooses a location from which to *start the OS* (if at least one is available), and finally starts it. The **BIOS** and **UEFI** SW are located on a chip *on your* **motherboard** (the **Printed Circuit Board (PCB)**), on which all your HW is connected – hard disks, USB controllers, graphics cards, and more.

**BIOS** is the older one and stands for **Basic Input/Output System**.

**UEFI** is the newer one and stands for **Unified Extensible Firmware Interface**. Initially, it was only **EFI**.

Nowadays, practically any such chip and SW is called BIOS. UEFI is only an interface specification; thus, any modern BIOS is based on UEFI.

Typically, if your system has an old **BIOS**, then you will have to work with **MBR**, while if it has **UEFI** support, then you should choose **GPT** (especially if you have hard drives over 2 TB in size!).

Most **UEFI** systems support **legacy BIOS** and **Compatibility Support Module (CSM)**, as well as **MBR**.

Rufus normally provides three modes (and, unfortunately, is the only SW that provides such options):

- **MBR+BIOS**
- **MBR+BIOS** or **UEFI-CSM** (most UEFI systems also support CSM booting, which can be compatible with Legacy BIOS boot mode to start the OS)
- **GPT+UEFI** (non-CSM)

For my *old 2013 fanless PC*, I have BIOS, but it supports **EFI**. Rufus has set **GPT+UEFI** (non-CSM), and the machine booted without problems after I set the correct BIOS settings. This will be described in the next section.

*Figure 2.12* shows how BIOS and UEFI are related to any HW connected to your PC's motherboard PCB:

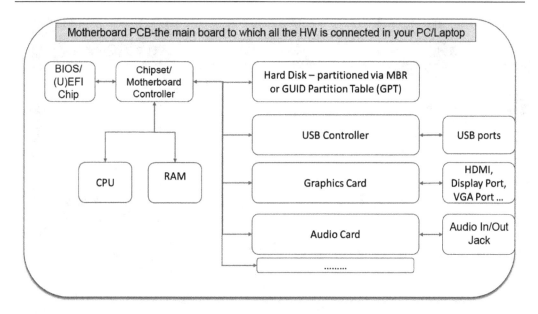

Figure 2.12 – Basic diagram of motherboard HW and connections

Now that we have learned how to install Majaro from a USB stick, let's learn how to set up the BIOS/UEFI.

## BIOS/UEFI setup for installation on a PC

This step is obligatory and not complex, so go through it no matter your level. You can do this via your PC or laptop's **BIOS/UEFI** setup, which runs before you start the OS. Usually, it consists of one or several messages with simplistic graphics before you see the Linux or Windows logo, if those messages are enabled at all. If they are disabled with an option such as **Fast BOOT Enabled** set to **True**, then despite **BIOS/UEFI** inevitably running, you might just see a black screen for a few seconds before your computer directly shows your OS's loading screen.

Different computers have different setups, so you may need to do some experimenting. You might have a hint on your screen to access a special menu, such as *Press Del to enter BIOS/Setup*. Depending on your machine's HW, the key might also be *F2, F12, F8, F9, F10*, or *Esc*. You must press those keys *right after you power up your machine* and *before* it starts loading an OS (if one is present on your machine). Inside **BIOS**, often, you can only navigate the menus and options *using the arrow keys on your keyboard*. Since a few years ago, many modern UEFI setups *also support a mouse. Figure 2.13* shows what an older **BIOS** might look like:

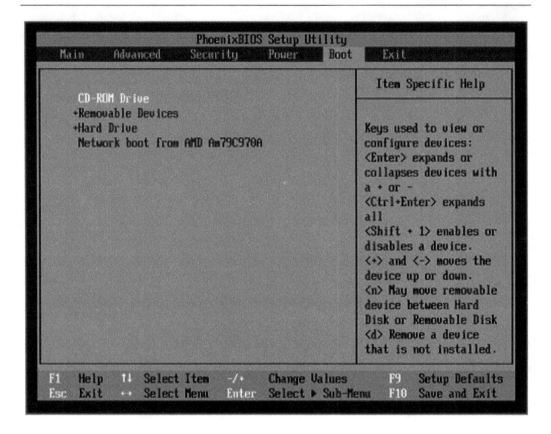

Figure 2.13 – An old classical BIOS view

*Figure 2.14* shows what a more modern **UEFI** setup with *mouse support* might look like:

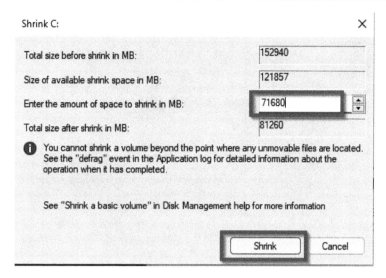

Figure 2.14 – A modern BIOS/UEFI view

Sometimes, when you activate the system settings, you get a menu, as shown in *Figure 2.15*, where each option is provided alongside the corresponding keyboard key. Here, you can see that the **BIOS** setup (named so despite supporting UEFI) can be entered via the *F10* key:

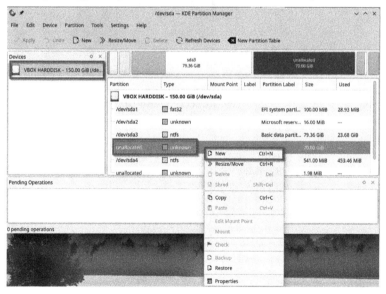

Figure 2.15 – Another modern BIOS/UEFI view

*Figure 2.16* shows another option from another manufacturer:

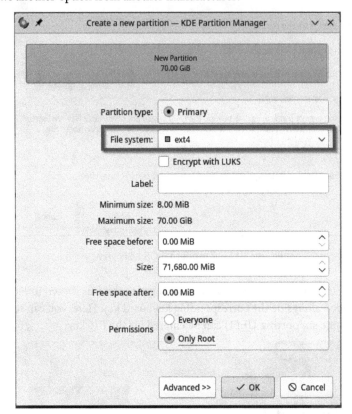

Figure 2.16 – Another modern BIOS/UEFI setup with a rich graphical menu

We won't include too many screenshots as there are hundreds of variants depending on your machine's manufacturer and model. Thus, whether you're using a keyboard or a mouse, you must navigate to the **boot sequence settings page**, and, from there, set the **USB Hard Disk**/*Removable devices* or *USB stick* to be the **First Boot Device/Boot Option #1**. Often, the setup will show you the name of the stick manufacturer (for example, Samsung USB Flash, Toshiba, ADATA, or others). If you cannot find the right options – as *there are hundreds of setups* – you must check which one is used on your machine and find the manual for it, or look for assistance in forums and web sites online.

It is important to note that *in 2007*, **Intel**, along with **AMD**, **AMI**, **Apple**, **Dell**, **HP**, **IBM**, **Lenovo**, **Microsoft**, and **Phoenix Technologies**, finally agreed to use **UEFI** (the improved **EFI**) as the universal replacement for old **BIOS**. Nowadays, practically almost no computer can be found with old-style **BIOS**. It might look like old **BIOS** (without a mouse and nice graphics), but it implements at least an early version of the **UEFI** specification. Only very old (from over 10 years ago) or special HW for non-common purposes uses pure **BIOS**.

### Secure boot

In addition to setting the USB stick as the **first boot device**, you must set one more option. Most UEFI setups have a **Secure Boot** option. Usually, it can be found under the **Security** or **Advanced** tab of the UEFI menu. **Secure Boot** was implemented as a request from Microsoft, but after a lot of noise from the FOSS community, it was changed from a firm feature to an explicit option. You must deactivate it to install and run Linux and other OSs.

### Compatibility Support Module

This optional tool that's included in the **UEFI** firmware allows **legacy BIOS** compatibility. It is abbreviated to **CSM** and offers backward compatibility by booting the machine as if it's running a legacy **BIOS** system with an **MBR** partitioned disk. In this scenario, booting is performed in the same way as on legacy BIOS-based systems – that is, by ignoring the **GPT** partition table and relying on the content of a *boot sector*. It also allows you to use older OSs *that don't support UEFI*.

> **Important**
>
> Manjaro and almost all modern Linux distributions support UEFI.

**UEFI** creates legacy **BIOS** compatibility by emulating a BIOS environment that's compatible with a given *BIOS-only* OS. If your computer is relatively new and came with Windows pre-installed, *CSM would have most likely been disabled by default.*

Usually, you don't need to enable it unless you want to install an older OS that doesn't support UEFI. The other scenario is *if you're trying to boot from an old storage drive* you recently hooked up, which already has an OS installed in **MBR** mode.

Finally, booting a legacy BIOS-based system from a **GPT** disk is also possible (though rare), and such a boot scheme is commonly called **BIOS-GPT**.

> **Note**
>
> If you wish to learn more, here is a good article on MBR, GPT, UEFI, the EFI boot partition, and a bit on GRUB: `https://www.happycoders.eu/devops/manjaro-tutorial-linux-bios-uefi-mbr-gpt-grub-sed-luks/`.

## How to understand whether your computer is BIOS/EFI or UEFI-based

Once you run the BIOS/UEFI setup with *F2, F8, F9, F10, Esc, Del*, or some other keyboard key, go through the menus, and if you see **EFI, UEFI, CSM, Compatibility Support**, or **Secure Boot**, *then it is at least EFI.*

## What needs to be done for (U)EFI setup

You have to make sure you meet the following requirements:

- **Secure Boot** should be *off*.
- **(U)EFI mode** should be *on* (if there is such an option).
- **CSM** (compatibility mode) should be *disabled* (if there is such an option).
- *Disable* **legacy mode** (if there is such an option).
- Of course, the USB stick that contains the installer must be *set as the first boot device*. On some machines, next to the name of the USB stick, it will also be noted that it will boot in **(U)EFI** mode.
- I also recommend *disabling* **fast boot** (if there is such an option), and if you have a setting for the initial timeout, if it is 0 seconds, set it to at least 3 seconds *to be able to see your boot screen*.

## What needs to be done for a pure BIOS system

Check and meet the following requirements:

- Only the USB stick with the installer should be set as the first boot device
- I also recommend disabling fast boot (if there is such an option)

## What needs to be done for a (U)EFI system to boot in BIOS mode

Check and meet the following requirements:

- **Secure Boot** should be *off*
- **(U)EFI mode** should be *off* (if there is such an option)
- **CSM** (compatibility mode) should be *enabled* (if there is such an option), and **legacy mode** should be *enabled* (if there is such an option)
- Of course, the USB stick that contains the installer should be *set as the first boot device*
- I also recommend *disabling* **fast boot** (if there is such an option), and if you have a setting for the initial timeout, if it is 0 seconds, set it to at least 3 seconds *so that you can see your boot screen*

# The installation itself

This section has the following subsections:

- Pure installation
- GParted and partition management

- Manual installation for EFI-based computers (for advanced users)

- Manual installation for BIOS-based computers (for advanced users)

- Installation on a virtual machine

- Installation in parallel with Windows (also known as Dual Boot)

- Installation on ARM platforms

The first part, Pure installation, describes the full Calamares GUI Installer options and explains installation basics. It is necessary for everyone and assumes one of the following cases:

- you install Manjaro on a clean machine without any other OS

- or you have Windows installed and want to install Manjaro alongside. In this case, just read it and then perform the actions from the sixth section about Dual Boot. Return to the first if necessary.

The second section provides hints for manual partitioning in advance (before starting the installer) and is essential for anyone not knowing how this is done.

The third section, Manual installation for EFI-based computers, is the typical case for an advanced user willing to do manual partitioning with Calamares, as all modern computers support at least EFI.

The fourth section is when you install Manjaro on an old machine with BIOS. If it is also a weak machine – any OS will be slow on it.

The fifth presents virtual machine installation. It is the easiest for a beginner with enough space on a macOS. It is also great if you have a Windows machine with enough HW resources but don't want to make a dual boot.

The sixth, Installation in parallel with Windows, is the most common case. I have used Dual Boot for over a decade.

The last contains a short description and links for ARM platforms.

Here are the tips for beginners:

- If you work with a separate machine – read only the first section.

- If you work on Windows, you can resize your partitions and Dual Boot

- If you work on macOS, working on a virtual machine is perfect.

- If you are on Windows and don't want to change your HDD partition – working on a virtual machine is again perfect. Once you know more, you can install Manjaro with Dual Boot or invest in a non-expensive second PC.

## Pure installation

Once you have the stick ready and have booted successfully, it is recommended that you have an *internet connection* so that you get additional and the latest SW packages from Manjaro's servers. However, this is *not obligatory; you can install* Manjaro *without an internet connection.*

Once the system boots, you will see the *initial Manjaro boot menu*, which, at the time of writing, has a 10-second timeout. It allows you to choose the installation system and keyboard input languages. What is more important is that if you don't move with the up/down arrows, this boot menu will time out and continue directly with the default option – **Boot with open source drivers**. The **Boot with proprietary drivers** option is helpful in two cases. The first is if you have some special HW with **proprietary** drivers upstream (already released to the FOSS Linux servers and Manjaro) you know about. The second is if the **Boot with open source drivers** option doesn't boot normally. Typically, the default option works well. *Figure 2.17* presents this menu:

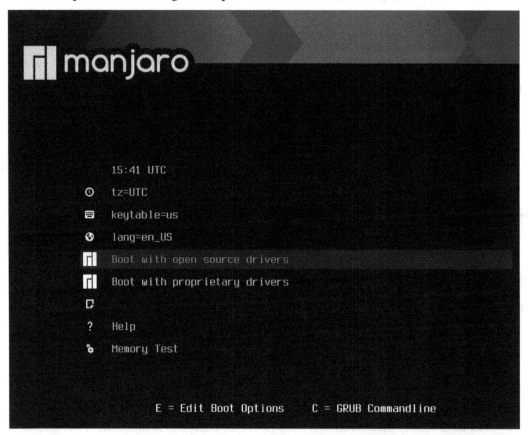

Figure 2.17 – Manjaro installation – first view after successfully booting Manjaro live

Next, Manjaro will print a long list of status messages. Normally, all of them *start with a green* **OK**, as shown in *Figure 2.18*:

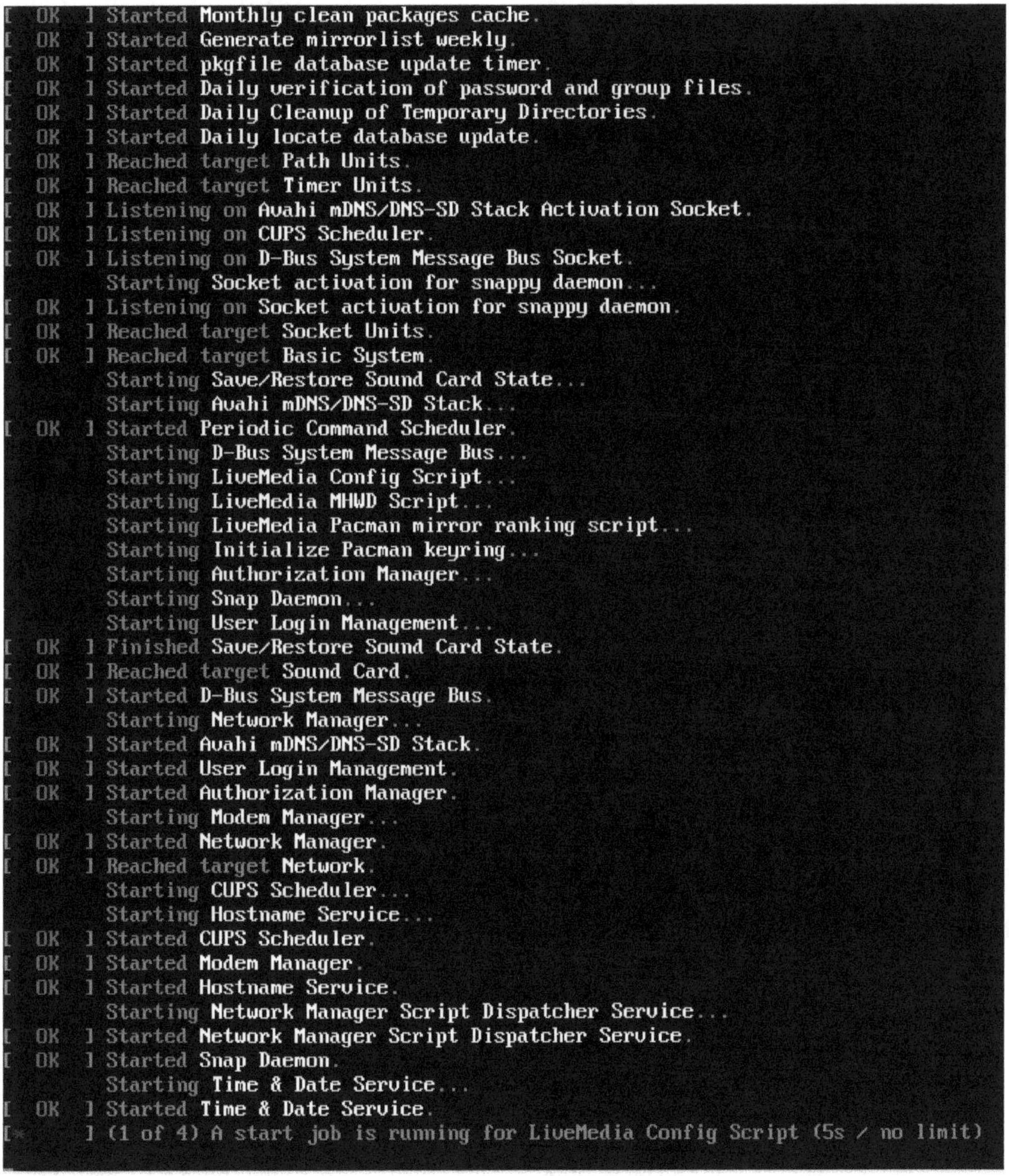

```
[ OK ] Started Monthly clean packages cache.
[ OK ] Started Generate mirrorlist weekly.
[ OK ] Started pkgfile database update timer.
[ OK ] Started Daily verification of password and group files.
[ OK ] Started Daily Cleanup of Temporary Directories.
[ OK ] Started Daily locate database update.
[ OK ] Reached target Path Units.
[ OK ] Reached target Timer Units.
[ OK ] Listening on Avahi mDNS/DNS-SD Stack Activation Socket.
[ OK ] Listening on CUPS Scheduler.
[ OK ] Listening on D-Bus System Message Bus Socket.
       Starting Socket activation for snappy daemon...
[ OK ] Listening on Socket activation for snappy daemon.
[ OK ] Reached target Socket Units.
[ OK ] Reached target Basic System.
       Starting Save/Restore Sound Card State...
       Starting Avahi mDNS/DNS-SD Stack...
[ OK ] Started Periodic Command Scheduler.
       Starting D-Bus System Message Bus...
       Starting LiveMedia Config Script...
       Starting LiveMedia MHWD Script...
       Starting LiveMedia Pacman mirror ranking script...
       Starting Initialize Pacman keyring...
       Starting Authorization Manager...
       Starting Snap Daemon...
       Starting User Login Management...
[ OK ] Finished Save/Restore Sound Card State.
[ OK ] Reached target Sound Card.
[ OK ] Started D-Bus System Message Bus.
       Starting Network Manager...
[ OK ] Started Avahi mDNS/DNS-SD Stack.
[ OK ] Started User Login Management.
[ OK ] Started Authorization Manager.
       Starting Modem Manager...
[ OK ] Started Network Manager.
[ OK ] Reached target Network.
       Starting CUPS Scheduler...
       Starting Hostname Service...
[ OK ] Started CUPS Scheduler.
[ OK ] Started Modem Manager.
[ OK ] Started Hostname Service.
       Starting Network Manager Script Dispatcher Service...
[ OK ] Started Network Manager Script Dispatcher Service.
[ OK ] Started Snap Daemon.
       Starting Time & Date Service...
[ OK ] Started Time & Date Service.
[*   ] (1 of 4) A start job is running for LiveMedia Config Script (5s / no limit)
```

Figure 2.18 – Status of modules during Manjaro's live initial load

After this, you will see the default view of the chosen flavor, along with the **Manjaro Hello** screen, as shown in *Figure 2.19*. This one is for **Xfce**. For the others, the screen will change a bit, but all the links and any further installation options will be the same:

Figure 2.19 – Manjaro live initial loaded view

It provides buttons with links for *basic information about the Manjaro project* (such as **DOCUMENTATION**, **SUPPORT**, and **PROJECT**). It also has the **Launch installer** button at the bottom, which you should click.

The first three steps are simple, and require you to:

- Choose your language.
- Choose a region and zone.
- Choose a keyboard layout. If your exact layout is missing, you can install it from an additional package after completing the system installation.

The next step is **the partitioning and boot setup**. If you install it on a machine with a clean hard disk (that is, no other pre-installed OS already exists on the hard drive), and there is no **MBR** or **GPT** scheme already set up, you will get the view shown in *Figure 2.20*:

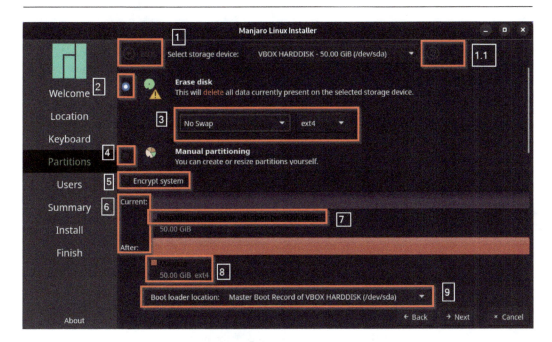

Figure 2.20 – Manjaro Linux Installer – Partitions

We have marked all essential points to review with *red rectangles*. They are as follows, according to the *white numbering* shown in *Figure 2.20*:

1. The current **BIOS/EFI** setting. This *cannot be changed*. It is based on what the **Calamares** installer is detecting based on the **HW** and **BIOS/EFI** setup. It will write **BIOS** there if it detects a pure BIOS scheme/system. If it detects **(U)EFI HW** and **SW**, it will write **EFI**. On the right-hand side of **Select storage device**, next to **1.1**, you can see *a question mark* since this example comes from setting up a *completely clean hard disk*. However, if Calamares detects a hard disk with an **MBR** scheme, it will mark it there with **MBR**. If the hard disk has a **GPT** table, it will say **GPT**. Later, I will show you how to change this, but for now, it is essential to note that even on older systems, **GPT** installation is possible since **Manjaro** supports it. The other important point is that with **GParted** or **KDE Partition Manager** applications (included in any Manjaro live installer), you can clean older **MBR** disks and later convert them so that they're **GPT**-based. We'll learn how to use it in the following subsection.

2. As this example is with an empty disk, we have only two options, marked as **2** and **4** – **Erase disk** or **Manual partitioning**. If you are a beginner, using the **Erase disk** option and setting up only point **3** is recommended.

3.   The **swap space** is the amount of space on the hard disk that's used to *store data temporarily* when the **RAM** is insufficient. Linux is smart and will typically not use it unless the **RAM** is full. When it is full, Linux will *move rarely used data to this space*. If you want to select the amount of **swap** by yourself, you have to choose **Manual partitioning**, but we will not consider this here. In addition, regarding the next dropdown, which shows **ext4** in the preceding screenshot, this is the recommended filesystem for *root* and *home*, so keep it as-is.

The swap menu gives you *four* options:

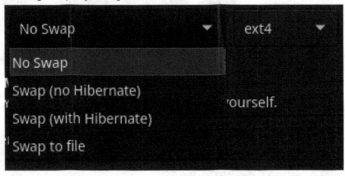

Figure 2.21 – Manjaro installer setup for the swap menu

Before we explain them, **Hibernate** means that your machine goes into complete shutdown but *keeps the OS's current state on the hard disk* (opened windows, SW, a browser with tabs, and so on). Going into **hibernation** is almost like a normal *shutdown* in terms of time. However, waking up and providing the machine exactly as it was is much *faster compared to full OS loading* and *manually opening* all user applications again. To highlight the difference, going to *sleep* will keep the data in RAM while causing limited power consumption. As you move your mouse, it will light up for several seconds. Loading from **hibernation** will take more time, but again, a lot less time than loading everything yourself from a normal power-up. Let's look at the four swap options in more detail:

A.   **No Swap** is not needed in either of the following situations:

   • You have a lot of RAM and expect it to *never fill up*. You plan *never to hibernate* and to use only sleep mode eventually.

   • If you have a minimal amount of hard disk space (and you understand that in this way, you will *not be able to hibernate*).

B.   **Swap (no Hibernate)** will usually allocate a swap space of approximately 0.5 times the RAM size if it is at least 6 GB. This option will reserve two times the RAM size if you have less than 6 GB. This option is usually for people with a small amount of RAM *who don't want to hibernate*. Nowadays, a regular PC or laptop will have, as an absolute minimum, 4 GB of RAM.

C. **Swap (with Hibernate)** will reserve as much memory as the amount of your RAM. If you have a *small amount* of RAM (less than 4 GB), it will reserve double the amount of RAM to serve during normal operations. For 8 GB of RAM, it will reserve 8 GB of swap space. For 16 GB of RAM, it will reserve 16 GB of swap space.

D. **Swap to file** allows you to keep swap data and/or hibernate if you don't want a dedicated part of your hard disk to be a separate swap partition. It can be (but not necessarily) a bit slower than a dedicated swap partition (options *B* and *C*).

Here is a helper on how to choose which option is best for you:

- If you have more than 100 GB of HDD space, it is recommended to choose **Swap (with Hibernate)** and don't bother further.

- If you have less than 100 GB but more than 40 GB of HDD space and less than 4 GB of RAM, it is recommended to choose **Swap ( no Hibernate)** or **Swap (with Hibernate)**, depending on your choice and the results in points **6**, **7**, and **8**.

- If you have less than 40 GB and at least 4 GB of RAM, you are better off choosing **No swap** or **Swap (no Hibernate)**.

- However, if you have 40 GB or less *but more than 4 GB of RAM* and *want to hibernate*, choose **Swap to file**.

4. We will discuss point **4** for **Manual partitioning** later.

5. Point **5** is essential – if you choose it, *your whole hard disk, along with your installed SW and personal data, will be encrypted.* This is appropriate for machines that contain **sensitive information**. However, if you don't plan to keep such information and your machine is quite *weak*, *slow*, or *old*, it's better *not to enable it* as encryption always requires some HW resources and would make the PC *slightly slower than without encryption*. Linux is quite secure by itself. On the other hand, my office workstation has **Intel Core i7** and **8 GB of RAM** installed *with encryption*. As the **ext4** filesystem is effective, and Linux is a highly effective OS, *I don't feel any delays*. In the end, it's your decision.

6. When you select the options from points **2** and **3**, you will see the results for your HDD partitioning before and after the installation in points **6**, **7**, and **8**.

7. Finally, point **9** provides information on where the bootloader will be installed if your installation is based on the **BIOS** scheme. If it is **EFI**-based, this point will not be depicted!

Once you've done this, on the next screen, you have to fill in your *name, login name, computer name* (for example, how it will be seen on the network), and, finally, *the passwords*. I have a few critical recommendations here:

- If you don't plan to do some special management with several users, select the option for the **administrator account** to have the *same password* as your **account**.

- Unless you will only use the computer for common browsing in a public area, *never select the option to log in automatically without a password*. Though Linux, particularly Manjaro, is secure and no serious action is allowed without a password, leaving your account unprotected is never a good practice.

- Always include *lowercase* and *uppercase letters*, some *numbers*, and *at least one special symbol*, such as ", !, %, #, @, &, ^, *, (, ), ~, _, <, >, $, or ", in your password. **It must be at least 12 characters**. This is because even with modern Linux and Manjaro's high level of security, a weak password can be hacked in seconds.

- A good and not difficult-to-remember password tip is to take something from a beloved book, movie, or song. Add a number and, for example, the & sign. An example is "WinniePooh135&". If it is something known to you, it will be easy for you to remember. Another excellent example is BilboLOTR324^.

Once you've set everything up, you will see something similar to the following:

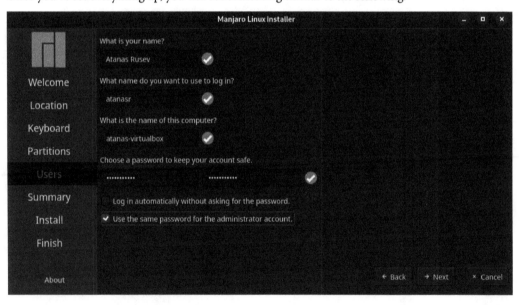

Figure 2.22 – Manjaro installer user and admin setup

Now, click **Next**; at the end, you'll see a summary of what will be done to your system, as shown in *Figure 2.23*:

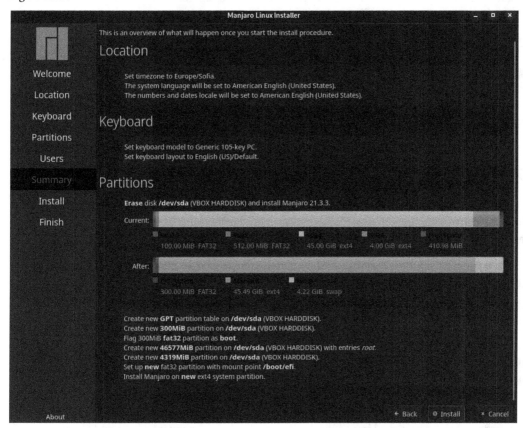

Figure 2.23 – Manjaro Linux Installer – full summary of the installation setup

Finally, click **Install**. If you're working on a fast machine, your PC will be *ready in a few minutes*. When the installer is ready, it will write the words **All done**. When you restart your machine, remove the USB installer stick, and the PC will boot into your new Manjaro. *After the first restart*, when you confirm the installation is working well, changing the settings back to the first boot device's HDD or SSD (if it hasn't happened automatically) is also recommended.

> **Important note**
>
> On some (U)EFI machines, this first restart leads to an unbootable state. Though this is rare and happens occasionally, if it does, there is a solution. In such occasions, the first thing to try is a HW restart, which you can do by pressing the power-on button for 10 seconds until the PC shuts down completely. Then, wait 10 more seconds and press the power button again. This time, the BIOS will find the GRUB bootloader path correctly, and the problem will never reappear. If this doesn't help, on many modern BIOSes, there is an option to activate a (U)EFI boot selection mode. In this case, the BIOS scans the available bootloaders on all connected mass storage drives and provides a suggested list of options. You can try the one that has Manjaro in its name. Again, once you've done this, the problem won't appear after your next restart.

## GParted and partition management

If we want some particular partitioning scheme and don't want to use the **Manual partitioning** option of the **Manjaro Installer**, we can use a partition manager *before the installation*. Remember, the **installer's manual partitioning** is enough for regular work, including *creating separate additional partitions*.

**GParted** is the default **GUI** tool for **HDD** partition management on Manjaro Xfce (and many other Linux distributions). With it, you can change from **MBR** to **GPT** and vice versa and *configure any partitions you want*. It is *included in the live installation* version of Manjaro (like Firefox and some other regular SW).

For Manjaro **KDE** Plasma, the default partition manager is **KDE Partition Manager**. It has a different interface but *practically the same general options*. We will not cover it here. If you need it and cannot figure it out yourself, check out https://docs.kde.org/trunk5/en/partitionmanager/ partitionmanager/partitionmanager.pdf or search the web for a PDF named *The KDE Partition Manager Handbook*. It is old, but I have thoroughly checked it, and it is still relevant. The last sentence proves how the Linux community doesn't change many interfaces and SW without a serious reason.

The information in the **KDE Partition Manager** is relevant for many general cases, so I recommend reading it if you need to work with partitions, *even if you use* **GParted**. Also, remember that many general warnings for GParted are valid also for **KDE Partition Manager**.

On Manjaro GNOME, the default partition manager is Disks. I will not cover it, as I find the two mentioned partition managers a lot better. No matter which partition manager you use, the general recommendations and actions are the same, as the task is the same. If GNOME is your choice, read this section to know the basics, then check out the KDE Partition Manager handbook. It will be easy for you to understand how to use the Disks partition manager GUI.

To start **GParted**, open the main OS menu, write partition, and click the icon shown in *Figure 2.24*:

Figure 2.24 – GParted Partition Manager icon

*Figure 2.25* shows its main interface:

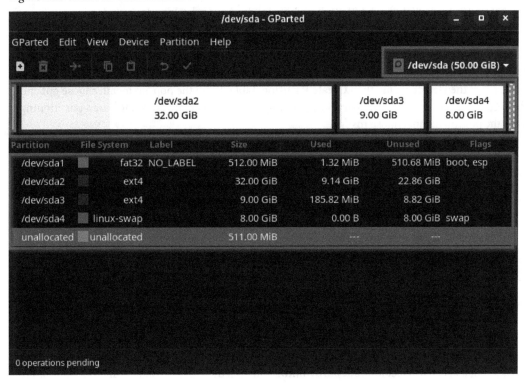

Figure 2.25 – GParted Partition Manager main GUI view

At the top right, we can see the name of the currently depicted **HDD**. On the light-colored row, we can see each partition. In the table below, we can see details of each of those partitions.

The following two warnings are also valid for KDE Partition Manager:

- Each *applied* deletion is irreversible

- Changing from **MBR** to **GPT** and vice versa means losing the opportunity to read any old information from any partition on the target HDD

To delete a partition, select it in the table or the rectangular representation and press *Del* on your keyboard. Once you've deleted all the partitions you want to, click the **Apply all operations** button, which is depicted with a checkmark: ✓ .

To create a partition, right-click on unallocated space and select **New**. Normally, you will keep the default options, caring only about the **size** in **MB** and the **Files system,** which is recommended to be **ext4**. *1 GB is equal to 1,024 MB*, so consider this when calculating. More on filesystems can be found in *Chapter 10*. Then, click the **Add** button and, again, the **Apply** checkmark symbol.

When changing from **MBR** to **GPT** or *vice versa*, you don't need to delete any partition. Changing this setting will wipe out all partitions on the given **HDD**. We know that **GPT** is always preferred (and Manjaro can boot in BIOS mode from a GPT partitioned HDD). To create a new **GUID partition table**, select the **Device** menu, then **Create Partition Table**. From the pop-up menu, choose **gpt** and click **Apply**. Then, create all the partitions you might need. To create an **MBR**-based partitioning from the pop-up menu, choose **msdos**.

---

**Interesting note**

Years ago, before GPT became the most used scheme, Linux distributions used extended and logical partitions to go around the limitations of the MBR scheme. Nowadays, with GPT, we don't need those in typical cases.

*Another note*: If you want to create an MBR partition, choose **msdos**.

You can find more information at `https://gparted.org/`.

---

## Manual installation for EFI-based computers (for advanced users)

Many people might be unhappy with the setup provided by the automatic installer (though it is excellent). In this case, there are two options: do the partitioning yourself in advance (with some partition manager) or select **Manual partitioning** and do it during the installation. Both options are OK, but let's see what happens if we choose **Manual partitioning**.

Ensure you have checked all BIOS-related settings listed in the *What needs to be done for (U)EFI setup* section before you continue.

As we are working with a clean installation, if you have some old partitions you don't care about, please delete them in advance by selecting them and clicking **Delete**, as shown in *Figure 2.26* (*be sure you know what you're deleting*):

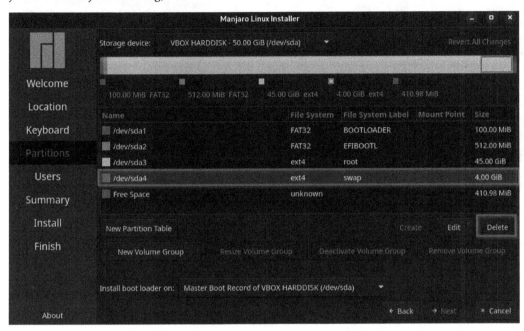

Figure 2.26 – Manjaro installer manual partition setup – deleting old partitions

Once you have deleted all partitions (or if you have no partitions at all), you will see only **Free Space** in the list, as shown in *Figure 2.27*:

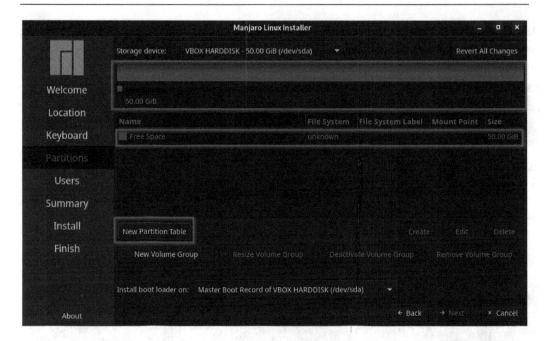

Figure 2.27 – Manjaro installer manual partition setup – new partition table

Then, you must click the **New Partition Table** button. Here, you can choose whether you want to create an **MBR** or **GPT** scheme. Always prefer **GPT** as it provides the following:

- *Bigger partitions* (**MBR** is up to *2 TB*, while **GPT** reaches up to *millions of TB*, depending on the filesystem)

- *More partitions – up to 128*, while the MBR scheme provides *only up to 4 primary* partitions

- If you want **EFI support** (recommended for Manjaro, almost all modern Linux distributions, and Windows), you need a **GPT**-based partitioning scheme

So, let's select **GUID Partition Table**, which will work even for BIOS-based computers. Then, we need several partitions. In the case of the BIOS scheme, we must create them in this order:

- An **EFI** partition that's **512 MB**, has the **FAT32** filesystem, /boot/efi as the mount point, and the **boot** flag

- A **Manjaro root partition** that's **at least 30 GB** and **ext4** as the filesystem – the bigger the better. In this case, as I'm working with 50 GB of free space, I will set it to 32 GB (32,768 MB). Use an **empty slash** / as the mount point and the **root** flag.

- A **home partition** *that's as big as possible* – in this case, **9 GB** (9,216 MB), with **ext4** at the filesystem, **/home** as the mount point, and **no flag**. Why do we need a home partition? Usually, if we reinstall/update or make some errors in the OS and need to reinstall, we can keep our data in the *home* partition and later simply attach it to another new installation so that we can keep, for example, our documents, pictures, and clips.

- An **optional swap partition**. As I am working with a small amount of RAM, I will set a swap space of 8 GB, **linuxswap** as the filesystem, **no mount point**, and the **swap** flag.

Here are some points to keep in mind:

- *1 GB is 1,024 MB, not 1,000. So, 8 GB is 8,192 MB.*

- For space distribution, we can consider the example shown in *Table 2.2* (for a machine with *2 to 8 GB* of **RAM,** and if you also want to have the opportunity to hibernate):

| | 50 GB HDD | 80 GB HDD | 100 GB HDD | 200 GB HDD | 500 GB <br><br> **or** <br> **Bigger HDD** |
|---|---|---|---|---|---|
| **boot** | 512 MB | 512 MB | 512 MB | 512 MB | 512 MB |
| **root** | 32,768 MB | 35,840 MB | 46,080 MB | 81,920 MB | 81,920 MB |
| **home** | 9,216 MB | 36,864 MB | 47,104 MB | 113,664 MB | 429,056 MB |
| **swap** | 8,192 MB | 8,192 MB | 8,192 MB | 8,192 MB | 8,192 MB |

Table 2.2 – Example distribution of space by partitions for installation

- Reminder: **Swap** shall only be skipped *if we have at least 4 GB of RAM* and *don't want to hibernate* to swap. In this case, the only possible hibernation is *hibernate to file*, which can be a bit slower.

For **swap**, we can use *Table 2.3* to define it, noting that this is *for regular usage* and *not for special or server needs*:

| RAM Size | Swap Only for Work | Swap for Hibernating |
|---|---|---|
| 1 GB | At least 4 GB – max 8 GB | The same as for work |
| 2 GB | At least 4 GB – max 8 GB | The same as for work |
| 3 or 4 GB | At least 4 GB – max 8 GB | The same as for work |
| 5 GB | 3 GB at least | 5 GB |
| 6 GB | 2 GB at least | 5 GB |
| 8 GB | Not more than 4 GB, can be skipped | 8 GB |
| 12 GB | Not needed | 12 GB |
| 16 GB | Not needed | 16 GB |
| 32 GB | Not needed | 32 GB |
| 64 GB | Not needed | 64 GB |

Table 2.3 – Swap space for work and hibernation based on RAM size

To set these settings for *each new partition*, select **Free Space** from the list and click **Create**, as shown in *Figure 2.28*:

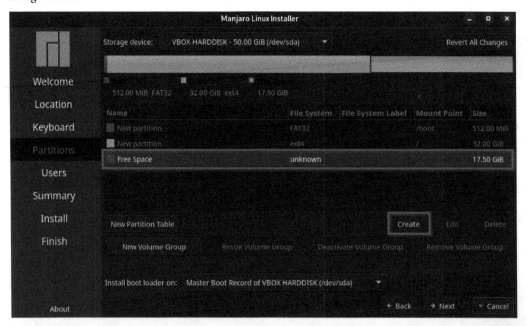

Figure 2.28 – Manjaro installer manual partition setup – creating a new partition

This will open the **Create a Partition** page. For /home, for example, we would have the setup shown in *Figure 2.29*:

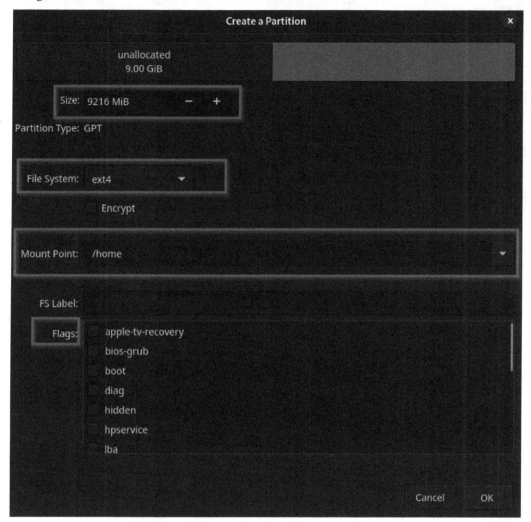

Figure 2.29 – Manjaro installer manual partition setup – setting up a new partition

In the end, we'll get the results presented in *Figure 2.30*:

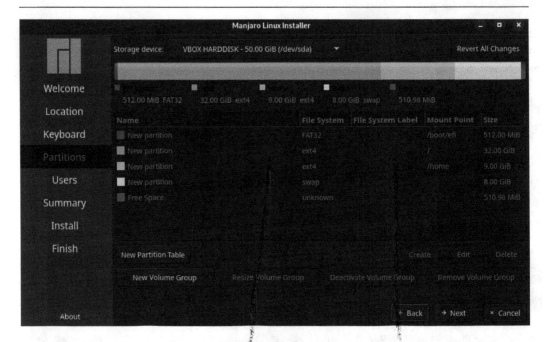

Figure 2.30 – Manjaro installer manual partition setup – new partitions overview

Then, fill in your username, password, and so on while keeping in mind the recommendations from the previous subsection. Finally, click **Next** and **Install**.

At the end of the installation, you will see **All done** in big letters. Now, you can restart your PC. As you have *already set up the first boot device to be the USB stick*, once your computer restarts, you must remove it to load from disk. After the first successful (re)start, once you've confirmed the installation is working well, changing the settings back to the first boot device HDD or SSD (if this hasn't happened automatically) is also recommended. Also, please consider the *Important note* point at the end of the initial part of the *The installation itself* section regarding potential restart issues.

## Manual installation for BIOS-based computers (for advanced users)

The booting scheme is the main difference between **BIOS**- and **UEFI**-based computers. So, all the points from the previous two sections are valid for *passwords, swap space*, and *the settings for root and home partitions*. If you haven't read them, please do so before going any further.

Ensure you have completed all the steps provided in the *What needs to be done for a pure BIOS system* or *What needs to be done for a (U)EFI system to boot in BIOS mode* section before you continue!

Usually, such cases will crop up when we work *with older PCs*. Nevertheless, if it is a **GPT**-formatted HDD, you will *inevitably lose everything when switching to* **MBR**. If it is already an **MBR** formatted HDD, the boot and root partitions will be overridden, and you will need most of the primary partitions, which leaves no space for another distribution or Windows on the drive.

Years ago, there was one workaround for this – as **MBR** gives only four primary partitions, each primary partition could be split internally into logical partitions by **GParted** and any other Linux partition manager. Search on the web for *dual boot of Windows and Linux on an MBR drive* if you need this solution.

For now, we must *delete all old partitions* and leave *only free space*, as shown in *Figure 2.31*:

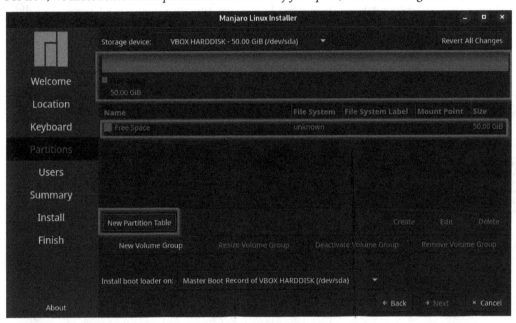

Figure 2.31 – Manjaro Installer manual partitions setup for BIOS-based computers

Then, you must click the **New Partition Table** button and choose the **msdos / MBR** scheme. Further, we must create the new partitions in this order:

- **Unallocated space of 8 MB** or a **fat32** partition with **no mount** point and **no flags** set – *required from Manjaro to write the basic bootloader* to the **Master Boot Record** for the drive

- **Bootloader** partition – **512 MB**, with **fat32** as the filesystem, **/boot** as the mount point, and the **boot** flag

- Manjaro **Root** partition – **at least 30 GB**, with **ext4** as the filesystem (the bigger, the better), an **empty slash /** as the mount point, and the **root** flag

- **Home** partition – *as big as possible*, but at least **10G B**, with **ext4** as the filesystem, **/home** as the mount point, and **no flag**

- **Optional swap partition** – as I'm working with a small amount of RAM, I will set a swap space of **8 GB**, **linuxswap** as the filesystem, **no mount point**, and the **swap** flag

To set these settings for *each new partition*, select **Free Space** from the list and click **Create**, as we did in the previous subsections.

In the end, we'll get the result shown in *Figure 2.32*:

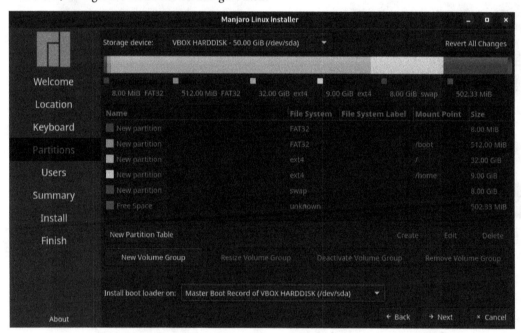

Figure 2.32 – Manjaro Linux Installer – partitions summary for BIOS-based computers

If you get the **info message** presented in *Figure 2.33*, simply click **OK** as we have just created the necessary partitions with the initial 8 MB one needed at the beginning of the hard drive:

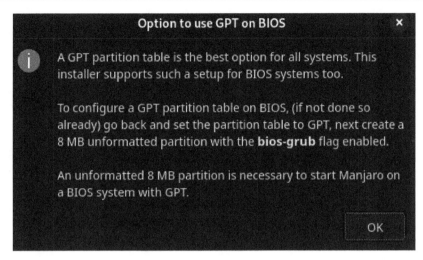

Figure 2.33 – Manjaro Live Installer with BIOS – warning for the boot sector

Then, fill in your *username*, *password*, and so on. Finally, click **Next** and **Install**.

At the end of the installation, you will see the words **All done** in big letters. Now, you can restart your PC. As you have already set up the first boot device to be the USB stick, once your computer restarts, you must remove it to load from disk. After this first (re)start, once you've confirmed the installation is working well, changing the settings back to the first boot device's HDD or SSD (if this hasn't happened automatically) is also recommended. Also, please consider the *Important note* point at the end of the initial part of the *The installation itself* section regarding potential restart issues.

## Installation on a virtual machine

For installation on a **virtual machine** (**VM**), you must set up a few basic **VM settings** and provide the **location** of the OS image. Then, once you have started it successfully, it will work exactly the same as on a regular PC from a USB stick.

I have used **VirtualBox** from **Oracle** for years. I have tried other solutions for this book's purpose, but **VirtualBox** is the best since it's the most robust, free, and updated frequently. Hence, I strongly recommend it and will present it in further detail here.

Here's what you need to do:

1.  Download and install it from `https://www.virtualbox.org/wiki/Downloads`.
    From the list on the *VirtualBox X.X.X platform packages* page, you have to choose your host
    (**macOS** or **Windows**). Keep in mind that, for a MacBook with an **M1** or **M2** processor, in
    June 2023, the provided version was still a development one, but it is expected that by the end
    of 2023, it might evolve into an official one.

2.  Install **VirtualBox** on your machine.

3.  Download **VirtualBox**'s **Extension Pack** *for all platforms (there is no other option)* from
    here: `https://www.virtualbox.org/wiki/Downloads`.

4.  Follow the numbers presented in *Figure 2.34*. Open VirtualBox, select **Tools** on the left, then
    **Preferences** on the right (as indicated by **2**). Go to **Extensions** and click the green plus icon
    on the right to *add the extension pack*:

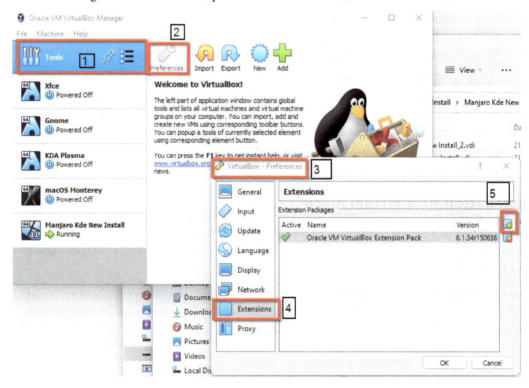

Figure 2.34 – Setting up VirtualBox

5.  You can install the Extension Pack by simply *locating the freshly downloaded file* for **All platforms**.

6.  Download one or a few Manjaro images of your choice from `https://manjaro.org/download/`.

7.  Now, in VirtualBox, click the **New** blue sun icon.

8.  Select **Expert Mode** from the bottom-left button – *this will be easier for you*.

9.  In the new pop-up window, choose a **name,** then pay attention to the **Machine Folder** area – the VM will also need a hard drive, and for it, Manjaro says you need **a minimum of 30 GB**, but it's better to have *at least 60 GB of space* on your hard drive. If your computer has separate drives or partitions, locate the VM in *a data drive with more free space*. For Windows, for me, this is **D: drive**. Keep *all the data* for one VM in *a separate directory*.

10. Select **Linux**, then set **Version** to **Arch Linux (64-bit)**. **Manjaro** is **Arch**-based, and **VirtualBox** doesn't have separate settings *for all possible distributions, only for the major* and a few specific ones.

11. Allocate **RAM Memory Size** according to *Table 2.4*. Rest assured that VirtualBox will only take as much as it needs, and this is the upper limit. To know the RAM size on Windows, open the **Start menu** and write `System`, then run **System Information**. The other option is to search for `memory` in the **Settings** application or to open the Task Manager. On macOS, open the **Apple menu** and run **About this mac**:

| Your Machine's Amount of RAM | Allocated RAM for the Virtual Machine |
| --- | --- |
| 4 GB | 2 GB |
| 6 GB | 3 GB |
| 8 GB | 4 GB |
| 12 GB | 6 GB |
| 16 GB or more | 8 GB |

Table 2.4 – Amount of RAM to be allocated for a VM

12. In the **Hard disk** section, choose the **Create a virtual hard disk now** option. The settings will look as follows:

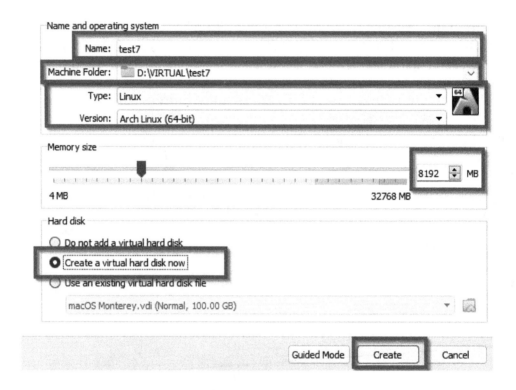

Figure 2.35 – VirtualBox – basic setup of a VM for Manjaro

13. Clicking **Create** will open an additional popup. Allocate the size you want – for example, 60 GB (use the box, not the slider – *the slider is uncomfortable*), then keep it **dynamically allocated** *so that it takes as much space as necessary and not the full maximum size*. Keep the type as the default of **VDI** (**VirtualBox Disk Image**).

14. Click **Create**.

15. You will now have the **VM** on the left under **Tools** in the main window of **VirtualBox**. Select it and click the yellow **Settings** wheel from the top-right area.

16. As shown in *Figure 2.36*, in the pop-up window, select **System** on the left. On the **Motherboard** tab, *disable* **Floppy** and *enable* the **Enable EFI (special OSes only)** option. Then, if your PC or MacBook has four processors or more, from the **Processors** tab (shown in *Figure 2.37*), change **CPU** to at least 2, but no more than 4, to indicate the number of CPUs to be allocated to the VM. **VirtualBox** will depict the total amount of CPU cores detected, so you don't need to look at the system characteristics to know how many cores you have:

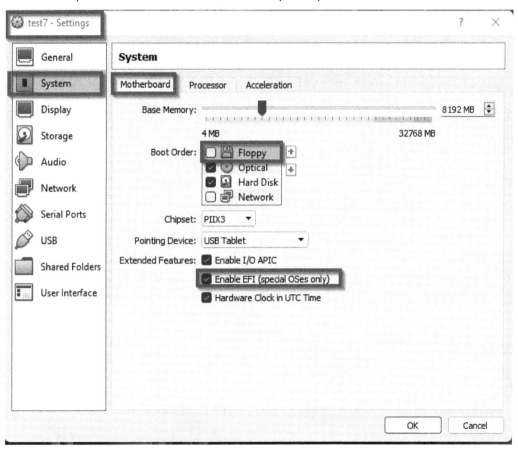

Figure 2.36 – VirtualBox – system setup for a VM

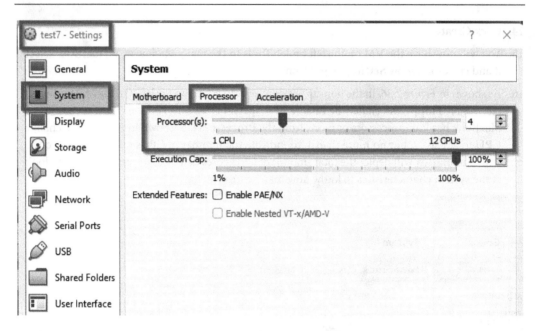

Figure 2.37 – VirtualBox CPU setup for a VM

17. Further, in the **Display** settings, make the memory *at least* **32 MB**. I tend to have a minimum of 64 MB. Set **Graphics Controller** to **VBoxVGA** since, as of March 2023, there is an issue related to both **VirtualBox** and *multiple Linux graphical drivers*, and the other options (**VMSVGA**, **VboxSVGA**, and **None**) lead to strange errors during execution. Of course, this might be solved in the future.

18. Then, as shown in *Figure 2.38*, in the **Storage** section, select **Empty** and then **Live CD / DVD** and, on the right, select the image of your choice by selecting **Choose a disk file…**:

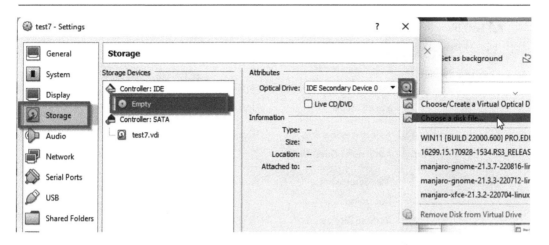

Figure 2.38 – VirtualBox – storage and ISO image setup of virtual machine

19. When you find the `.iso` image of the **Manjaro** flavor you chose, click **OK** and start the machine by clicking the big green right arrow icon from VirtualBox's main view.

20. Further, follow the steps provided in the *Pure installation* section, or if you want some special partitioning, from *Manual installation for EFI-based computers (for advanced users)*. You don't have to do any specific **BIOS** setup as **VirtualBox** will work from scratch without special settings.

21. In any case, if you want to do something more for **BIOS** settings, reset the VM and immediately keep pressing the *Esc* key to enter those settings. You will not find the regular BIOS settings, though, and with tens of installations behind my back, **Manjaro** will start from the ISO immediately.

22. When it starts, it is essential to note that sometimes, the resolution will be **800x600**, which is insufficient for any modern OS. Most of the time, VirtualBox will manage the window size automatically, but only after the guest OS graphical mode is started. If issues arise, you might need to change it manually, as shown in *Figure 2.39*:

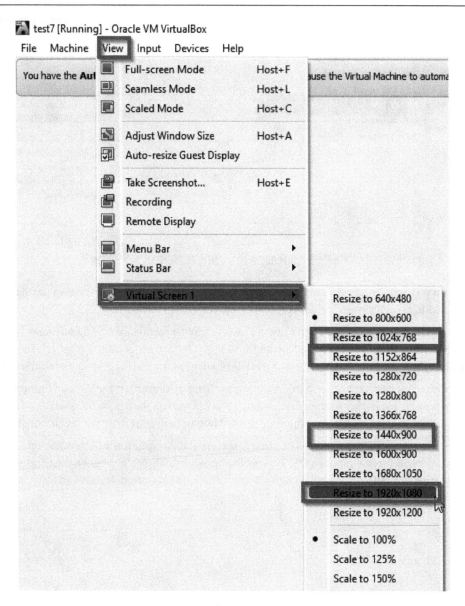

Figure 2.39 – VirtualBox – virtual screen setup

23. Usually, this changes it automatically, and you don't have to bother with setting up **Manjaro**. Also, note that *the mouse integration is excellent* – and as you would probably expect, when your mouse is over the screen, you work with the VM. When your mouse is outside the virtual screen area, your host OS controls are in charge. Despite this, *when the VM is in text-only mode during initial loading, if the control of the mouse is captured, releasing it is a setting currently set to the right-hand Ctrl key*.

24. That's all – everything should now be working. If you experience any issues, search the web – **VirtualBox** is widely used, so solutions for common problems are easy to find.

## Installation in parallel with Windows (also known as Dual Boot)

Here, we will discuss **Windows** versions **10** and **11**. Some tricks might be necessary for older versions, and if you look on the internet, you'll be able to find various guides.

Here's a hint: **Windows 7** might already be installed in **BIOS** mode, so in this case, you will have to look for how to make logical partitions and (most probably) manually install it.

I will not even consider **Windows 8** as it was, by common opinion, a weak and not widely used OS, just like **Windows Vista**. *Both were seriously avoided by colleagues, customers, and big companies I have worked with over the last 12 years.*

**Windows 10**, on the other hand, has been quite good and stable for many years, and so far, **Windows 11** looks good to me.

Whether you install Windows **10** or **11**, know that **Windows** creates more partitions than **C** and **EFI**. One is a small **Microsoft Reserved Partition**, and many times there is a **Microsoft Recovery Tools** partition on many laptops. Some laptop manufacturers add even more recovery partitions.

> **An important hint**
>
> If you see guides that provide instructions on how to use **EasyBCD**, *don't use it on Windows 10 or 11*. It will most probably *ruin the basic GPT table*, resulting in you losing all the HDD information. *Often, recovery is impossible after using* **EasyBCD!**

Manjaro is set up *automatically* for installation along with **Windows 10** and **11**. However, you should always make space for it first.

Open your Windows **Start menu**, type `partition`, and open the **Create and format hard disk partitions** Control Panel service. This will start the **Disk Management** tool. Here, you must first understand which is your *primary hard drive* (in case you have more than one, like me). It will be the one that has **C: drive** in it. Select it from the vertical list and see which partition is marked, as shown in *Figure 2.40*:

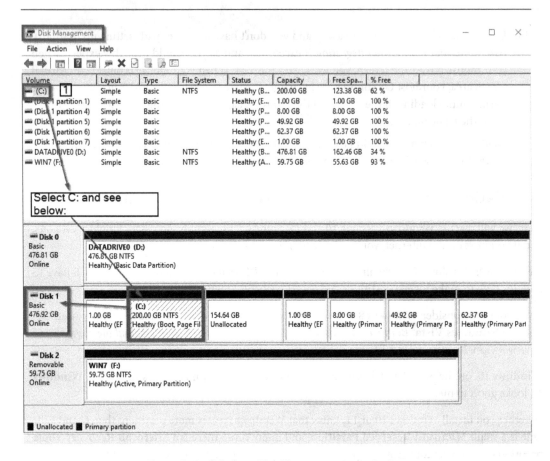

Figure 2.40 – Windows Disk Management – basic view

Now, on the right of **C**, I already have *unallocated space*. But, if I didn't, I'd have to select **C** and *shrink it*. What is **shrinking**? As an example, if your C drive is **200 GB** in size, you can reduce it to **100 GB**, leaving **100 GB** *free/unallocated*. Thus, you then have 100 GB for your Linux installation.

It is essential to leave *an absolute minimum of* **80 GB** on the **C drive** for **Windows** so that if you install a more space-hungry piece of SW, you will have enough space. The pure Windows 11 Pro installation takes about *30 GB minimum*. On my main Windows machine, which runs **Windows 11** and has a lot of SW installed, I have already used 74 GB.

It would be better to have at least **100 GB** for **Windows** and **100 GB** for Linux.

For Manjaro, we know it needs at least **30 GB**, but when we install more SW and modules, for an average user, **Manjaro** will need a minimum of **60 GB**.

*Any partition (especially a primary OS partition) also needs some free space to operate well* – that's why I say Windows needs **100 GB**.

Here's another example: **Disk Management** shows me **C** with a size of **149.36 GB** on this machine. I want to install Linux along with it. I have already used **30 GB** from the fresh **Windows 11** installation I have performed. So, I will leave **80 GB** for Windows and allocate **70 GB** for Linux. How can I do this? In the **Disk Management** area, select **C**, right-click on it, and select **Shrink Volume…**, as shown in *Figure 2.41*:

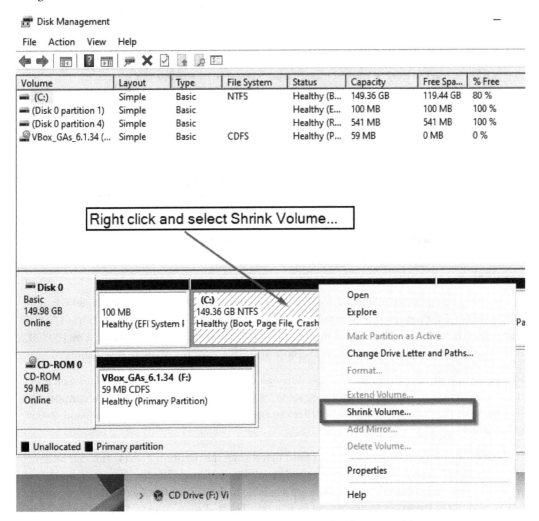

Figure 2.41 – Windows Disk Management – shrinking a partition

Then, after a few seconds of analysis, Windows will tell me the *size before shrinking* and the *available size to shrink to*. As the size is in MB, I will enter **70 GB** (70 x 1,024 = **71,680** MB) and click the **Shrink** button, as shown in *Figure 2.42*:

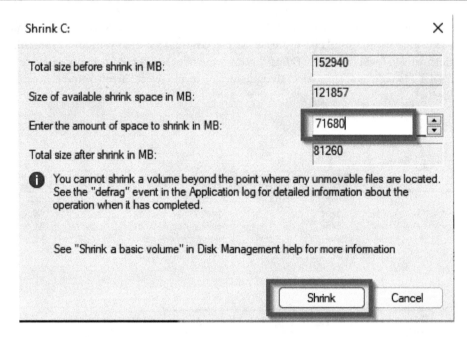

Figure 2.42 – Windows Disk Management – shrinking the partition

And that's all we need to do.

Suppose I have one primary drive, and someone has already shrunk **C** to, for example, **80 GB**, and has created **D** or **E** for data *on this same primary drive*, with a **size of 400 GB**. Let's say **D** has **170 GB** free and **230 GB** used. In this case, I will shrink not from **C**, but from **D**. I would shrink it by **100 GB**, which leaves me with **70 GB** more free on **D**. Those **100 GB** are for my Manjaro installation.

Good – now that we have the space ready, go through the *BIOS/UEFI setup for Installation on a PC* section – in particular, the *What needs to be done for (U)EFI setup* subsection – then connect the **USB stick** with the prepared **Manjaro** installer (according to *Installing on a USB stick*) and restart your computer.

Once your computer boots successfully, you must first format the partition to **ext4**. This can be done *easily under Linux* (under Windows, this can only be done with special tools – *it is strongly advised to avoid it*). Thus, run **GParted** (instructions are available in the *GParted and partition management* subsection) or **KDE Partition Manager** and format the allocated space as **ext4**. To find the partition tool, open the main menu, type `partition`, and run the corresponding application. In it, select your **primary hard drive** with the *unallocated space*, right-click it, and select **New**. How will you know which drive is correct? It will have *the same sizes for partitions* as the drive *you were shrinking* in Windows. In this example, I have **70 GB** of unallocated space that I've prepared on **Windows**. In **KDE Partition Manager**, it looks like this:

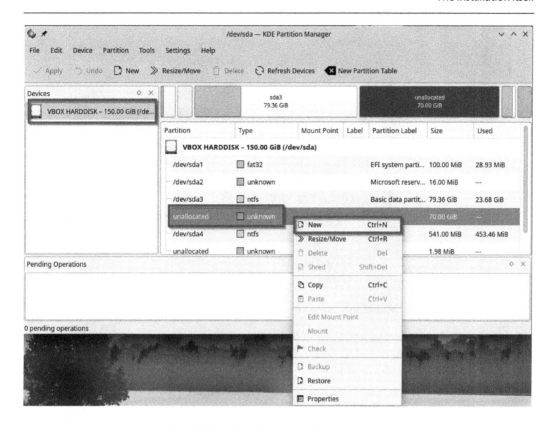

Figure 2.43 – KDE Partition Manager – allocating a new partition

Then, in the pop-up window, select **ext4** under **File system** and leave the other options as-is, as shown in *Figure 2.44*:

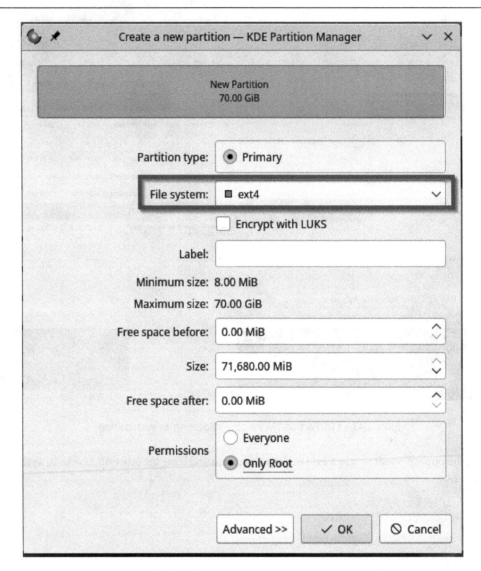

Figure 2.44 – KDE Partition Manager – setting up a new partition

Now, click **OK**. This will close the window. Then, click the ✔ **Apply** button in the top-left part of the window and click **Apply all operations**.

Before you move on, if you haven't read the *Pure installation* section, please do so. Though it is not for dual boot, the main points on *passwords*, *sizes*, and so on are still valid.

Next, start the installer from the **Hello menu** and select your *language*, *zone*, and *keyboard*. When you reach the *main partition window*, select **Install alongside**.

In this example, I have prepared a 70 GB partition for Manjaro, so I will not make a separate **home** partition (as the size is too small). I will also leave the installer to do everything automatically as it works well. Later, from Manjaro itself, I can create a home partition if I want.

Here is my screen and some explanations:

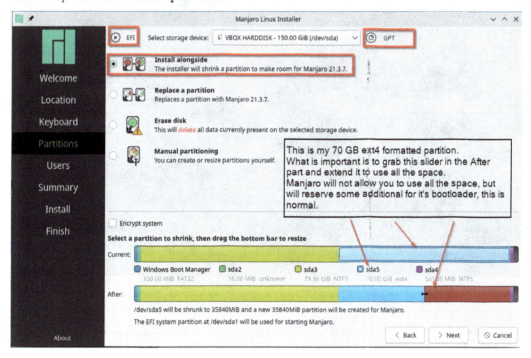

Figure 2.45 – Manjaro Linux Installer – dual boot setup

As explained, once you choose **Install alongside**, select your part of the **HDD**, then *extend the slider* so that *the red part becomes as big as possible*.

Then, go through the other regular screens and click **Install**. You might have noticed that most of my guides are created based on a **VM**. The reason I've done this is so that I can provide screenshots. Taking them with a camera would cause a lot of issues. Apart from on real machines, I have installed Manjaro on VMs on *four different systems*, including this very laptop I'm currently writing on. Everything presented on the VMs *has been tested beforehand on real machines*.

*There is only one last special hint*: I have experienced peculiar behavior a few times. After installing Manjaro, *one restart is not always enough*. Sometimes, you have to *enter the BIOS/UEFI machine's boot menu* and, from the **EFI boot options**, manually choose **where to boot from**. Once you boot from Manjaro's EFI partition successfully, on each (re)start, the bootloader screen of choice for **Windows** or **Manjaro** will appear for you. So far, it has worked rock-solid for me on all four machines. It looks as follows:

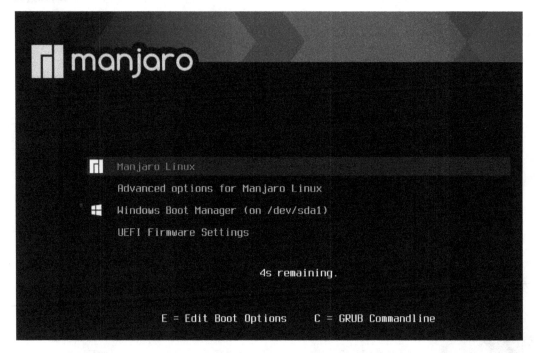

Figure 2.46 – Boot menu of the installed system with Manjaro and Windows

Be fast – *the default timeout is 5 seconds* and *loads your last choice*. Once you start moving with the arrow keys, the timer will be disabled. At this point, you can boot to the OS of your choice.

When you start using the machine, you might notice one strange thing – *the clock might have issues when you switch between Windows and Linux*. The reason is that, by default, in Windows, time is stored on the local machine *in local time*. Linux assumes the time is stored in *UTC time*. This can be frustrating as internet browsers use this time to *validate web page certificates and validity*. In general, the timestamps can be critical for both OSs and may lead to peculiar problems. So, you must tell **Windows** *to use UTC* or tell **Manjaro** *to use local time for* **Real Time Clock (RTC)**.

I prefer to change Windows. This is as easy as doing the following:

1. Open the **Start menu** and type `regedit`. From the search results, open the **Registry Editor** application. Click **Yes** when you're prompted for system changes.

2. When the program opens, either navigate manually to the following path on the left or copy and paste this path into the address field:

```
Computer\HKEY_LOCAL_MACHINE\SYSTEM\CurrentControlSet\Control\
TimeZoneInformation
```

You will see the following:

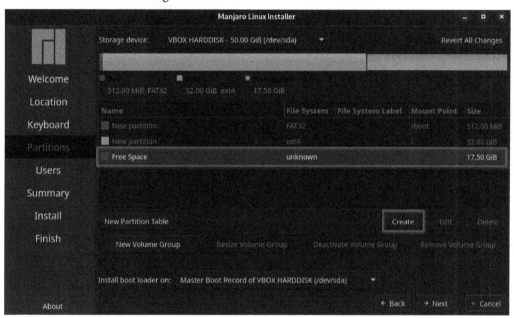

Figure 2.47 – Setting up the time/clock in Windows for dual boot systems

3. The **RealTimeIsUniversal** key (one of the listings on the right) will *most probably be missing*. Right-click on the white spot under the listing on the right, select **New | DWORD (32-bit) Value**, and name it `RealTimeIsUniversal`. Double-click on it and set its value to `1`.

4. *Reboot* Windows.

5. This will fix it. If you experience more problems, search the web for more information, though this solution is quite common and has never failed me.

## Installation on ARM platforms

This book is centered on the **desktop** versions of **Manjaro**, and its volume doesn't allow me to write sections on **Raspberry Pi** and other **ARM** platforms. It would take *several hundred pages more* to add, for example, **RPI4B**. Still, it is important to mention that many articles and guides exist on **Manjaro** for **ARM** platforms. One of the reasons to like **Manjaro** is that the amount of supported **ARM** platforms has grown steadily in the last 4 years. It started with only three, and now it **has grown to over eight**, each of which has support for several modifications. This results in *18 special distributions for platforms*! One of them is **Phosh**, which is a special mobile platform for the **PinePhone**.

The ARM versions are challenging since *ARM microcomputers* are significantly weaker than desktops and have different HW and drivers. They require severe modifications, such as removing as many unnecessary services as possible, adding the corresponding custom HW and drivers layer, making a custom filesystem snapshot, and more.

A few years ago, I wrote a series of articles on Manjaro on an RPI4 B+ with a touch screen (`https://triplehelix-consulting.com/manjaro-arm-linux-setup-on-raspberry-pi-4/`). It explains the *display* I got, *HW considerations*, *heat dissipation*, and the **Xfce** flavor *installation*.

Around 1 year ago, I ported the **IMGUI** library and a basic application with almost no effort to **ARM**. I could *cross-compile* and also compile *directly on the RPI4B+* under **Manjaro Xfce**. The **Manjaro ARM team** did a magnificent job making, supporting, and extending Manjaro versions for ARM.

Here is the Manjaro forum for ARM: `https://forum.manjaro.org/c/arm/100`. The blog section at `https://blog.manjaro.org/` always has separate posts for Manjaro ARM.

Searching through the web, you can easily find several guides, such as the following links:

- `https://9to5linux.com/hands-on-with-manjaro-linux-arm-on-raspberry-pi-4-a-gem`
- `https://linuxhint.com/install_manjaro_raspberry_pi_4/#:~:text=Once%20you%20have%20Raspberry%20Pi,downloaded%20and%20click%20on%20Open`
- `https://www.zdnet.com/article/hands-on-manjaro-linux-on-the-raspberry-pi-4/`

Apologies for not providing more links, but again, this book is for the desktop edition, and covering such a wide variety of technologies would take up another book.

## Summary

In this chapter, we covered all possible topics regarding installation, including the various flavors, the preparation steps, USB stick preparation, detailed BIOS and UEFI guides, and finally, the various ways to install Manjaro.

In the next chapter, we will cover the basic settings of the three official flavors: **Xfce**, **KDE Plasma**, and **GNOME**.

# 3

# Editions and Flavors

Now that we have installed **Manjaro,** it is time to see what each of the official flavors offers. The graphical environment brings with it numerous settings and specifics. This chapter provides short descriptions of the most important ones. However, before we step into this, I will give you a few hints on a few good things to do after any fresh Manjaro installation.

The following topics will be covered in this chapter:

- A few important steps after installation
- The Xfce edition and settings
- The KDE Plasma edition and settings
- The GNOME edition and settings
- The other editions – Cinnamon, Lxde, Mate, and so on

Here are some general points:

- The settings we will cover here are based on the **graphical user interface** (GUI). This means these are clickable options and refer to how the OS is set up with the mouse.
- There are thousands of GUI options in each Manjaro flavor. Here, we will cover as many basic points as possible.

## A few important steps after installation

### For beginners

Before we move on, it is essential to note what a **root** (also known as a **superuser**) account is. This is the main **administrator account** on Linux. On all distributions, it is the one that starts the OS, mounts filesystems, loads the graphical environment, and has full privileges for managing and changing the system. After this, you can log in with a *regular* **user account** such as *MaryJ* or *JohnS*, which has fewer privileges. Even if a user account has elevated privileges, you might need to provide the **root** account **password** for some specific commands and **software** (**SW**) with low-level system access.

If you have followed my advice during the installation, you have set the **Use the same password for the administrator account** option. As a result, you will not need to be concerned with whether you need the **root** or **user** account **password** when requested. If you haven't done this, know that regular SW installations and all **sudo** actions require the user password since your user is added to several administration-related groups. To learn more, please read *Chapters 7* and *12*.

The first stop here is the **Manjaro Hello** menu. You saw it just before the installation. It is enabled by default on any new Manjaro installation and looks like the installation launcher menu. On an installed system, it will have a button called **Applications** instead of **Install**. It allows you to apply changes to the currently installed SW directly. If you want, you can skip this step for now and run it later. However, keep in mind that it will make your life easier because instead of installing or uninstalling each app separately, you have a great set of tens of applications to choose from to be added or removed in bulk. *Figure 3.1* shows what this looks like for **KDE Plasma**:

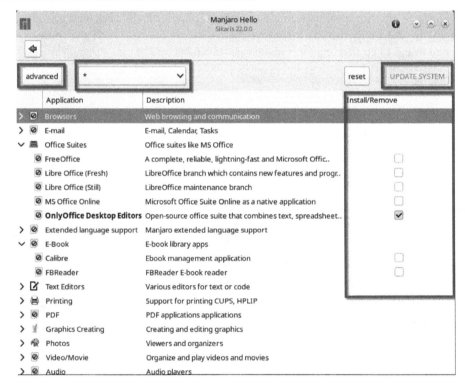

Figure 3.1 – Manjaro KDE Plasma Hello – default applications install/remove view

Select any application you want to add or remove, then click the **UPDATE SYSTEM** button. The **advanced** button provides a broader set of applications.

Remember that you can also open the **Manjaro Hello** system modification app later, even if you deactivate it on startup (by turning off the **Launch at start** option).

In **Manjaro Hello**, especially for **GNOME**, we also have a **GNOME Layout Switcher** button. It provides a complete modification of the default GNOME layout. This can also be executed separately later.

Though Manjaro is great and has already been tweaked a lot, there are always a few steps to take after installing any OS on any PC.

Before we move on, knowing about the **Meta** and **Super** keys is essential. Both originated in the 1960s. Due to a complicated history, what Windows users know as the **Windows key** is referred to in different environments as the *Meta* (**KDE**) or *Super* (**Xfce**) key. It normally opens the main menu (also known as the **application launcher**).

The following are a few general steps to complete after a fresh installation:

1.  Change *Manjaro Mirrors* so that it's using the ones closest to you, not global ones (for the fastest updates and SW download).

    To do this, open the application launcher, write Add, and open **Add/Remove Software** (it's the same for KDE Plasma, GNOME, and Xfce – its name actually is **Pamac GUI**). Click the three vertical dots ⠿ or lines ≡ to open the options, and from there, choose **Preferences**. You must provide your user password here as this changes the OS settings. In the default **General** tab, scroll to **Use mirrors from** and choose your country or a close one that you have the fastest connection to. Manjaro has servers on all continents but not in all countries. Then, click the button below this dropdown named **Refresh Mirrors**. *Figure 3.2* shows an example from Xfce:

Figure 3.2 – Pamac GUI – setting mirrors via Preferences | General

2. Close the **Preferences** menu, and again, in the **Add/Remove Software** area (the Pamac GUI), go to the third tab, **Updates**. Manjaro is constantly updated – the older the release you have installed, the more updates you will have. Even if you have a newer release after a week, it might have at least some pending updates.

If you have update issues with a message stating `invalid or corrupted package`, open the three vertical dots ⦚ or lines ≡ menu again and choose **Refresh databases**, then try the update again.

3. Again, in **Preferences**, you can choose how frequently you want to be notified about updates. Let's set it to *every day* or *every week* and not enable **Automatically download updates**.

4. Depending on your distribution, go to the **Preferences** option of the Pamac GUI again, select the third tab (**Third Party**), and enable **Flatpak** support. This allows us to download and install applications built with such packages so that we will be able to use more SW.

5. Add the **Microsoft TrueType fonts** if you are working with Microsoft Office documents. For this, you must first enable **AUR support** (from the **Third Party** tab of **Preferences**), refresh the databases again, and then search for `ttf-ms-fonts` via the looking-glass icon.

6. By default, Manjaro comes with multiple applications. Although we will look at most of them, you should know that we always have **Mozilla Firefox** for web browsing and **Mozilla Thunderbird** as a mail client. If you want **Skype**, **Viber**, **Spotify**, **WhatsApp**, or any other application, we will look at how to install them in *Chapters 5* and *6*.

7. To finish this section, I will give you a first glimpse of the **Terminal**. Once you have installed Manjaro (or any Linux distribution), you should know that the Terminal is super easy to use. Open the application launcher (with the *Super* or *Meta* key), type `terminal`, and press *Enter*. With the following two commands, you will get detailed information about your Linux kernel and your distribution, along with its version:

```
uname -a
lsb_release -a
```

With the installation setups succesfully completed, let us now see the various Xfce editions and their settings.

## Xfce edition and settings

As you may recall, Xfce is one of the lightest Linux GUI environments. It is old and mature, initially released in 1997. Despite being light, it has an excellent set of options. Manjaro 23.0.0 comes with **Xfce4** version 4.18. The official documentation page is `https://docs.xfce.org/`.

## Xfce desktop and bottom panel bar right-click settings

To start, on a freshly loaded and logged-in system, right-click on the desktop and select **Desktop Settings…**. The **Background** tab is self-explanatory. Check its options, but also check the second and third tabs – **Menus** and **Icons** – as shown in *Figure 3.3*:

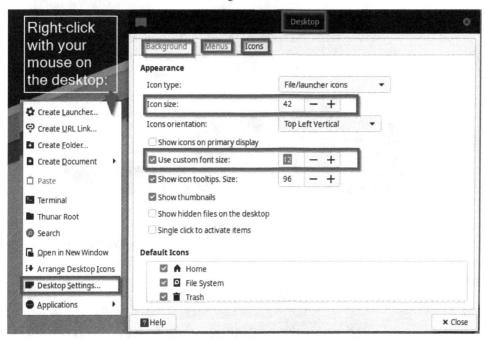

Figure 3.3 – Xfce Desktop Settings… and Icons tab

Regarding the **Menus** tab, it is interesting to check **Window List Menu**, which is currently enabled by default. When we middle-click the mouse on the desktop, it will show a listing of all **workspaces** (virtual desktops) and their opened applications. *"What is a workspace?"* you might ask. It is like having several desktops. An example of a workspace is provided in *Figure 3.4*:

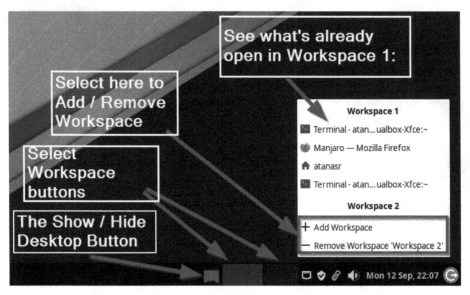

Figure 3.4 – Xfce workspace options and right-click menu

The next exciting buttons are on *the bottom-right part of the screen*. Clicking/double-clicking or right-clicking them will provide you with information or options, as shown in *Figure 3.5*:

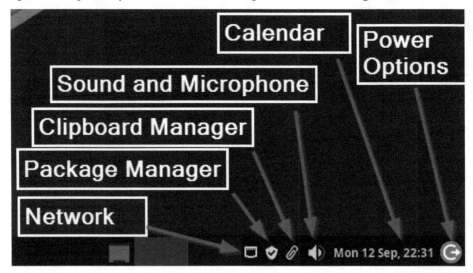

Figure 3.5 – Xfce default panel items

## Xfce application launcher and settings

Our next topic is the **application launcher** (the MS Windows name is the *Start menu*), which in Xfce is called the **Whisker menu**. It has a simple layout with a good integrated search. If you type any keyword, such as Network, Internet, Manager, or Settings, it will immediately give you the correct suggestions. It also has some basic categories, and on top, icons for **Settings** and session control (Lock Screen, Power Control, and Switch User), as shown in *Figure 3.6*:

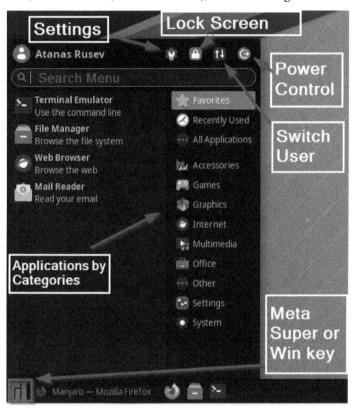

Figure 3.6 – Xfce Whisker menu view and functions, which can be opened with the Meta or Super key

All further settings are nicely arranged in the **Settings** menu, as shown in *Figure 3.7*. However, if you type any of the names from it in the **Whisker menu** *search field*, such as Desktop, Tweaks, or Panel, you will have a direct application to begin with. We can see them all in *Figure 3.7*:

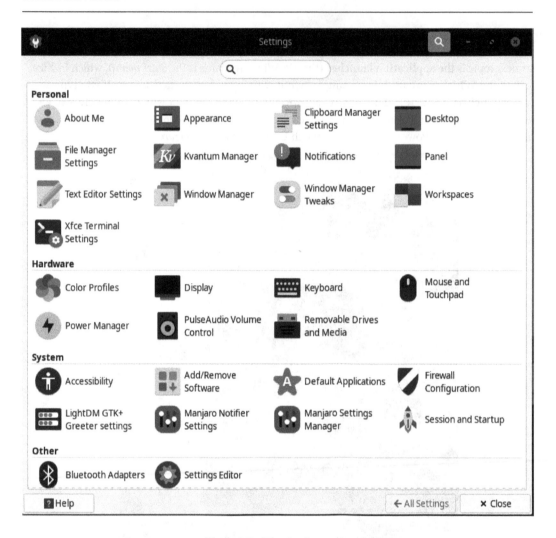

Figure 3.7 – Xfce Settings area

For most of them, as soon as we change some settings, we will see the results on the existing opened windows, icons, and elements.

Let's start with **Appearance** – again, whatever you change here will have an immediate visible effect. Here, we have a **Style** tab to change the window borders. Further, there is an **Icons** tab, which also has different styles. While the previous two tabs are nice, and you can easily choose from the available options, the **Fonts** and **Settings** tabs are essential for people with specific monitors or glasses. In the **Fonts** tab, we can change the **dots per inch** (**DPI**) setting of the current default fonts (using the keyboard arrows helps). This menu is shown in *Figure 3.8*:

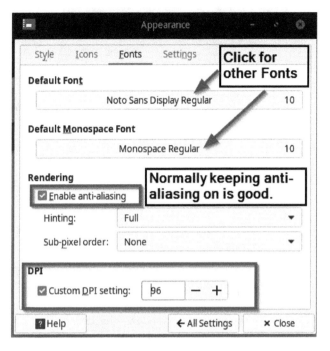

Figure 3.8 – Xfce Appearance window – the Fonts tab

In the **Appearance | Settings** tab, I always keep the current default image options as I believe they are helpful. Regarding the **Enable event sounds** option – it's up to you. The **Enable editable accelerators** option is for keyboard shortcuts and fast editing. I changed shortcuts by mistake when I tried it (as I generally click and type quickly), so I don't enjoy this one. Regarding **Window Scaling**, at the time of writing, only **2x** is supported. This menu is shown in *Figure 3.9*:

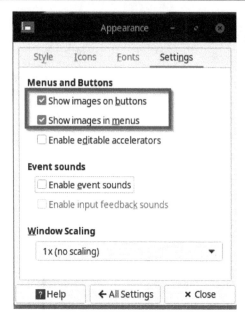

Figure 3.9 – Xfce Appearance window – the Settings tab

Another nice feature is pop-up tooltips while *hovering* (keeping the mouse over) on an option. For some menus and options, hovering over it often shows a small popup explaining more about the given option, as shown in *Figure 3.10*:

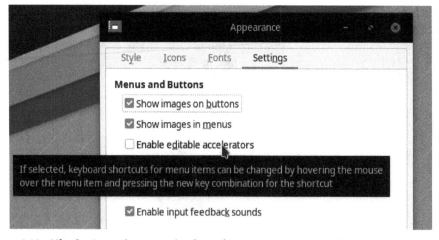

Figure 3.10 – Xfce Settings tab – example of an information popup when hovering over an option

Now, click the **All Settings** bottom to return to the main menu and choose **Window Manager**.

We have four tabs here. In the **Style** tab, the theme doesn't influence the view much. However, you can change the order of the standard buttons for *menu, title, minimize, maximize,* and *close* here. This can be seen in *Figure 3.11*:

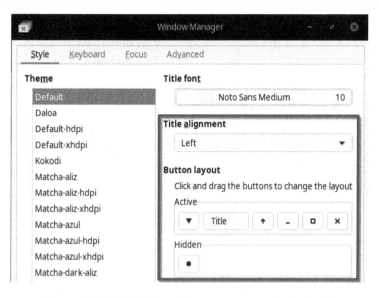

Figure 3.11 – Xfce Window Manager settings – the Style tab

The next tab, and one of my favorites, shows the default **keyboard shortcuts**. I love *Alt+F10* to toggle maximize and random size and *Ctrl+Alt+D* to show/hide the desktop. This can be seen in *Figure 3.12*:

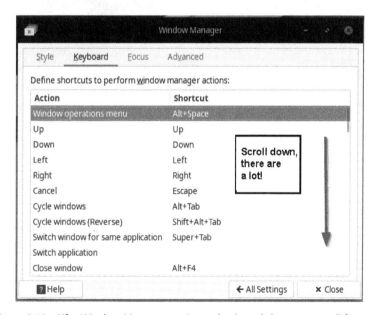

Figure 3.12 – Xfce Window Manager settings – keyboard shortcuts, scroll for more

It is essential to know that, at the time of writing, *Alt+Tab* will switch between different workspace applications by default. Hence, if you switch to another application in another workspace, this will switch the workspace as well. It is worth checking out all the shortcuts.

Next, we have the **Focus** tab. It defines what happens when we *hover* over different open windows. I like the defaults, but this is the place if you want to change them.

Finally, we have the **Advanced** tab. It contains the settings for *windows snapping* (automatically aligning one window to another) while moving them. I prefer snapping both to screen borders and other windows. We can also wrap workspaces and double-click action options.

For the following settings, click the **All Settings** button again and go to **Window Manager Tweaks**. On the **Cycling** tab, I like to disable **Skip** for **Skip Pager**, enable *hidden windows*, and *cycle through windows on all workspaces* (if I have enabled the workspaces). The **Workspaces** tab has the interesting option of using mouse scroll to change workspaces when hovering over their menu bar area. This is shown in *Figure 3.13*:

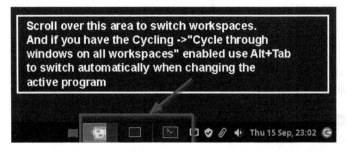

Figure 3.13 – Xfce workspaces – scroll to switch them or do it automatically via Alt+Tab

In the last tab of the **Window Manager Tweaks** area, called **Compositor**, we have some great settings for **opacity** and options for what happens with windows when switching. Opacity is the quality of lacking transparency, so this means transparency control. Check them out if you like.

The file manager application on Manjaro Xfce is called **Thunar**. When you run it (by looking for File Manager or Thunar in the **Whisker menu**), you can edit its settings from its menu via **Edit | Preferences**. The same menu is available from the **Settings** panel and is called **File Manager Settings**.

Once again, go back to the **Settings** main view and find **Notifications**. We have a **Do not disturb** mode option in the **General** tab, which means we're only notified of urgent notifications. In the **Appearance** tab, we have a **Show Preview** button. This preview is also activated when you change the notifications theme. In the **Applications** tab, we can control each app's notifications separately. The fourth tab is about notifications logging (this is disabled by default).

There are a few more essential **settings** that I recommend you quickly look at.

In the **Hardware** section, we have **Power Manager**, with crucial points divided into five sections: **General**, **System**, **Display** (when to sleep and switch the display off), **Security** (for locking the session), and **Devices** (for laptop charge monitoring).

Again, in the **Hardware** section, we have **PulseAudio Volume Control** with the **Playback**, **Recording**, **Output**, **Input**, and **Configuration** tabs. Further, in **HW**, we have **Display** (with the ability to save the current or selected settings related to some displays in profiles), **Keyboard** (also with several interesting general keyboard shortcuts); **Mouse and Touchpad**, and **Color Profiles**.

The last option in **HW** is **Removable Drives and Media**. It controls the automatic mounting of plugged-in storage devices (HDD, USB sticks, and others) and, despite being rare nowadays, a *burning CDs* control, **Cameras**, **Printers**, and **Input devices**.

The **System** section also has several important parts. We already know about **Add/Remove Software** (and it was reviewed in detail in *Chapter 5*). Here, you will also find the **Default Applications**, **Firewall Configuration**, **Lock screen** (in **LightDM Greeter Settings**), **Manjaro Settings Manager** (for the locale, language, user accounts, time and date, and so on), and **Session and Startup** (including applications started at system start) settings.

The last exciting settings app is **Bluetooth Adapters** in the **Other** section.

To close the **Xfce** subsection, we will look at the right-click menu of the application launcher (**Whisker menu**) and the bottom panel (or menu bar). The application launcher allows us to open the **Edit Applications** and **Properties** menus. While the first is more for advanced users (with options for launching each given application), the second is for everyone. It controls the Whisker menu's appearance, behavior, commands, and search actions. *Figure 3.14* shows how to access these two menus:

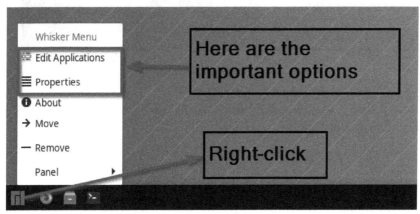

Figure 3.14 – Xfce Whisker menu right-click options

The last point in this section is **Panel Preferences**. You can use the **Add New Items** option for it, but I would recommend making any changes through **Panel Preferences**, as shown in *Figure 3.15*:

Figure 3.15 – Xfce Panel – right-click menu

You can select the panel's style, location, and appearance, add and remove items (launchers for apps, separators, and other elements), and reorder them easily. This can be seen in *Figure 3.16*:

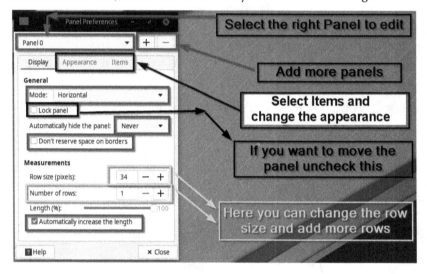

Figure 3.16 – Xfce Panel Preferences

Last but not least, here are two videos that provide an overview of some of the settings mentioned here, as well as a few other options:

- `https://www.youtube.com/watch?v=91G08NvJLQE`

- `https://www.youtube.com/watch?v=_o_cpZQIIpM`

# KDE Plasma edition and settings

The **KDE Plasma** environment is famous for having the richest settings. Manjaro KDE 23.0.0 comes with **Plasma 5.27.5**. Let's take a look at it.

**KDE** is an extensive ecosystem – apart from the Plasma desktop, it provides an application-building framework, toolkits, and many native applications (one more reason why it is rich!). Hence, the documentation for the whole project also includes tutorials on its many applications. The primary location is the **Tutorials** section (https://userbase.kde.org/Tutorials) of the **KDE UserBase** (https://userbase.kde.org/Welcome_to_KDE_UserBase). For the applications, check out https://userbase.kde.org/Applications.

To start, let's take a look at the default notification and control icons on the main menu bar, as shown in *Figure 3.17*:

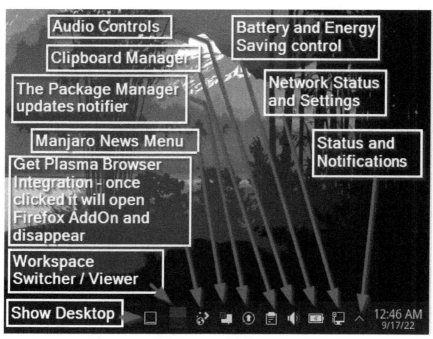

Figure 3.17 – KDE Plasma control icons on the panel

Again, as on Xfce, right-clicking on the icons provides more options and access to configuration menus. When *hovering with the mouse*, the *information popups* work here as they do in most menus and desktop icons.

## KDE Plasma desktop right-click menu

Right-clicking on the desktop gives us the menu shown in *Figure 3.18*:

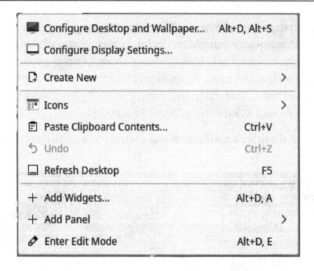

Figure 3.18 – KDE Plasma – Desktop right-click menu

The first menu option, **Configure Desktop and Wallpaper**, opens a rich configuration, which at the time of writing, by default, is set to **Folder View Layout**. I prefer it as it provides more configuration options compared to **Desktop Layout**. The **Wallpaper** options are pretty rich compared to Xfce – look at them for yourself. The **Mouse Actions** options are as shown in *Figure 3.19*:

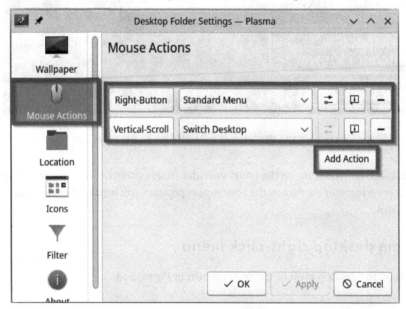

Figure 3.19 – KDE Plasma – configuring the mouse actions

You can add more actions if you have a mouse with more buttons. **Standard Menu** is the right-button option shown in *Figure 3.19*, but we can also choose **Switch Desktop**, **Switch Activity**, **Switch Window**, **Paste**, or the application launcher. The **Location** options are only interesting if you want to move the default desktop directory location. The **Icons** options are also rich – here, we have **Arrangement**, **Sorting**, **Size**, **Label**, **Hovering**, and others.

**Configure Display Settings** is the next option in the menu when right-clicking on the desktop. On the **Display Configuration** page, we have **Resolution**, **Orientation**, and **Scaling**. **Compositor** controls the smoothness of menus. Usually, the default settings work fine, but if you deactivate it, you will have more resources, which is good for weaker PCs.

When you are with the default **Folder View** desktop layout (set in **Configure Desktop and Wallpaper | Wallpaper | Layout**), the **Create New** option exists in the desktop right-click menu. It provides the standard options shown in *Figure 3.20*:

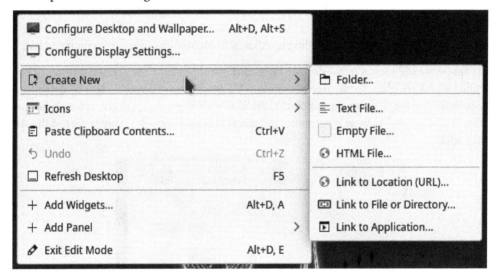

Figure 3.20 – KDE Plasma – desktop right-click – Create New

For non-Linux users, **Link to Application** might sound similar to creating a Windows application shortcut, but in the Linux world, this is a custom command with a lot more options.

The easiest way to create a shortcut icon on the main menu bar for an app is to run it, right-click on its icon, and select **Pin to Task Manager**. This will keep the icon even when you close the app. You can also right-click its icon inside the main menu (application launcher), which provides more options. This is shown in *Figure 3.21*:

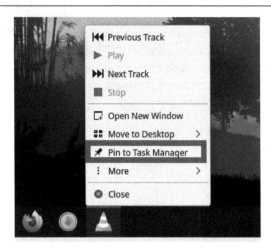

Figure 3.21 – KDE Plasma – Pin to Task Manager

You can remove it in the same way as adding it – that is, by right-clicking on the main menu bar.

The next option we will look at in the desktop right-click menu is one of the greatest features of KDE and is called +**Add Widgets**. At the time of writing, KDE Plasma provides *66 widgets* by default, and we can download more. We can choose clocks, calculators, timers, sticky notes, system monitors, calendars, HW controls (such as Bluetooth), and many others. *Figure 3.22* shows part of the **Add Widgets** panel:

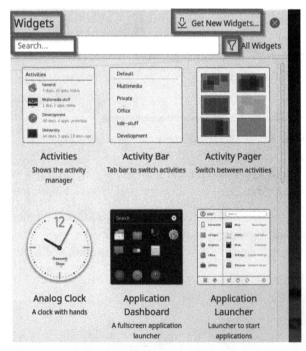

Figure 3.22 – KDE Plasma – Add Widgets panel

The final fast link in the desktop right-click menu is the **Add Panel** feature, which adds more menu bars. It is shown in *Figure 3.23*:

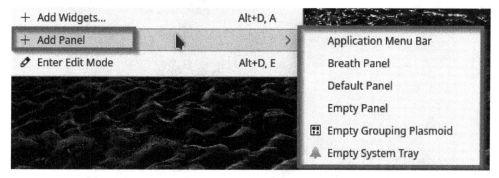

Figure 3.23 – KDE Plasma – Add Panel capabilities

When you add application menu bars, they will be added *clockwise, first on the left*; adding another one will place it on the top edge. Further, you can move, resize, or edit any of these panels by clicking on a panel empty part and choosing **Enter Edit Mode**.

The last desktop right-click option is **Enter Edit Mode**, which allows us to *move* and *resize* all widgets and panels. Once you have finished, right-click on the desktop and select the text that changed to **Exit Edit Mode**. Removing a panel is only possible in **Edit mode**, which is designated by additional margins around the desktop elements, as shown in *Figure 3.24*. If the margins are present, you have to exit it by right-clicking again explicitly:

Figure 3.24 – KDE Plasma – Edit mode with margins highlighted

## KDE Plasma application launcher

Finally, we reach the application launcher or main menu (named the *Start menu* in Windows). In KDE Plasma, it is related to the **Meta** or **Super** (that is, *Windows*) key. It is presented in *Figure 3.25*:

Figure 3.25 – KDE Plasma – main menu or application launcher with highlights

We have a window on the right for the category that was chosen on the left. In addition, the **Places** button gives us a direct link to multiple user locations, such as **Desktop**, **Documents**, **Downloads**, **Music**, and so on. Right-clicking on the Manjaro-branded launcher icon provides us with the menu shown in *Figure 3.26*:

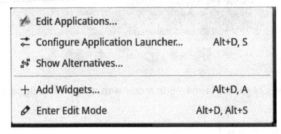

Figure 3.26 – KDE Plasma – application launcher right-click options

The first option, **Edit Applications**, opens the **KDE Menu Editor** area. It allows us to edit groups, menus, and individual program launch settings. You can also add custom commands and SW. It is shown in *Figure 3.27*:

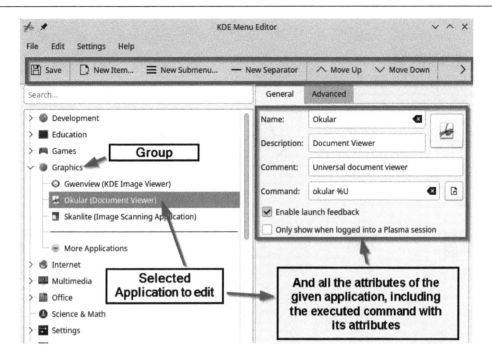

Figure 3.27 – KDE Plasma – KDE Menu Editor with application management

The next right-click menu button option is **Configure Application Launcher**. This allows us to change the application launcher's view and assign a custom shortcut.

**Show Alternatives** is a superb option in the launcher right-click menu that provides different application launcher looks and organization with a graphics preview. **Add Widgets** and **Enter Edit Mode** are the same menus that were presented earlier.

Regarding the basic OS settings, open the application launcher, type `Settings,` and look for **Manjaro Settings Manager**. This will provide you with basics such as **Locale Settings**, **Language Packages**, **Time and Date**, **User Accounts**, and a few others:

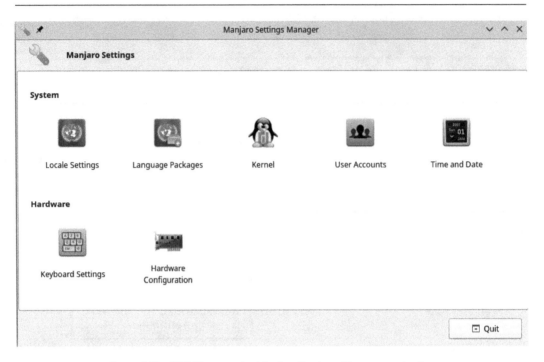

Figure 3.28 – KDE Plasma – the Manjaro Settings Manager main view

To open the same search as in the application launcher, press the *Alt*+spacebar global shortcut. This will open the **KDE Runner** direct search in the top-middle part of the screen. Type your search term (for example, `Settings` or `Network`) to open any related application.

## The KDE Plasma System Settings area

The final stop is the greatest of all – the KDE Plasma **System Settings** area. To open it, search for `System Settings` in the application launcher; its icon is shown in *Figure 3.29*:

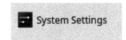

Figure 3.29 – KDE Plasma – the System Settings icon in the application launcher menu

This will open the one-stop shop for all Plasma settings *with integrated search* and tens of custom settings for the graphical environment. One of its nicest features is that you can preview the effects of the settings for all graphical options in the **Appearance** and the **Startup and Shutdown | Login Screen** areas. *Figure 3.30* shows its home page view:

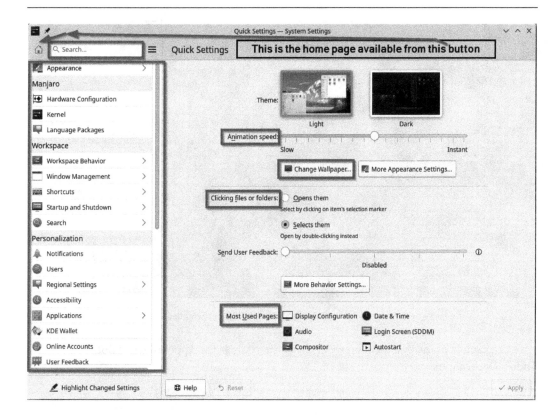

Figure 3.30 – KDE Plasma – the System Settings home page view with highlights

Pay attention to the left of the page for different sections such as **Manjaro**, **Worskpace**, **Personalization**, and others.

I recommend going through all of them at least once as you might find many useful and interesting options. We will look at a few of them.

In **Workspace | Startup and Shutdown | Desktop Session**, we have the **Session Restore** section. By default, it is set to *When session was manually saved*. The application launcher has a left arrow button for user and session control in its bottom-right corner. This opens a menu with a **Save Session** option, as shown in *Figure 3.31*. Suppose you have opened applications (for example, Firefox, the Terminal, File Manager, VLC player, and so on) when you press the **Save Session** option. All these apps will be *loaded automatically* upon the next logoff/login, instead of you getting an empty desktop:

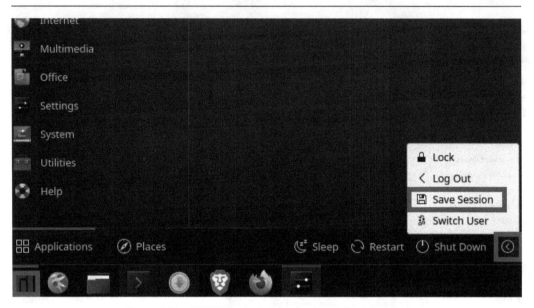

Figure 3.31 – KDE Plasma – System Settings – Save Session

To see an example of the graphical preview, check the settings for **Appearance | Global Theme | Window Decorations** and **Colors**, as shown in *Figures 3.32* and *3.33*:

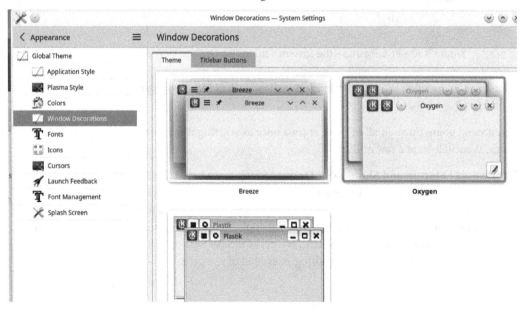

Figure 3.32 – KDE Plasma – Window Decorations – System Settings

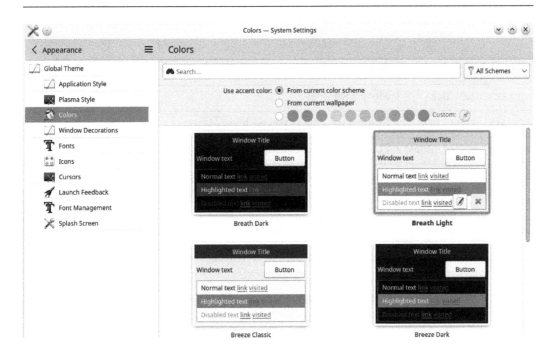

Figure 3.33 – KDE Plasma – Colors – System Settings

Check them all out to get the best of Plasma. Some settings will require your user password. In addition, some might require you to log off and on.

To close this topic, **System Settings** contains sections for **Kernel** version switch; **Hardware**, **Input Devices** (keyboard, mouse, touchpad, and game controller), **Display and Monitor**, **Audio**, **Power**, and **Bluetooth**; touchscreen configuration in the **Workspace Behavior** page, **Network**, and many other options. Apart from taking at least a brief look at each, use the search functionality to look for topics – this will speed up your navigation.

## KDE Plasma general keyboard shortcuts

The last thing we'll mention about the KDE Plasma environment is its excellent shortcuts set and hints. You can find a nice list of the most common shortcuts at `https://community.kde.org/KDE_Visual_Design_Group/HIG/Keyboard_Shortcuts`. However, Michael Tunnel explains a few tweaks in his article at `https://frontpagelinux.com/tutorials/17-kde-plasma-features-that-you-didnt-know-about/`. The article also contains a YouTube video that's worth checking out.

*Tables 3.1* and *3.2* show a small portion of the most important ones:

| Press What? | To Get This Effect |
|---|---|
| *Super* key + left-click | Move a window. |
| *Super* key + right-click | Resize a window as you wish. |
| *Super* key + = or - (minus) | Zoom in and out of areas of the desktop. |
| *Super* key + any arrow key | This tiles the windows in the desktop's top, bottom, left, or right half. Combining adjacent arrows tiles the window in one-quarter of the desktop. |
| *Ctrl + F1* (or *F2, F3,* or *F4*) | This switches the desktop – only up to four desktops. |
| *Alt* + spacebar | Opens the **KRunner** menu – that is, the application launcher search. |

Table 3.1 – KDE Plasma keyboard shortcuts 1

The next set of options provides live information – so you can watch the different windows in real time:

| | |
|---|---|
| *Ctrl + F10* | To show all currently open (including minimized) windows of applications in all workspaces and switch to any of them |
| *Ctrl + F9* | To show all currently open applications in the current workspace |
| *Ctrl + F8* | To show all workspaces tiled and allow you to move windows between them |
| *Ctrl + F7* | To show all instances of one program no matter the workspace – if you have three open Terminals in three different workspaces and use this shortcut, it will show you the three on one screen, and you can switch between them |

Table 3.2 – KDE Plasma keyboard shortcuts 2

For more customization, check out **System Settings** for the **Shortcuts** section.

> **Important note**
>
> There is an older, strange behavior of KDE, which, when present, is annoying. In the past, sometimes, KDE used to lose its *Super* key binding to open the application launcher. Other distributions also reported the problem. If this happens to you, in case it is not fixed yet, for **Manjaro**, the solution is simple. Right-click the application launcher, open the **Configure Application Launcher** area, choose **Keyboard Shortcuts**, and click the **None** keyboard shortcut. Then, use the *Super+A* combination. Finally, click **Apply** and close the **Application Launcher Settings** area. Now, KDE will open the menu, even if only the *Meta* key is pressed.

Luckily, apart from this problem, KDE Plasma remains one of the greatest and most versatile graphical environments. While writing this chapter, I extensively tested all the options presented in this version. I also have a machine that has had Manjaro KDE Plasma installed for a few months, so I can say that the distribution is stable and magnificent.

For a short overview, I would recommend going to YouTube and searching for `Manjaro KDE Plasma`. Always look for newer videos, such as these:

- `https://www.youtube.com/watch?v=rYHOCcP0JCE`

- `https://www.youtube.com/watch?v=MGeTUUlQEUg&t=84s`

- `https://www.youtube.com/watch?v=hIDh2HiV_dg`

# GNOME edition and settings

The last official release is **GNOME**. It is also an old and mature system that's been in development since 1997. The community initially wanted to avoid the Qt libraries, but they support it now. The **Manjaro** GNOME release we will look at is *23.0.0*, while the **GNOME** version is *44.1*. Though some screenshots are from **Manjaro** *21.3.7*, all descriptions have been checked with **GNOME** *44.1*.

**GNOME** is a modern-looking graphical environment. Before we start, if you remember **Manjaro Hello** and have already disabled its launch at startup, when you open it for GNOME, it will feature an additional button called **GNOME Layout Switcher**. Clicking it will provide you with four options for the layouts of the graphical environment: **Manjaro**, **Traditional**, **Tiling**, and **GNOME**. The layout switcher is also available as a separate application – press the *Super* key and search for `Layout`.

Manjaro and GNOME layouts are pretty similar. In *Figure 3.34*, we can see a description of the top information bar and the Dash bottom menu:

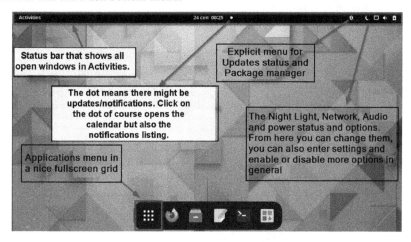

Figure 3.34 – GNOME flavor main view with highlights

The **GNOME layout** shows the **Dash** bottom menu onscreen *only* if you press the *Super* key. The **Manjaro layout** hides it *only* if you maximize the currently opened application. That's why I will keep the **Manjaro layout**. It is essential to point out that many online sources use the name **Dock menu** for the **Dash** bottom menu.

The **traditional layout** is a regular bar below, with a launcher menu on the left and status and control icons on the right – some people say it looks like Windows 7.

In the **tiling layout**, all the windows are tiled automatically on the left or right, depending on the space (and there are settings for them, of course). This mode can be handy for big screens with multiple opened windows but is rarely comfortable for regular daily work.

**GNOME** has had one primary purpose for a long time – it maintains a kind of minimalism (a bit like macOS), together with nice graphics with beautiful overlays and effects. Additionally, the GNOME community provides a nice set of applications that share the same values.

To compare – while GNOME strives for simplicity, providing *a limited set of GUI tools and options*, KDE Plasma provides *rich desktop settings* and applications with *rich menus and options*. This doesn't mean that GNOME lacks options. It is functional and good.

This section will cover essential points about the *Manjaro GNOME layout* since it is more attractive compared to the **KDE**, **Xfce**, and traditional looks.

The GNOME desktop environment is officially named **GNOME Shell** since the desktop environment usually is an overlay (a shell) above the OS. Regarding documentation, most of the GNOME community web articles date back to 2014 and 2015, so keep this in mind if you look online. On the other hand, there is a help menu – press the *Super* key and type `Help` to find the app. Unfortunately, *it is not up to date either*, so many contemporary options are not described in it. The *latest release announcement* and the three before it can be found at `https://release.gnome.org/`. You can find a general features description there. The official GNOME user help can be found at `https://help.gnome.org/users/`, but it is ancient (for GNOME 3 from 2008).

I like the article at `https://news.itsfoss.com/gnome-42-features/`, which provides a basic GNOME 42 overview. In general, finding such articles is helpful when you're looking for an updates overview. I would also look at the Manjaro forum GNOME tag for help: `https://forum.manjaro.org/tag/gnome`. Here is where you'll find the GNOME forum itself: `https://discourse.gnome.org/`.

Before we start, remember that in **GNOME**, the default *close window* key combination is *Super+Q*.

We looked at the top bar in *Figure 3.34*. The next point is the **Dash** menu (again, emphasizing that many guides and people call it the **Dock** menu). It has pinned applications in a macOS style and a grid to open a full menu to search for or launch an application. It is shown in *Figure 3.35*:

Figure 3.35 – GNOME Dash/Dock menu

When you open an application, it will be added to the right-hand side, so you can *right-click* on its icon on the **Dash** menu and select **Pin to Dash**, as shown in *Figure 3.36*:

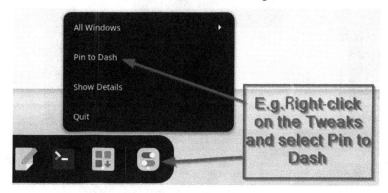

Figure 3.36 – GNOME – the Pin to Dash option in the right-click menu

When one program is pinned, it will go to the left of the vertical line, while all unpinned open programs will go to the right-hand side. The **Pin to Dash** option is also available from the **Application** menu search itself (without the program being started) – simply right-click and select it.

After you've pinned an application, the **Dash** menu *might* disappear – this is a known issue. If this happens, press the *Super* key twice, and the **Dash** menu will reappear. However, the default behavior is an application window to cover the menu to hide it automatically, which explains why it disappears in the first place. There is a setting for this – we will check it out later.

Also, whenever you press the *Super* key, you will enter an open applications overview, which resizes all open windows to smaller windows. At this point, you can move them to other desktops with the mouse. This view is shown in *Figure 3.37*:

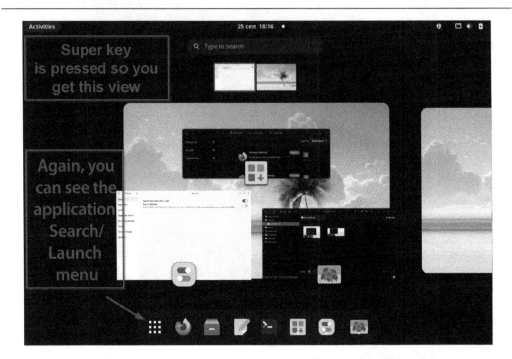

Figure 3.37 – GNOME application and workspace switcher with Dash visible via the Meta/Super keys

In this view, we can drag and drop an application to the free desktop on the right. This can be done via the main big center field or the small upper workspace screens listing. GNOME always adds a new screen to the right of the top screen listing. This effect is presented in *Figure 3.38*, where we have seven separate workspaces:

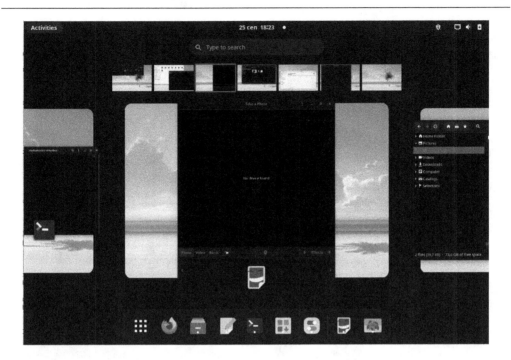

Figure 3.38 – GNOME switch workspace view

With *Alt+Tab*, you can switch between all applications from all workspaces (virtual desktops). This will switch the workspace accordingly. If you want to go to the left or right workspace using the keyboard, use *Super+Alt* and the left or right arrow. Clicking on the bottom icon of an already open application will bring you to the given desktop. However, if you want two separate Firefox instances, right-click the Firefox icon and select **New window**. *Shift+Alt+Super*+the left/right arrow will move the currently opened window one workspace to the left or right. More keyboard shortcuts are provided in the settings menu; we will look at these later.

The applications menu looks like this:

Figure 3.39 – GNOME Dash application menu view with highlights

You can access the **Dash** settings by right-clicking on the application menu and selecting the only option that appears: **Dash to Dock Settings**. This will open a powerful menu.

On the first tab, you can edit the **Dash** screen position, icon sizes, and monitor(s) to show the *Super* key dock view. The first page also has the critical **Intelligent autohide** option, which, when disabled, will always keep the **Dash** on top, shrinking your active screen area. I chose to disable **Intelligent autohide** and keep the **Dash** on the left while also shrinking its size a bit so as not to lose too much of my screen.

The third tab, **Behavior**, contains click and scroll action controls and the custom ability to activate apps via shortcuts with *Super*+a digit (*0-9*).

The fourth tab, **Appearance**, contains settings related to shrinking, opacity, colors, and themes.

The second tab, **Launchers**, apart from several **Show** options, includes important options for showing the trash can, volumes, devices, and more at the bottom. This is shown in *Figure 3.40*:

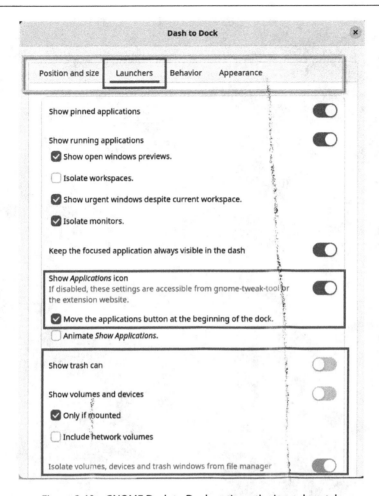

Figure 3.40 – GNOME Dash to Dock options, the Launchers tab

Our next stop is the applications. You might have already noticed that some of the applications are grouped, and if you click on a group, you get a popup containing the group's contents. To scroll, use the mouse or click the dot page marker at the bottom, as shown in *Figure 3.41*:

Figure 3.41 – GNOME application menu group view

Looking into all the folders might not be convenient, so I usually use the application menu search function at the top. In any case, of the multiple tools and applications, I recommend checking out the ones listed in *Table 3.3*. We will check out some of them briefly later in this section:

| Icon | Name | Usage |
| --- | --- | --- |
| ☼ | **Settings** | This is the main settings application and includes **Network**, **Bluetooth**, **Appearance**, **Displays**, **Power**, **Sound**, **Printers**, **Region**, **Language**, and various other options. It is simplistic but powerful. |
| S | **Tweaks** | Tweaks is the editor for additional **Appearance**, **Startup**, **Bar**, **Windows**, and other graphical fine-tuning settings. |

| Icon | Name | Usage |
|------|------|-------|
| | **Extensions** | Extensions are additions to GNOME for multiple purposes – from custom settings to nice animation effects, tools, menus, and others. Manjaro GNOME comes with some preinstalled. |
| | **System Monitor** | As you'd expect, with this tool, you can check CPU, memory, processes, and filesystem usage. |
| | **Qt5 Settings** | This area contains appearance settings for any **Qt5-based applications** – check them out. I think the default theme is quite impractical. |
| | **Qt6 Settings** | The same as above, but for **Qt6** – check them out as well. |
| | **Main Menu** | If you see it, it's only useful if you add a regular **Manjaro main menu**; otherwise, skip it in GNOME. |
| | **Fonts** | Here, you can check out all available system-installed fonts. |
| | **Screenshot** | A great feature is that it supports not only pictures but also video, so you can easily create captures of your actions. |
| | **Touche** | An extremely rich touchpad driver configuration that includes 2, 3, and 4-finger configurations. |
| | **Disks** | Here, you can manipulate disk **partitions** with a tool different from **GParted** (also installed). |
| | **Manjaro Settings Manager** | The default Manjaro settings for **Locale**, **Language**, **Kernel**, **Users**, **Time**, **Date**, **Hardware**, and **Keyboard**. |

Table 3.3 – GNOME's important default applications

We'll start with **Settings** . At first sight, there are plenty of categories, but in contrast to Xfce, Plasma, and some other graphical environments, they are relatively limited options. In addition, the search engine looks for more items and keywords only in menu titles, but not in sub-menus. Though not as functional as KDE's, the available options are enough for all basic OS features, as mentioned previously. It is worth noting the following:

- In **Network**, we have settings to connect to a VPN server easily.

- A separate menu for **Sharing** makes your life easier (compared to Xfce and KDE Plasma).

- Of course, there is a detailed **Shortcuts** section in **Keyboard**.

- Unfortunately, I couldn't find options for **auto-mounting** hot-plug disk drives, but regular USB-based devices work perfectly.

- The **Accessibility** menu is worth checking out as it has some text, alerts, typing, and pointing options that aren't supposed to be located there.

- While Xfce has no online accounts, and for KDE, you have only four options, GNOME offers integration for **Google, Nextcloud, Facebook, Microsoft, Flickr, Foursquare, Microsoft Exchange, Last.fm**, IMAP / SMTP for any mail, **Enterprise Login**, and **Media Server**.

- There is an interesting **Privacy** section.

- What puzzles me is that there is a section called **Applications**, but although there are many default applications, this section is empty. I installed some games and a few applications (Gimp, Slack, Solitaire, Aqueducts, and Atomix), logged out, and logged back in. Even then, the section remained empty. The single **Install some...** button didn't do anything, so I guess it will be a future development.

The next stop will be **Tweaks** 5 . As explained earlier, this tool allows you to *modify multiple appearance and behavior settings*, located in Xfce and KDE Plasma in the OS settings. As GNOME is minimalistic, it removed them from the **Settings** menu, and we have a separate application for them. Remember that some of these require an explicit logoff and logon to become effective. Let's take a look at some interesting parts of the different tool sections:

- In **General**, there is a control for whether you wish to suspend a laptop when its lid is closed. It also has a setting to allow audio over-amplification.

- In **Appearance**, we have themes for the cursor (mouse pointer), icons, and legacy applications. We also have desktop selection, positioning control, and lock screen background options.

- **Fonts** offers some generic font controls.

- In **Keyboard & Mouse**, I like that there is a **Show Extended Input sources** option. This is useful by allowing us to use, for example, German input letters (umlauts) on a US keyboard. On the other hand, in the past, this option has caused a crash upon adding a new regular keyboard layout, so search the web for any keyboard-related strange behaviors. Under **Mouse**, the *middle-click* paste is a nice addition. There is also a setting to disable the touchpad while typing and a few options for the touchpad mouse click.

- The startup applications are in **Tweaks**, so if you want a browser, mail client, Slack, Spotify, or something else to start automatically upon logging in, add it here.

- For the top bar, we have some clock and calendar options. **Weekday** is a nice addition.

- **Window Titlebars** provide several generic click options and the default min and max buttons.

- Finally, we have an interesting option in **Windows** called **Attach Modal Dialogs**. It is on by default. I like it turned off since, in many applications, this controls whether pop-up dialog windows are locked to the parent – that is, you cannot move the secondary window independently. There are also **Focus** controls here that control whether to switch focus on clicking, hovering, or secondary clicking.

Our next stop will be **Extensions**   . These are crucial for additional functionality. This is like the many plugins for **Mozilla Firefox** or **Google Chrome** – without them, the two browsers would not be the same.

Manjaro GNOME has several of them already integrated, such as **ArcMenu** for applications, **Dash to Dock**, which we have viewed in *Figure 3.35*, the **Pamac** updates indicator, which tells you about pending updates, and **X11 Gestures** for multi-touch features.

Once you open **Extensions**, you will see which extensions are installed and enabled. They always provide a website in the list, and many provide settings. They are famous and massively used. Try a web search for `the most famous extensions for Gnome shell` to see some collections of the best ones.

The next stop is the **Qt5** and **Qt6** settings   . I have covered them together as they look exactly the same inside. We have two versions since the Qt library has significant internal differences between versions 5 and 6. From the user's perspective, it doesn't matter. We simply have the same settings related to menus, fonts, icons, and other user settings. Whether or not they will be applied in a given Qt-based application depends on which Qt version it uses.

Here is my recommendation for the Qt settings. The current default theme is, in my experience, dysfunctional. Everything is dark, making my eyes feel strained just by looking at it. So, in the **Appearance** tab, choose your style and palette. I have chosen the **Fusion** style, with an **airy** palette, and then, in the **Icon Theme** tab, the **Papirus-Light** theme. Remember that you can change as much as you want, but once you do, click **Apply**, wait 1-2 seconds, and switch to different tabs two times. Then, go back to **Appearance** to see the result. The framework needs some time to apply and render the new view.

Next is the **Screenshot** tool   . Apart from creating a snapshot of a custom region, the whole screen, or some window, it can also create **WebM** videos. You can choose whether or not to show the mouse pointer, and later, you can play them with **VLC** or the default player, **Videos**.

The last stop is **Manjaro Settings Manager**   . We reviewed it in Xfce and KDE Plasma, and there is no difference here. Just remember to check it for locale, language, kernel, and so on.

Having checked many of the preceding options, I can say that sometimes, there are instabilities and strange choices regarding the default settings. The instabilities were three crashes at the beginning of 30 hours of extensive usage, one wrong relationship between the **Dash to Dock** and **Material Shell** settings, and two minor applications not starting. The **GNOME** flavor became stable after the system was configured according to my preferences, and apart from the issues in the beginning, I enjoyed it a lot.

Apart from this, GNOME provides a different and excellent user experience, and once you get used to it, the system is quite user friendly. The learning curve is a bit steep, but tweaking it is worth the effort. The default applications are simplistic and perfect for regular daily usage. If you want more options, richer menus, and applications, go for **KDE Plasma**. **GNOME** is your best choice if you like minimalism and a modern view.

## The other editions – Cinnamon, Lxde, Mate, and so on

It is impossible to cover all possible editions in one chapter. As you have seen, the preceding three have different and rich settings, but most mature projects have these too. From my experience, as I have used both **Mate** and **Cinnamon**, they are also good desktop environments. What I can recommend is, when looking for a potential change (or wondering whether to get an official or a community edition), look for online video presentations of the given graphical environment and consider whether you want to check it out. In addition, you could always try it on a virtual machine. You have to consider that sometimes, community editions might not get updates as fast as the official ones.

## Summary

In this chapter, we reviewed a few basic steps to do after a fresh Manjaro installation. We also looked at the basic settings of the three official flavors of Manjaro – Xfce, KDE Plasma, and GNOME. We looked at their menus, basic applications, setup, and graphics capabilities.

In the next chapter, we will continue learning how to get more information about any features, news, potential issues, ideas, and troubleshooting. This is important for any OS. After all, since many people invested time in writing them, we'd better use their tips and guides to make our lives easier and to save time. They have often stumbled on questions that may arise in our daily use of Manjaro.

# 4
# Help, Online Resources, Forums, and Updates

Now that we have seen the official **Manjaro** flavors, it is time to check out where we can find all the information we might need about Manjaro. The web is rich in official resources, and we will look at them in a structured way. The last part of the chapter explains Manjaro's Rolling Release Development Model, together with generic points on Linux backward compatibility and the SW updates.

The sections in this chapter are as follows:

- Help
- Online resources
- The forum – the greatest collection of knowledge
- The Rolling Release Development Model
- Updates

## Help

**Manjaro**, like many other good distributions, is rich in online resources. The best source of help for any issues is the **Manjaro forum**. However, before you approach it, there are a few points you need to know.

If you are looking to clarify some system settings, such as how to set up desktop properties or configure a network or Bluetooth device, they will be directly related to your Manjaro flavor. **Xfce** has one specific set of menus, while any other edition will often have at least slightly or even completely different menu interfaces. As a result, when seeking help, you should always add your flavor name. Here are a few examples: `Manjaro Xfce Bluetooth settings`, `Manjaro KDE Plasma system settings`, and `Manjaro Gnome application launcher options`.

In addition, some applications might have integrated help. See, for example, the KDE Plasma **File Manager**, shown in *Figure 4.1*:

Figure 4.1 – KDE Plasma Dolphin File Manager Help menu

Not all applications provide a *manual* or a **Help** menu, but it's always worth checking. Remember that the manual, *how-to* guide, and basic or detailed instructions might be available online.

No matter the Manjaro flavor you use, in the main menu (the *application launcher*), when you open and type in its search (e.g., `Network`), it will show you all applications associated with this keyword.

The **Manjaro wiki** (`https://wiki.manjaro.org/index.php/Main_Page`) is a good place to find information about essential topics, such as software management, hardware, networks, booting, and desktops. It is a good collection of articles with integrated search.

The concise **Manjaro FAQ** at `https://wiki.manjaro.org/index.php/Manjaro_FAQ` is always worth reading to know more about it.

## Troubleshooting

I have included this subject as a separate section for those who might look specifically for it. Troubleshooting typically means solving standard configuration issues or specific known problems. These are usually related to specific software or subsystems. Thus, we may have a troubleshooting guide, menu, or options for an app or a subsystem only if there are such standard issues for it. For example, the **Pacman** software package manager has such a page: `https://wiki.manjaro.org/index.php/Pacman_troubleshooting`. If you don't have a specific reason to look for a troubleshooting guide, you can directly look for an instruction manual, how-to-use articles, or forum posts about specific issues. For advanced hardware troubleshooting, we will review how to do it in Chapter 14, as it requires more knowledge.

# Online resources

There are a lot of online resources for Manjaro, listed in *Table 4.1*:

| Main website | `https://manjaro.org/` |
| FAQ | `https://wiki.manjaro.org/index.php/Manjaro_FAQ` |
| Manjaro download page | `https://manjaro.org/download/` |
| Blog | `https://blog.manjaro.org/` |
| Officially supported software, with an online search engine | `https://software.manjaro.org/applications` |
| Manjaro forum | `https://forum.manjaro.org/` |
| Manjaro basic wiki | `https://wiki.manjaro.org/index.php/Main_Page` |
| Manjaro tutorials | `https://forum.manjaro.org/c/contributions/tutorials/40` |
| Manjaro packages status | `https://packages.manjaro.org/?query=%23manjaro` |
| Manjaro repository and database update status | `https://packages.manjaro.org/status/` |
| Manjaro firmware update | `https://docs.manjaro.org/updating-firmware/` |
| Manjaro cheat sheet | `https://wiki.manjaro.org/index.php/CheatSheet` |
| Manjaro team | `https://manjaro.org/team/` |
| Manjaro partners | `https://manjaro.org/partners/` |
| Manjaro sponsors | `https://manjaro.org/sponsors/` |
| Manjaro hardware | `https://manjaro.org/hardware/` |
| Donations | `https://manjaro.org/donate/` |
| Videos | `https://manjaro.org/videos/` |

Table 4.1 – List of Manjaro online resources

The Manjaro web content is growing steadily; there is a lot to read, as you can see from the preceding table. The main website and its pages were reworked entirely in 2022. As a result, the **blog** section has entries since May 2022, when the site went online. The blog documents monthly highlights about Manjaro. The forum is also frequently updated.

From the preceding list, the **forum** will be reviewed as the primary source of knowledge in the next section.

Before that, here are a few interesting points:

- The wiki might look like a short set of 60–70 articles on main topics; however, each contains subsections and separate sub-articles. If you delve into it, use the embedded search and/or any web search engine.

- The hardware page is interesting – there are more than a few Manjaro-approved laptop and mini PC offers for enthusiasts. As Manjaro works on the PinePhone smartphone as well, there are also offers for that.

- The firmware update page contains a link to the **FWUPD** utility, which supports over 1,000 devices. These include laptops, computers, minicomputers, and workstations. Check out the complete list here: `https://fwupd.org/lvfs/devices/`.

## The news

All Manjaro-related news can be found in the **forum** and the *Online resources* subsection listed the links. We will look into it in the next section. The Manjaro team also has several official channels on social platforms. As the core developers are busy, some don't have frequent updates. Here are the social platforms, with a few words on each:

- **Twitter** – At the moment, this is the most frequently updated, with two main accounts to follow: `https://twitter.com/ManjaroLinux` is the main one and `https://twitter.com/ManjaroLinuxARM` is for ARM

- **Discord** – There is a public server with 4k members

- **Mastodon** – Currently frequently updated: `https://mastodon.social/@manjarolinux@masto.ai`

- **YouTube** – Rarely updated: `https://www.youtube.com/@manjarolinux8985/videos`

- **Facebook** – No update since 2020: `https://www.facebook.com/ManjaroLinux`

## The forum – the greatest collection of knowledge

We have now reached the primary source of help, discussions, news, and other information about Manjaro. The **forum** is well structured, and we will dive deeper into all the topics it explores. What is important is that you can find out *everything about Manjaro here*. Each section has drop-down filters for *tags*, *all*, *solved*, or *unsolved* issues. We have, as expected, a search, or we can use a web search such as this: `some search topic site:https://forum.manjaro.org/`. Each forum post listing orders, by default, from newest to oldest (including when filtered). To start with, here are a few points on its user interface.

We have the three-line menu button at the top left for showing/hiding the left sidebar. It contains links for **Topics (latest)**, **Manjaro**, **Categories**, **Tags**, and **Messages**. Below is a button to add a custom section to your private forum view and another to show all keyboard shortcuts. The **Manjaro** main links are essential for any beginner.

The looking-glass search icon is in the top-right corner, next to our profile icon button.

In each section (except the first three), we have *subsections* introduced on the right, at the top of the page. We have all these elements highlighted in *Figure 4.2*, presenting, in particular, the **Announcements** category. To access it, on the left, in Categories select All categories, then choose Announcements:

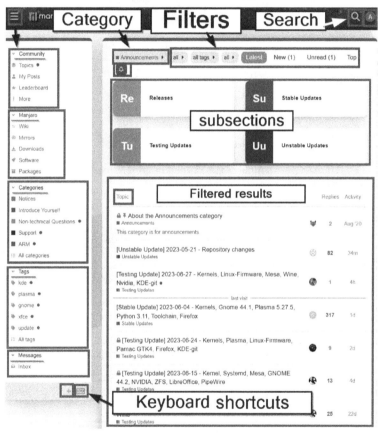

Figure 4.2 – Manjaro forum with menus and subsections for the Announcements category

*Figure 4.2* also presents the top screen filters for the *current* **category**, **subsections**, and **tags**, and a choice of whether to see **all**, **solved**, or **unsolved** issues. Further, depending on the section, we may have filters for **Latest**, **New**, **Unread**, and **Top** posts. In *Figure 4.3*, you can see the different tag filters in the **Support** category:

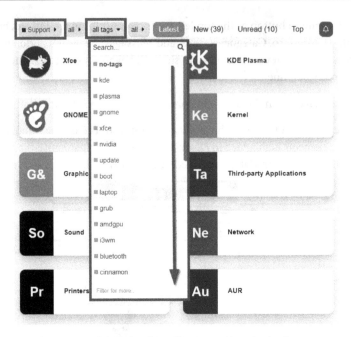

Figure 4.3 – Example of tag-filtering options in the forum

All the top-level forum Categories are described in *Table 4.2*:

| Notices | Every forum has rules. They are pinned here in the **Top** posts. |
|---|---|
| **Introduce Yourself** | If you wish – please be brief. |
| **Non-technical Questions** | Questions about the project and the distribution; don't ask support questions here. |
| **Support** | Support from community members, nicely separated into categories, which we'll discuss in more detail later in the chapter. |
| **ARM** | Everything for ARM – updates, versions, releases, news, discussions, and technical assistance – again, nicely separated. |
| **Announcements** | Releases, stable updates, testing updates, unstable updates, and news. |
| **Manjaro-Development** | For developers. |
| **Contributions** | For any contributor with some additional requirements for them – not everyone can post here. |
| **Languages** | Topics discussed in Português, Deutsch, Français, Español, Русский, 中文, Türkçe, and Suomi. |
| **Feedback** | Suggestions on improving the user experience; here, you can explain any usage difficulties you encounter. |

Table 4.2 – All top-level categories in Manjaro's forum

Let's look at each top-level section.

## Notices

If you want to post anything in the forum, please read the **Top** posts *Welcome to the new Manjaro Forum* and *Strict Forum Rules and guidelines* carefully. They are essential; the second clearly describes how to write in the forum.

## Introduce Yourself

Whether you write an introduction is entirely up to you. Usually, people here write a few words about themselves and explain how they started using Manjaro.

## Non-technical Questions

This section covers topics such as the frequency of software updates, packages, mirrors, kernels, UI, features, plans for future development, and more. Many of the topics are technical, but not exactly from the developers' point of view. Any generic discussion not explicitly formulated as a support question is here, hence the category title.

## Support

This section currently contains 14 subcategories:

- **XFCE**
- **KDE Plasma**
- **GNOME**
- **Kernel**
- **Graphics & Display**
- **Third-party applications**
- **Sound**
- **Network**
- **Printers**
- **AUR**
- **Virtualization**
- **Software & Applications**
- **Gaming**
- **Snap/Flatpak**

You can find thousands of already solved issues and ask about new ones here. The following are a few examples of existing topic titles:

- Bluetooth won't "turn on"

- USB tethering is super-slow

- How to make Pamac stop doing updates when packages are installed

- Formatting a 4 TB HDD

- Help to install a Wi-Fi USB dongle driver

This is the place to search for whatever question you have. The answers come from the community. I have asked questions, answered a few, and searched hundreds of times through this section for advice and solutions. Manjaro has a big community; as more members join, more questions will be covered.

## ARM

Manjaro currently supports 15 non-PC ARM platforms from 9 categories. For all of them, the **CPU architecture** is **ARM**. It is entirely different from **x86**, **x64**, **Intel**, and **AMD**. Most of these platforms are usually not as powerful as regular PCs and laptops; hence, they share specifics on how **OSs** and **packages** are developed and structured. The generic **Manjaro ARM** edition serves as a basis for most Manjaro ARM releases. For all these reasons, we have a united Manjaro-ARM sub-community.

The subcategories here are as follows:

- **Stable Updates**.

- **Testing Updates**.

- **Unstable Updates**.

- **Lomiri** (one more desktop environment such as Plasma and Xfce).

- **Phosh**: A GNOME-based desktop environment that works well with smartphones – it's like an Android phone but completely Manjaro/Linux-based! It runs on the **PinePhone**.

- **Plasma Mobile**: This is also for the PinePhone.

- **Sway**: This is a super-lightweight Wayland-based edition that is fast and thus appropriate for ARM devices (especially those with fewer resources, such as a weaker CPU and less RAM).

- **Releases**: This is mainly for the generic Manjaro ARM release, but some other releases have also been announced here.

- **General ARM Discussion**: This is quite a vivid section on the common topics for software support on Manjaro ARM. There are discussions on hardware, boot, custom ARM hardware (not from the officially supported platforms), tools, and so on.

- **Technical Assistance**: Here, you will find technical questions about all platforms.

## Announcements

**Announcements** contains four main subcategories – **Releases**, **Stable Updates**, **Testing Updates**, and **Unstable Updates**. It is used only to announce releases, updates, and testing results.

## Manjaro Development

This section is for developers. If you are interested, you can start contributing by **testing**. Of course, you would have to be at least an advanced user. Without development experience, most of the tasks would be beyond your capabilities. Here are the main subcategories in this section:

- **Pamac**
- **MSM (Manjaro Settings Manager)**
- **Calamares** (the unified installer)
- **Architect**
- **Packaging**
- **Layout Switcher**
- **mhwd (Manjaro Hardware Detection)**
- **Others** (mirrors, kernel modules, and others)
- **Dev Tools**
- **QA (Quality Assurance)** (requests for and results from testing)
- **Community Editions** (for all non-official Manjaro flavors)
- **Translations** (for any language)

## Contributions

This section covers Manjaro community contributions. We have four subcategories here:

- **Tutorials**: A section with multiple how-to guides. The topics cover Manjaro installation, forum features, Flatpak, troubleshooting, TigerVNC, AUR, PKGBUILD, kernel headers, and so on.
- **Design/Art**: This is the source for anything related to graphics, such as desktop wallpapers, desktop environment theme mods, and so on.
- **Spins**: These are unofficial, community-made Manjaro images. You will often find Manjaro flavors different from the official ones, such as Cinnamon, LXQt, and Mate.
- **Software**: In addition to the officially supported rich list of software, here we have additional scripts, drivers, and also applications.

## Languages

This section provides discussions in languages other than English. It is helpful for people worldwide who do not speak or are not fluent in English.

## Feedback

This section is for feedback, discussions about the Manjaro community organizational points, and a few general user experience discussions. It is not exactly pure feedback, such as whether a particular feature is working or not working. The subcategories are as follows:

- **Participation Systems**: A section purely for developers and contributors.
- **software.manjaro.org**: A section related only to Manjaro software servers, discover servers, GitLab, and packages.
- **Wiki**: For incorrect, outdated, or broken link issues at `wiki.manjaro.org`.
- **Forum**: For anything related to the forum itself.
- **manjaro.org**: For any issues and points about the main website.
- **Accessibility**: Reports and discussions related not only to helping people with disabilities but also ways to simplify and ease the experience of regular users.
- **Feature Request**: This is the place to request a feature. Remember that **Manjaro** is free and open source, and people will not always be available to develop a new feature. If your request is related to some hot topic and would improve something in Manjaro significantly, it might be treated as a high priority.

# The Rolling Release Development Model

Two general SW and OS release concepts exist: **Fixed/Point** and **Rolling Release Models**.

**Fixed** or **Point Release** distribution updates are delayed until a new version of the *whole* OS is released, typically every *6 to 18 months*. Thus, *fixes* for them can be *pretty slow*, and users must often *manually apply workarounds* and *patches* for current issues. The advantage of Fixed Release distributions is that this allows extensive testing. Package conflicts and issues are not excluded completely, as any Linux distribution contains thousands of packages. On the other hand, **Debian** is an example that is renowned for the quality of its Fixed/Point Release distribution. Their **Stable** branch is often called *super-stable* and has rarely had any issues at all.

In contrast, in a **Rolling Release Model** distribution, every package is updated when the developers are ready with its new version. This means that they constantly work on *one single current version*. As a result, the system is *always up to date* with the latest *SW, drivers, packages, fixes,* and *features*. Some examples are **Arch**, **Manjaro**, and **OpenSUSE**. You need to *update frequently* (which is *easy*), as when installing a new SW, it is presumed that you have an up-to-date system. The theoretical disadvantage of Rolling Release distributions is that they are more vulnerable to potential package conflicts and issues. Practically, in four years of permanent Manjaro usage, I have never experienced any severe issues, except once with the Timeshift application. It was handled by the Manjaro team quickly, and they offered a fast OS workaround update.

If you have to block a package or application update, the Pamac GUI offers an easy way to *mark it* to be skipped during updates. You also need to know that no other SW that depends on this SW or package would require it to be updated.

### Disabling updates for chosen packages and enabling downgrade

For any flavor of **Manjaro**, when you open the **Pamac GUI** (i.e., the **Add/Remove Software** application), in **Preferences | Advanced**, you have two important options: **Enable downgrade** and **Ignored upgrades**. The first one will only set the Pamac settings to be able to downgrade but will not show you older versions of SW packages. The reason is that they reside in another repository. We will see how to use it in *Chapter 8*. The second option will open *an incremental search* through the currently installed packages. Find the package you want to stay at its current version and click **Choose**.

### Backward compatibility

Some of you may raise this topic. The support for *older technologies/HW/drivers* is the generally accepted development model for *many Linux-related packages*. For example, **ext2** and **ext3** filesystems, although rarely used for years, are still supported by the **ext4** implementation *for reading*.

However, to allow normal development of *new* and *improved technologies*, the community periodically *suspends support for rarely used older ones*. Thus, if you have some specific use case scenario, you must search the web and the forum. *If there is no information*, feel free to *ask in the forum*. Keep in mind that, usually, there are *improved/better* newer options.

### New stable and unstable updates for SW (what is a branch?)

A **repository** is a *server* where we keep the current state of the developed software. It always supports some *version control* to allow proper tracking of changes. A **branch** is a copy of a repository with a different purpose from the original. Thus, if we develop an infotainment system for **Ford**, which will be integrated into three different models, we will probably have *different branches* for each.

Manjaro has three branches – **Stable**, **Testing**, and **Unstable**:

- **Stable** is the one from which we all get our SW when we update our OS.

- **Unstable** is the *playground for new SW and OS features* – the Manjaro team integrates the *latest SW packages* for initial *testing* and *checks*. As *Manjaro* is an *Arch*-based distribution, the set of chosen and verified packages from the *Arch user database* is transferred here several times daily. A limited number of users work here, as when an issue arises, it may require advanced knowledge and skills to be fixed.

- **Testing** is the intermediate state – once a particular SW *has passed at least basic checks* in the **Unstable** branch, it is transferred to this one *for several weeks*. Since more people use it, more users may report any residual potential problems with the various packages. Usually, once a package *has passed a couple of weeks of successful testing* in **Unstable** and then in **Testing**, it is ready to move to the **Stable** branch.

News and information on what is being tested in branches are announced in the **Announcements** forum category. Switching the branch is unnecessary for any regular user, but if you want to know how to support the team and do some tests, it is easy and described here: `https://wiki.manjaro.org/index.php/Switching_Branches`.

## Updates

By default, Pamac's **Add SW** GUI has the **Check for updates** option in the **Preferences** menu *enabled upon a fresh installation*. There is an additional option for how frequently to check for them. For third-party **AUR-** and **Flatpak**-based packages, you can enable update checks if you enable the given sources. We will explain more about them in *Chapter 5*.

How do **Updates** work? All the SW is located on **servers**, and the Manjaro team periodically *updates the packages*. Triggered automatically or manually, our Manjaro OS refreshes the *local OS databases* with the *latest versions of all packages on the servers*. Each installed application is checked to determine whether the latest version from the server is newer than the currently installed one. If so, you will *get a notification* in your panel bar. If you want to trigger the check manually (even though it is done upon each login), open the **Pamac GUI** and click on **Refresh databases** from the **Options** menu. That's all – just like the app stores for **macOS**, **Windows**, and **Android**, you are notified and can install updates with a single click. For **official repositors/official SW**, you can also enable **Automatically download updates** in the **Preferences** menu. Still, I prefer instead to be notified, as I'm not too fond of the system performing actions without my knowledge.

## Summary

In this chapter, we have covered all possible sources of help, manuals, news, and information about **Manjaro**. As it is freely and voluntarily supported by its vast community, you cannot expect an answer immediately if you have complicated questions; sometimes, it can take a week. On the other hand, this community has already posted thousands of hints and solutions for many topics. We also covered the basics of the Manjaro Rolling Release Development Model and its updates. New SW and fixes are frequently provided; getting them is effortless, as for any other major OS.

In the following two chapters, we will continue our journey by covering all the general classes of applications we may need. There are over 10,000 officially supported **Manjaro** applications. We will start with the **Pamac GUI** for adding and removing applications, libraries, and modules. It is easy to use and, in addition to the official ones, supports **Flatpak** and **AUR** packages.

If you find the content so far valuable, I will be eternally grateful to you for a short book review on the platform you purchased it from. With this, you will help people interested in learning more about Manjaro and Linux.

# Part 2: Daily Usage

In this part, we cover all essential topics for installing any type of application – official, third-party, office, gaming, Windows-based, and so on. It starts with GUI installer specifics and the supported application containers, adding essential notes on different packages. For the most important types of software – office tools, mail clients, and browsers, there is a detailed review explaining which are the best and why. In *Chapter 6*, we add short lists of the tens of free applications in any major category, emphasizing messaging applications and text editors. We then have a dedicated review of the gaming category, as its development in the Linux world has exploded in the last six years. As `.deb` packages are the most commonly used, a short section explains when and how conversion is possible.

*Chapter 7*, the last in this part, drives into the absolute power of terminals and commands. It is an excellent primer to lay the base for most of the rest of the book. The commands are presented with short but clear explanations, each with examples. The gradual approach to their intricacies makes the content digestible for any person of any age, regardless of their level.

This part has the following chapters:

- *Chapter 5, Officially Supported Software Part 1*
- *Chapter 6, Officially Supported Software Part 2, 3D Games and Windows SW*
- *Chapter 7, All Basic Terminal Commands – Easy and with Examples*

# 5
# Officially Supported Software – Part 1

For years, the Linux **free and open source software** (**FOSS**) community has provided a massive number of free applications. Many of them have also been ported to macOS and Windows. A few examples of such, available for all desktop operating systems, are **VLC** (the best free video player); **GNU Image Manipulation Program** (**GIMP**) and **Inkscape** for graphic designers; **Audacious** to play music collections; **Audacity** to edit audio files; **Python** to write scripts and **software** (**SW**) for ML, AI, and thousands of applications; **WordPress** to make a website of your own for free; **Mozilla Thunderbird**, a mail client; **Shotcut**, a video editor; and the **Brave** browser.

In this chapter, we will learn all the general points about user application installations. We will then switch to the basic applications everyone uses – office tools, browsers, and image viewers. As this is an extensive topic, we will continue with other application types in *Chapter 6*.

The sections in this chapter are as follows:

- Pamac – the Add/Remove SW GUI application
- The Flatpak, Snap, and AppImage containers
- The Manjaro repositories
- Office tools, calendars, and mail clients
- Browsers
- Photo, video, image, and graphics

## Pamac – the Add/Remove SW GUI application

To start, in any Manjaro flavor, type the word software in the main menu and open the **Add/Remove SW** application. Its other name is **Pamac** – remember it, as it is the official **Manjaro** package manager. It is equivalent to the Play Store on an Android phone, the App Store on an iPhone, and the Microsoft Store on Windows. **Pamac** also provides a **command-line interface** (**CLI**), which we will cover in detail in *Chapter 8*.

The **SW** categories in Pamac are the same in all flavors:

- **Featured**
- **Photo & Video**
- **Music & Audio**
- **Productivity**
- **Communication & News**
- **Education & Science**
- **Games**
- **Utilities**
- **Development**

**Pamac**'s integrated search looks by default in the official Manjaro repositories. They provide thousands of applications to install. We can also install a few other application package types from the **Flatpak**, **Snap**, and **Arch User Repository** (**AUR**) repositories. To enable them, open the three vertical dots (kebab menu) dropdown at the top-right of the screen, and choose **Preferences**. In the **Third Party** tab, we can enable **Flatpak** support (which is strongly recommended). **Snap** support is available by default only for **Xfce**, as most of the Linux community dislikes it. For advanced users, we have **AUR** support for all flavors. These additional package types will be added to **Pamac**'s search results, allowing us to install thousands more applications. We will look at them later.

There is brief information for each package when we click on it, often including an official web page, license, and repository. When searching the web for more information on some SW, always add the words `for Linux` to get relevant results. `for Manjaro` might also work, but most of the SW presented here is available for almost all Linux distributions. After all, this is why the Linux community is great. Conversely, adding `for Manjaro Linux` will provide results directly related to Manjaro. Adding `KDE`, `Xfce`, or `GNOME` is also often necessary.

Remember that some installations might require a restart, for which you will usually be notified.

Additionally, the *downloads* might be *slow* if you haven't *switched to local mirrors*. If, for some reason, the mirrors you work with are down, you should switch to others.

**Pamac** depends on *a large list of databases*. If those are *not refreshed*, we might get errors. To update them, open the drop-down menu from the three vertical dots, and choose the **Refresh databases** option. This is usually done automatically after changing the mirrors server or making a system-wide update. Sometimes, we may need to refresh manually, such as when enabling **Third Party** repositories.

We will continue with regular updates of the already installed SW. Open the **Pamac** GUI, and check the options at the top. Select **Updates**, and then you will see the results in the left-hand column, as shown in *Figure 5.1*:

Figure 5.1 – The Updates view of the Pamac GUI

You will either see pending updates or the words **Your system is up to date**. In general, updating frequently is strongly recommended. At the bottom, you will see the total download size, and clicking **Apply** might require your password. After the updates have finished, you will be notified if a restart is required.

In the **Installed** tab, we can see everything already installed. While there or in the **Browse** tab, pressing *Ctrl+F* will open the search applications menu. When you click on the downward arrow to the right of a package result, the arrow changes to a checkmark. Then, you can select **Apply**, as shown in *Figure 5.2*:

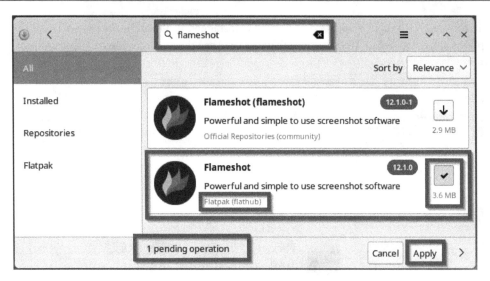

Figure 5.2 – Installing an application

Installing a *Manjaro official repository* or **AUR** application often requires additional packages and libraries as dependencies. This is great as, this way, many applications share common resources, avoiding duplications. Conversely, although rarely, it is possible to have collision points, as one application may require one version of a library and another may require a different one. While testing the hundreds of applications described in this and the following chapters, such issues happened to me only once. In such cases, find a **Flatpak** version of the application. **Snap** and **AppImage** are also an option. We will discuss them all in detail in the next subsection.

Often, you also have optional dependencies responsible for some special/additional functionalities. I usually choose some of them based on my needs. Reading the application description on the home page and checking the online documentation helps if we are unsure. For example, the **snapd** program includes the optional bash-completion package. It is harmless and required by a few other applications. As I know it, I'm happy to install it.

When a few applications are part of one bundle, or the one we choose depends on another, we might be forced to install additional ones. Let's see a simple example. I want only to install the spreadsheet program **Calc** from the **LibreOffice** suite. This is impossible, so the **Pamac GUI** automatically selects all the applications from the suite, even though I have selected only **LibreOffice Calc**, as shown in *Figure 5.3*:

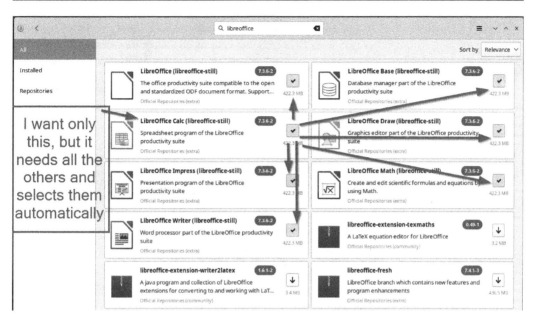

Figure 5.3 – Installing a bundled application

From the **Pamac GUI**, you can also delete any SW or package you don't want. Open the Pamac GUI, go to **Installed**, select the trash-can image next to the package, and click **Apply**.

The last important point here is that if you have serious trouble with updates or installations, and a database refresh and PC restart don't help, there is a solution. Open the application launcher, type `terminal`, and launch the Terminal application. In it type the command (without the dollar symbol) `$ sudo pamac update -- force-refresh`, and press Enter. This command will ask for your password; provide it, and press *Enter again*. We will learn more about commands in *Chapter 7* and this one in particular in *Chapter 8*. For now, know that any such command is executed like this in the Terminal, and that we never type the dollar sign.

## The Flatpak, Snap, and AppImage containers

Many applications are available for any Manjaro graphical environment. However, others are made especially for **GNOME, KDE**, or another Linux distribution. Thus, there is a difference in the available applications for the different Manjaro flavors.

The Linux community has three technologies to solve this issue – **Flatpak**, **Snap**, and **AppImage**. All these are **SW containers**, a way to pack a given application written – for example – for the **GNOME** environment with all their dependencies and provide it to Linux users, using any graphical environment on any distribution. In addition, all three containers also offer an **isolation layer**, so the given application cannot access your OS internals.

In addition, let's imagine the following situation. Say one app needs the **GTK** library version 4.6.6 but another needs 4.7.1. Though there are ways to keep both versions, in some rare cases collisions can happen. When each of the two applications are packed in separate containers, collisions cannot happen.

The drawback is that *containerized applications are bigger and sometimes slower*. Currently, **AppImage** is not offered by the Manjaro application manager, but it is provided for direct download via your browser (for example, on the Viber website).

> **Info**
>
> These two links from TechHut (Brandon Hopkins) provide interesting results from application containers performance testing and features explanation: `https://www.techhut.tv/flatpak-vs-snap-vs-appimage/` and `https://www.youtube.com/watch?v=OftD86RgAcc`.
>
> The article at `https://bytexd.com/differences-between-snap-appimage-and-flatpak/` provides more general details on all the formats.

All of the formats have their own online store to check the thousands of offered applications – **Flatpak** at `https://flathub.org/home`, **Snap** at `https://snapcraft.io/`, and **AppImage** at `https://www.appimagehub.com/`.

**Flatpak** is the preferred container in Manjaro and most Linux distributions as it is an independent technology. **AppImage** is also independent but has fewer applications and is not that stable yet. **Snap** is developed by **Canonical** (**Ubuntu**) and, despite being **FOSS**, is not preferred.

Manjaro has integrated **Flatpak** support in all three official flavors. **Snap** support is provided only for **Xfce**. When you enable any of them, you will see SW packaged in such containers, along with the official repositories packets in **Pamac**'s results. Just open the three vertical dots' drop-down menu, select **Preferences,** and go to the last tab, **Third Party**.

Remember that if a SW is available as both a **Flatpak** and native application, *the native application is always a lot smaller and potentially faster*. Thus, to not overload your system with hundreds of containerized apps, or if you have less hard disk space, it is always recommended to use the native application version. Size difference examples are provided in *Figure 5.4*:

Figure 5.4 – The size difference between Flatpak and Official Repositories applications

Conversely, if you have a regular PC/laptop with enough space (for example, 100+ GB free) and don't plan to install very big applications, the Flatpak, AppImage, and Snap versions are fine.

> **Important note**
>
> If you enable Flatpak support in **Preferences**, it is possible for Flatpak packages to not appear in Pamac's results, even after refreshing the databases and restarting the Pamac GUI. In this case, restart your computer. Generally, as with any modern OS, Linux is complex and in constant development. Sometimes, a restart of the system helps, as it reinitializes its services.

For **KDE Plasma** or **GNOME**, you must install `snapd` and then the **Snap Store GUI** application to access **Snap** applications (instructions later in this section). Although marked for advanced users, even a beginner can do this.

For **AppImage**, download the SW and change the file permissions to allow execution. For example, **Viber** provides an **AppImage** file here: `https://www.viber.com/en/download/`. Currently, the file is named `viber.AppImage`. Download it, and it will save automatically in your `/home/UserName/Downloads/` directory. Create a folder named `APPimages` (or something similar) in your home directory. Then, move the AppImage file from **Downloads** to the directory. Right-click on it, select **Properties**, and go to the third tab, **Permissions**. All three flavors have a checkbox option to mark the file as executable. In **KDE**, the option is written as **Is executable**; in **Xfce**, **Allow this file to run as a program**; and in **GNOME**, **Allow executing file as program**. When this option is marked, the file is executed upon double-clicking.

> **Important note**
>
> AppImage can sometimes require additional settings. I have tried numerous apps. While Viber was no issue for me, several simple applications could not start due to a missing library, not being able to get access permissions, and so on. This is usually an issue with the AppImage file itself.

## For advanced users

Here are the command-line instructions to install **Snap** on Manjaro **KDE Plasma** and **GNOME**. On the second one, some instabilities are possible. In particular, the Snap Store GUI currently doesn't work on GNOME. I will not provide detailed explanations of what each command does. `pamac` is reviewed in detail in *Chapter 8*, `systemctl` in *Chapter 13*, and `ln` in *Chapters 7* and *9*.

To install **Snap** on **KDE Plasma**, open the **Terminal**, type the following and press Enter:

```
$ sudo pamac install snapd
```

If you are asked for the optional `bash completion support` dependency, I recommend allowing it by typing 1 and pressing *Enter*. Then, type and execute each of these two commands:

```
$ sudo systemctl enable --now snapd.socket
$ sudo ln -s /var/lib/snapd/snap /snap
```

Restart your system to ensure the **Snap** paths and **AppArmor** are initialized and updated correctly. To test `snapd`, install the `hello-world` application, considering that the first line is the command, and the second is the result you shall see printed in your Terminal

```
$ sudo snap install hello-world
hello-world 6.3 from Canonical√ installed
```

When finished, type its name directly in the **Terminal**, press Enter, and you will get the `Hello World!` string as a result:

```
$ hello-world
Hello World!
```

Then, install the **Snap Store** by executing the following:

```
$ sudo snap install snap-store
```

Restart again. Finally, open the main menu, type `Snap`, start the Snap Store GUI application, and search for programs.

## AUR

AUR is full of more experimental and stable applications and packages, but it is not recommended for beginners. If enabled, **Pamac** searches include over 40,000 additional packages from it. In addition to downloading them, **Pamac** *will build and install them for you*, relieving you of the need to know all developers' details for a given package. Consequently, there is a risk for of people not knowing what they do, the worst-case scenario being breaking your Manjaro installation. Of course, if you find a big group of users who already use the given package and see no issues, *in my experience, you are 100% safe*.

Most **AUR** packages need the `base-devel` Official Repositories package to be built. Thus, *find it in Pamac and install it*, as most times, you will not be warned it is required. This is because **AUR** is used mainly by developers, who always install it in advance.

Some SW is available only as a **Deb** or **Rpm** package. Deb is a format from **Debian**, while Rpm comes from **RedHat** and **Fedora**. There are tools to convert/install them on Manjaro in *Chapter 6*, although this is strongly discouraged. Currently, conversion is mostly successful for **Deb** and better be avoided for **Rpm**.

## The Manjaro repositories

The **Manjaro repositories** are **Manjaro servers**, which provide both OS and SW updates. They work globally, 24/7. The full list with status, location, supported protocols, and updated timestamps is available at `https://repo.manjaro.org/` and `https://wiki.manjaro.org/index.php/Repositories_and_Servers`.

Most SW packages *come from trusted* **AUR** *users* but not directly from Arch repositories. Instead, they are *transferred to Manjaro servers*, and to ensure OS stability and security, *any system-critical SW is tested before being released*. With standard applications, any creator team tests their SW before releasing it.

A useful page in the Manjaro wiki gives more information on mirrors and how they help: `https://wiki.manjaro.org/index.php?title=Manjaro_Mirrors`. The easiest way to switch to mirrors from countries you have the fastest connection to is a command that tests and reorders your mirror list. To do this, open the Terminal and execute the following:

```
$ sudo pacman-mirrors --fasttrack
```

This command will test your current connection, and if it is slow, it will start a test on all possible servers (133 at the time of writing). It will then define those with the best connection speed. Depending on your internet connection and current settings, this can be slow.

Enough of the technicalities – let's start with the available SW.

# Office tools, calendars, and mail clients

We will start with a **PDF** and any **document viewers**.

Manjaro **Xfce** and **GNOME** flavors come preinstalled with the **Evince Document Viewer** GNOME application, which supports **PDF**, **PS**, **EPS**, **XPS**, **DjVu**, **TIFF**, and **DVI** files (with SyncTeX) and comic book archive files (**CBR**, **CBT**, **CBZ**, and **CB7**).

**KDE Plasma** has preinstalled the quite famous **Okular**. It supports, among others, **PDF**, **EPub**, **DjVU**, and **MD** formats for documents; **Postscript** (**PS**) documents based on **libspectre** / **CHM**; **JPEG**, **PNG**, **GIF**, **Tiff**, and **WebP** for images; **CBR** and **CBZ** for comics; and **DVI**, **XPS**, and **ODT**.

**Xfce** opens **PDF** files by default in the **Firefox** browser. Nowadays, all modern browsers support it. I prefer changing this in the **Default Applications** configuration module. To do so, type `Default Applications` in the main menu search, open it, go to the third tab (**Others**), find **MIME/type application/pdf**, and choose **Document Viewer** or **Okular** (if you have installed it). I prefer **Okular**, as it has more advanced features, but both offer highlighting and notes in a saved copy.

## Mail clients with calendars

The following are available directly on any Manjaro flavor:

- **Thunderbird** (Mozilla): We will start with the most famous one. Apart from **SMTP**, **IMAP**, and **POP3**, Thunderbird also offers **MS Exchange** support. It has a great calendar, which can sync with **Google Calendar**. It offers a built-in **RSS** reader and multi-account support. In addition, *hundreds of plugins are available for it*. It is *stable*, *reliable*, and *mature*, as the project started in July 2003. Finally, it has integrated **PGP** *encryption support*, an address book (you can also import one), and mail filtering and folders support. I have used it for years and think it's great! As it requires a bit more resources, it can be a bit slower on weak machines.

- **Evolution** (GNOME project): This is the direct competitor of **Thunderbird**, as it has all its features. Interestingly, some users tend to switch from Mozilla's application to **Evolution** due to issues with PGP encryption and the UI. I tried it, and adding a **Google** account to it took *a matter of seconds*, which also added my calendar events automatically. The contacts took around 15 seconds to load. It was a great experience. This application is also old and mature, started in May 2000, and also requires a bit more resources like Thunderbird.

- **Claws Mail**: A *lightweight mail client*, Claws Mail is also very popular among Linux users. It is the default mail application on some lightweight distributions, but not Manjaro. Comparing it with the previous two, it has fewer features. Automatic integration with **Gmail** is not so easy, requiring a few specific Gmail settings to be changed. It supports *encryption*, but additional features, such as an **RSS** aggregator, *calendar*, and others, might require additional plugins. It is generally lightweight, and emails are displayed as simple text. You cannot send HTML or other rich formatted emails with it. Claws is used mainly to read mail and rarely for anything else. If you want more features, use **Thunderbird** or **Evolution**. This might be your choice only if you want a simple and limited lightweight mail reader.

- **Geary**: This is another simple and lightweight mail client. Connecting to **MS Exchange** might be an issue for it. Again, depending on your system, **Gmail** might require additional steps to connect.

- **Kmail**: Another rich mail client considered an alternative to Thunderbird. It is **KDE**'s default mail application. I haven't tested it, but some reviews say it is good enough. If you want a powerful mail client and don't like the first two, **Kmail** is worth trying.

**Mailspring** and **Blue Mail** are only provided for Manjaro KDE as Flatpak.

I recommend trying, in this order, **Evolution**, **Thunderbird**, and finally, **Kmail**. If your system is too slow or weak, or you have an explicit reason to try another client, check out **Geary** or **Claws Mail**. The rest of the available mail clients are neither better nor stable, so I would not try any others.

## Office

The modern, rich office suite SW comes historically from **Microsoft Office**. At a minimum, this means **Word**, **Excel**, and **PowerPoint**. Its additional applications (**Access** for databases, **Outlook** for mail, **OneNote**, the cloud, and communications services) are rarely necessary for everyday users. So, I will review here the best alternatives, focusing on editing `.doc` and `.docx` documents, Excel sheets, and presentations.

Our alternatives list is rich – **LibreOffice**, **SoftMaker Office/FreeOffice**, **OnlyOffice**, **WPS Office**, **Apache OpenOffice**, and **Calligra**. The other option is to use the online services of **Google Docs** or **Microsoft Office365**, but they are both limited.

The main issue for me is compatibility. In 2014, I was sending multiple versions of my CV as an **MS Word** .doc file created with **LibreOffice**. When opened in **Microsoft Word**, it had completely different content ordering, which was devastating for me. Back then, I wanted to switch entirely to **FOSS** office. Eight years later, there is no such issue, as the alternatives have improved significantly. Despite this, to be sure that an **MS Office** document created with a **FOSS** application will look the same when opened with an **MS Office** application, ask a friend to check this for you. The other option is to export in **PDF**, as it keeps the content format and order. **HTML** is not an option, as its interpretation depends entirely on the browser/SW that will open it.

*Table 5.1* lists all **MS Office** alternatives. We're interested in the **FOSS** ones in this book, so I will review only them.

| Suite | FOSS | Free version |
| --- | --- | --- |
| **LibreOffice** | Yes | |
| **SoftMaker Office** | No, paid – owned by SoftMaker Software GmbH, Germany | No |
| **FreeOffice** | No, closed source – owned by SoftMaker Software GmbH, Germany | Yes |
| **ONLYOFFICE** | No, paid – owned by Ascensio System SIA, Latvia | Yes |
| **WPS Office** | No, paid – owned by the Chinese company, Kingston | Yes |
| **Apache OpenOffice** | Yes | |
| **Calligra** | Yes, KDE-based | |

Table 5.1 – A comparison of office tools for Linux distributions

## LibreOffice

The **Libre Office** suite is the official winner. It provides the best **MS Office** compatibility and includes the text editor **Writer**, the spreadsheet program **Calc**, the presentation builder **Impress**, **Draw** for drawing (both raster and vector graphics), **Math** for editing formulas, and **Base** for databases.

I opened drafts of this book's chapters, full of comments, with **Writer**. All the document information, annotations, and contents were presented correctly. **Writer** supports more formats than **OpenOffice** for saving, including the .doc, .xml, and .docx **Microsoft** formats. **LibreOffice** also provides a lot of guides online.

**Calc** opened my complex Excel sheets without problems. It also opened the book .xlsx schedule with dates and periods and calculated them correctly. There are *draw functions, separations, filters,* and everything you might need.

The presentation application **Impress** provides all you need for a good presentation, including *shapes*, *animations*, *arrows*, and *templates*.

The only thing to remember is that to have **Microsoft** and other fonts, you must install additional font packages from **Pamac**. Otherwise, you should stick to the default ones.

There are *thousands of templates online* for all **LibreOffice** modules. Check out `https://extensions.libreoffice.org/?Tags%5B0%5D=118` and `https://www.libreofficetemplates.net/`.

Regarding documentation and supported file formats, open the latest *Getting Started* guide at `https://documentation.libreoffice.org/en/english-documentation/`. In it, search for `File formats LibreOffice can open` and `File formats LibreOffice can save to`.

**LibreOffice** is available for **Linux**, **Windows**, **macOS**, **iOS**, and **Android**. It supports many languages, some available as extensions, and you can change the proofing/spellcheck language runtime.

Yes, it looks like the old **MS Office** 2003. Conversely, all the tools you might need are there. If you insist on a ribbon interface (the MS Office look for the last decade), you care for design, not functionality. Considering that it is **FOSS** and has everything you might need, we owe a big *thank you* to its authors and community.

## OpenOffice

**OpenOffice** is the parent of **LibreOffice**. It was started by **Sun Microsystems**, based on the older proprietary StarOffice suite. A community created the **LibreOffice** branch just before **Oracle** bought **Sun**, as they were afraid **Oracle** would make **OpenOffice** proprietary. Now, both suites are **FOSS**, and their modules are the same. Their licenses are different but compatible; both share many improvements and ideas.

**OpenOffice** has *several cons*, resulting in a firm *no* for it. First, it doesn't support most **Microsoft Office** formats for writing. Second, it *has a less frequent update model*, so **LibreOffice** has more and faster fixes and improvements. Third, **OpenOffice** is available only as a **deb** or **rpm** package and is not in the official Manjaro repositories. To install it, you must convert it from **deb**, which is covered in the next chapter.

## Calligra

**Calligra**, the last alternative on our list, is a **KDE** project. It offers a suite of applications, the main being **Words** – the word processor, **Sheets** for tables, **Karbon** for vector images, **Stage** for presentations, **KEXI** for databases, and **Plan** for project management. It also has the unique **Gemini** – a document preparation tool for touchscreen devices.

It has a unique interface distinct from MS Office, but its integration with MS formats is not yet good enough. It can save in .docx; however, opening complex .xlsx files resulted in incorrect data and formatting. **Calligra** favors **Open Document Formats (ODFs)** such as **ODT** (text), **ODS** (spreadsheets), and **ODP** (presentation), which are also used by **LibreOffice** and **OpenOffice**. The final verdict here is that **Calligra** is suitable only when MS Office formats are not necessary.

### The conclusion

The *official winner* is **LibreOffice**. It offers *the best* **MS Office** compatibility, excellent documentation, and hundreds of templates online. If you use only **ODF** formats, you can try **Calligra** – it is also a great suite. It is nice to know that **ODF** documents are accepted officially in some administrations. Both **Calligra** and **LibreOffice** are available in the official Manjaro repositories.

**Microsoft 365** *online* is an option only for basic functionality and if you explicitly need modern **MS Office** text styles (considering its limitations).

**Google Docs** *online* is great, but it is also seriously limited in options.

## Browsers

A modern browser for any user should cover the following requirements:

- Support *all modern web technologies* (e.g., HTML5, the latest CSS and JavaScript additions, video and audio content, notifications, etc.).

- Any rich website such as **Facebook** or **LinkedIn** must work *fast* and *flawlessly*.

- *Plugins* should be easy to add and *available*.

- It should have, by default, *at least mid-level security*, and *higher levels* should be *configurable*.

- *RAM and CPU use* should be acceptable, combined with stability, even when tens of tabs are open.

- Any potential *web-based activity tracking* should not be present in it. Thus, any proprietary browsers such as **Google Chrome** are not an option.

From all web browsers, I tested only those available directly in all official Manjaro flavors. They are in three categories – **Firefox**, **Chromium**-based (**Brave**, **Vivaldi**, and **Opera**), and others. I recommend only the first two. (**Tor** is the only good one from the last category but for a very special use case.) Let's see why.

**Firefox** is the browser that led the modern internet web content revolution and superseded the old **MS Internet Explorer**. It has had *high security and privacy standards* since its creation in September 2002. Until a few years ago, it suffered periodic issues with performance and quality, but they are now solved. It is still one of the best **FOSS** browsers, hence its *default* status *on many Linux distributions*.

**Google** released its proprietary **Chrome** browser for the first time in 2008, along with its **FOSS** version, **Chromium**. **Chrome** is based on it, and **Google** adds a proprietary SW layer for tracking and reporting. **Chrome** is also the default browser on **Android** devices. **Chromium** is essential, as creating a browser is a highly challenging task, and this one is **FOSS**, *powerful*, *efficient*, and *constantly developed for all the latest web technologies*.

As a result, **Google Chrome** has been the *top web browser* since 2011. Before this, **Firefox** was the leader, but then it steadily dropped its market share, stabilizing at around 10%. **Chrome** became faster, more reliable, integrated with **Google services**, and had *a better add-on system*. The downside is that **Chrome** collects users' data and sends it to **Google**'s servers.

The Linux community avoids **Chrome**, as for years it did not work well for Linux. Even in *June 2023*, it is still unavailable in the official Manjaro repositories, even though **Chromium** is. **Microsoft Edge** is now available for Linux and also good, as since *2019*, it has been based on **Chromium**. It is *officially available for many Linux distributions*, including Manjaro. Some reviews claim it has one of the highest privacy and security standards, but being a **Microsoft** and *not* **FOSS** product, the Linux community avoids it.

To conclude this limited historical overview, we have one final important point. If a browser doesn't support *the latest web technologies and standards*, it isn't good at all. **Firefox** and all **Chromium**-based browsers have kept up with all web updates since their creation.

## Test results

**Firefox**, **Brave**, **Vivaldi**, **Chromium**, and **Opera** have different user interfaces. They were all *stable with 25 opened tabs*, 15 of which had *complex, heavy, interactive content* such as online documents, sheets, the draft of a big chapter of this book, **IMDb**, **Amazon**, and **Maps**. No website crashed; all features worked *flawlessly*.

### Regarding resource usage

**Brave** and **Vivaldi** used the smallest amount of RAM. **Chromium** consumed 5 to 10% more, and **Opera** and **Firefox** required 15 to 20% more RAM than **Brave** and **Vivaldi**.

### Regarding ad blockers, security, and some features

The three **Chromium** derivatives have *integrated ad blockers*, all with good controls, although **Brave** and **Opera** have a few *better options for ad blocking*. **Brave** is known to have one of the highest levels of security against site trackers and ads (it even offers **Tor** mode), but the other two are *also well equipped*. **Vivaldi** has a great user interface and magnificent support. Keep in mind that **Vivaldi** is 99% open source, as its user interface SW is closed. Considering that enough people check and use it, it is widely considered private and sound.

**Firefox** doesn't have an integrated ad blocker, but plugins are available. Despite this, *its privacy and security are high*, with *rich settings* to track protection, block dangerous websites and links, and so on.

Bookmarks can be imported into all the browsers discussed so far. **Vivaldi** and **Opera** offer some additional features. Last but not least, Firefox, all **the recommended browsers** are available for **Linux**, **Windows**, and **macOS**. All of them, except Vivaldi, supports **Android** and **iOS** (if this matters to you).

If you want an alternative to **Firefox**, I cannot say which **Chromium**-based browser is the best. I suggest you try them out in this order – **Brave**, **Vivaldi**, and then **Opera**. Currently, I use **Brave** primarily on all my computers.

### The special addition

In the third category, I mentioned only **Tor**. Twenty years ago, two military researchers developed it to anonymize user activity on the web. It distributes the network packets so that your activity cannot be tracked. As a result of the traffic distribution, it is significantly slower than regular browsers. It is used by official agencies, military personnel, political activists, and regular people to access anonymously hidden websites (from the deep and dark web). You can find more information in this Kaspersky article: `https://www.kaspersky.com/resource-center/threats/deep-web`.

### The other browsers

From the other browsers, the only two relatively good ones were **Midori** and **Epiphany**, but both had issues with RAM consumption and slow loading of heavy websites. **Falkon**, **Pale Moon**, **Links**, **Seamonkey**, **NetSurf**, and **Otter** were awful for many reasons, related to performance, unloadable websites, and user interfaces.

# Photo, video, image, and graphics

This section will look into all well-known tools for video and imaging work. Many were initially developed only for Linux and would satisfy all your daily and special needs.

## Photo/images

Let's start with the simple photo and picture viewers. For **Xfce**, the default is **Viewnior**; for **KDE Plasma**, it is **Gwenview**; and for **GNOME**, it is **gThumb**. They all open images in multiple formats and are available for all three official Manjaro flavors. For simple viewing, I like **gThumb** and **Gwenview** the most.

The other alternatives we have available on all flavors are the following viewers:

- **Photoflare** is suitable for common effects such as blur, sharpen, color/channel modifications, and batch process.

- **Photos** (Pantheon) is a simple photo manager that automatically looks for images on your system and shows additional classification by date. It has hardly any editor options.

- **Photos** (GNOME) is a limited functionality viewer (press *Ctrl + E* to open the **Crop**, **Color Enhance**, and **Filters** options).

- **Ephoto** is a limited functionality simple viewer.

- **KphotoAlbum** is a good photo manager for people working with **Exchangeable Image File Format** (**EXIF**) data, **geo position**, **timestamps**, and other collection-related characteristics. It is not for simple viewing.

- **Deepin Album** is the most simplistic single-image file viewer. It has no options at all.

- **Shotwell** is a simple viewer with limited functions (**Rotate**, **Crop**, **Red-eye**, **Adjust**, and **Enhance**).

- **Showfoto** is a sophisticated manager with additional effects, transformations, enhancements, color modifications, and even face recognition.

I recommend **Gwenview**, **gThumb**, or **Shotwell** for a simple daily image view. For batch processing of photos and working with collections, use **Showfoto** or **Photoflare**.

## Graphics

In this subsection, we will review editing tools for images, 2D, 3D, and video files.

### Raster image editing

These are tools, such as **Adobe Photoshop**, that directly manipulate image pixels (dots):

- **GIMP**: The most powerful and famous alternative to **Adobe Photoshop**.

- **Krita**: A magnificent raster graphics editor.

- **Darktable**: A photography workflow application. It can manipulate raw images and apply filters on collections.

### Vector image editing

These are tools such as **Adobe Illustrator** and **CorelDraw**. They create and edit images presented in geometrical descriptions that are infinitely scalable (zoomable), not bound to a fixed number of pixels like raster graphics.

**Inkscape** offers a rich set of features. It is widely used for artistic and technical illustrations such as cartoons, clip art, logos, typography, diagrams, and flowcharts. Its vector graphics allow sharp printouts and renderings at unlimited resolutions. Inkscape uses the standardized SVG file format as its main format, supported by many other applications, including web browsers. Finally, it has been the most popular free vector imaging tool for many years.

**Karbon (Calligra)** is a highly customizable and extensible vector drawing application with an easy user interface. That makes it great for beginners to vector graphics and artists wanting to create breathtaking images. Whether you want to create clip art, logos, illustrations, or photorealistic vector images, look no further – **Karbon** can do it.

### 2D and 3D

For pure **2D**, we have only **Synfig Studio** to create and edit **2D** animations and compositions. The rest of the presented SW is **3D**.

**Blender** is a free and open source **3D** creation suite. Many users consider it *the best FOSS 3D SW*. It has a *complete 3D pipeline*, with *modeling, rigging, animation, simulation, rendering, compositing, motion tracking*, 3D interactive applications, multiple *visual effects, video editing*, and *game creation*. It also supports *skinning, texturing*, UV unwrapping, *smoke and fluid simulation, camera tracking*, and particle and soft body simulation. Advanced users employ Blender's Python scripting API to customize the application and write specialized tools, often included in **Blender**'s later releases. **Blender** is well suited to individuals and small studios benefitting from its unified pipeline and responsive development process. In addition, it is cross-platform and runs equally well on **Linux**, **Windows**, and **macOS**. Its interface uses **OpenGL** to provide a consistent experience.

Other **3D** apps to consider are as follows:

- **FreeCAD**: `https://www.freecadweb.org/downloads.php`

- **Wings 3D**: `http://www.wings3d.com/?page_id=84`; this can be downloaded as a Flatpak package

- **Bforartists**: `https://www.bforartists.de/download/`; this can be downloaded from **AUR**, as an AppImage or deb package

- **Sweethome3d** is an interior design application to place your furniture in 2D and get a 3D preview

### Video

The following list does not include all Linux video editing applications. Despite this, those presented are considered the best and are available in all official Manjaro flavors. Some YouTube vloggers solely use these tools to make professional videos.

**OpenShot** is easy to use and supports **4K** video. It depends on **Blender** 3D graphics SW, so *you must keep* **Blender** *updated accordingly*. It supports *many audio, video, and image formats* and has a good drag-and-drop feature. *Cutting, trimming, snapping*, and *cropping* are easily applied. It also supports *video transition, compositing, 3D effects*, and *motion picture credits*. **OpenShot** is a user-friendly SW, providing keyframe animation, easy *video encoding, digital zooming*, editing, and *mixing audio and digital video effects*. This video editor is a comprehensive tool for *high-definition videos* in the **HDV** and **AVCHD** formats.

**Kdenlive** has 4K support and no animated titles. It is part of the **KDE** project and is *one of the best open source alternatives* to **iMovie**. If you're migrating from **macOS**, this is what you want. Like **OpenShot**, **Kdenlive** is an all-purpose, multi-track, non-linear video editor that supports various video, audio, and image formats. It also supports tiles using texts and images, built-in effects and transitions, audio and video scopes for footage balance, proxy editing, autosave, and keyframe effects. Compared to **OpenShot**, it adds a *customizable layout* to make the editing process better fit your needs.

**Shotcut** is more advanced than **OpenShot** and **Kdenlive**, and it also supports 4K video. However, due to its higher number of features, it has a *steep learning curve* – in particular, audio mixing can be a complex task.

**Pitivi** supports simple features such as *snapping, trimming, splitting*, and *clip cutting*. It also has an *audio*-mixing feature and can use keyboard shortcuts and scrubbers. With this SW, video and audio can be linked, which is a great advantage. It is the first open source video editor that supports the **Material Exchange Format** (**MXF**). Pitivi's user-friendly interface provides drag and drop, direct manipulation, reduced complexity, and native theme navigation.

**Flowblade's** interface is *similar in layout* to **OpenShot**, as is the feature set. One of its highlights is the included *extension filter set for video, audio, and images*. Like **OpenShot**, it focuses on ease of use; you won't have a steep learning curve. **Flowblade's** bag of tricks includes drag and drop support, proxy editing, a range of supported formats (including video, audio, and image), batch rendering, watermarks, and video transitions.

Here are a few links to help you further with the choice:

- `https://www.fosslinux.com/6506/top-10-best-video-editors-for-linux.htm`
- `https://itsfoss.com/best-video-editing-software-linux/`
- `https://filmora.wondershare.com/video-editor/free-linux-video-editor.html`
- `https://www.lifewire.com/best-linux-video-editors-4176979`

# Summary

In this chapter, we covered adding and removing SW and the basics of the different *user application containers*. We looked at the most common SW for daily use – *mail clients, office suites*, and *browsers*. We explored *image* and *video*-related SW, showing the might of **FOSS**.

*Chapter 6* will continue with *music, audio, messaging*, and *text editors*. Further, we will see how many more application categories have been developed for Linux and investigate the already actively developed area of Linux *gaming*. The last part of the chapter is dedicated to the *conversion of deb and rpm packages* and the *usage of Windows SW*.

# 6

# Officially Supported Software Part 2, 3D Games, and Windows SW

This chapter covers the rest of the common **software** (**SW**) categories on Manjaro with the best available applications in them. They are mostly from the official Manjaro Repositories. We will also list many applications from non-common categories and look into gaming support. The chapter will end with the conversion of `.deb` and `.rpm` packages and how to start Windows SW on Manjaro.

The topics we will cover are as follows:

- Music and audio
- Teams, Zoom, Viber, Spotify, WhatsApp, Signal, and Telegram
- Text editors
- Drivers, tools, and simple games
- Advanced 2D/3D game support on Linux
- Creating application shortcuts and converting .deb and .rpm packages
- Using Windows SW on Linux

## Music and audio

There are three categories here: *media players*, *editors*/**Digital Audio Workstations** (**DAWs**), and *music servers*.

**PulseAudio** is the **Free and Open Source** (**FOSS**) audio package/driver on almost all Linux distributions. Thus, if you type `volume` or `audio` in your main menu's search, you will find application(s) for audio input and output control via **PulseAudio**.

In **Xfce**, the **PulseAudio Volume Control** GUI is installed by default. The sound control is integrated directly into the system settings on **KDE** and **GNOME**. In the last two, search for `pulse` in Pamac to add the **PulseAudio Volume Control** or **Equalizer** applications.

## Classic audio players

Xfce comes with **Audacious**, KDE with **Elisa**, and Gnome with **Lollypop**. They are all excellent for listening to music collections on your drive.

However, in all Manjaro flavors, we can install **Elisa, Audacious, Clementine, Rhythmbox, VLC media player** (also good for audio, not just video), and **Sayonara**. According to multiple online comparisons, other not-so-good options are **Amberol, cmus** (a console music player for advanced users), and **Strawberry Music Player**. Before making your choice, I recommend considering the streaming players listed in the next subsection, as they are the standard nowadays. **Clementine** and **Rhythmbox** are considered the best, according to reviews. These three links can help you with choosing a classical music player: `https://itsfoss.com/best-music-players-linux/`, `https://www.fosslinux.com/47158/mp3-players-linux.htm`, and `https://fossbytes.com/best-linux-music-players-ubuntu/`.

## Streaming players

Nowadays, music streaming is widespread. On Manjaro, you can install **Clementine, Spotify, Rhythmbox** (which provides connections to Last.fm, Magnitude, SoundCloud, Libre.fm, podcasts, and online radio stations), **Sayonara**, and **VLC**. There are even more, and new ones may be added.

Keep in mind that **Clementine** (`https://www.clementine-player.org/about`) provides connections to **Spotify, Grooveshark, SomaFM, Magnatune, Jamendo, SKY.FM, DI.FM, JAZZRADIO.com, SoundCloud, Icecast**, and **Subsonic** servers. It also plays Box, Dropbox, Google Drive, and OneDrive content and is also available for Windows and Android.

## Audio editors and DAWs

This is a whole special category for audio creators. Here we find **Audacity, LMMS, Ardour, Mixxx, Rosegarden, Qtractor, Hydrogen** drum machine, **Reaper, Muse, Guitarix, MuseScore, Bitwig Studio**, and **Waveform**. If you type `Midi` in **Pamac**, you can find more related applications and packages.

Available only as an **AUR** package, we have **Cecilia, ocenaudio, Traverso DAW**, and **Aria Maestosa**. Converting them from `.deb` is also an option. These reviews can help you with the choice: `https://itsfoss.com/best-audio-editors-linux/`, `https://www.fosslinux.com/50360/top-open-source-audio-editors-for-linux.htm`, `https://filmora.wondershare.com/audio-editing/linux-audio-editor.html`, and `https://www.tecmint.com/free-music-creation-or-audio-editing-softwares-for-linux/`.

## Music servers

These applications allow you to stream video and audio on multiple devices at home. We have here **Plex**, **Kodi**, **Emby Server**, **Subsonic**, **Gerbera**, **Jellyfin**, and **Mopidy**. Check out these links for more information: `https://www.linuxlinks.com/musicservers/`, `https://www.tecmint.com/best-media-server-software-for-linux/`, and `https://www.how2shout.com/tools/best-open-source-free-music-server-software.html`.

# Teams, Zoom, Viber, Spotify, WhatsApp, Signal, and Telegram

I've written a separate section for these apps as they are all famous and important to most of us. They are all available in **Pamac** from the Manjaro official repositories or as a **Flatpak**. For most of them, there are also **AUR** versions. I use them all.

For **Teams**, I have downloaded an unofficial **Flatpak** that connects to **Teams web**, and it works fine. Microsoft offered a **.deb** version of its software in 2022, but in July 2023, it was stopped. If it becomes available again, you can download it and follow the conversion and installation instructions at the end of the chapter (in the *Converting deb and rpm packages* section).

If you search the web for `Zoom download Linux`, you will find the official installation instructions for Zoom. Since 2022, there is a separate section for **Arch** (which works for Manjaro). A **Flatpak** version is also available, and I recommend it.

As of July 2023, **Viber** is available as a **Flatpak** on all Manjaro flavors and works perfectly. It is also available for download as an **AppImage**. Though there were previously issues and bugs, they are all now solved.

**Spotify** is directly available in **Pamac** as a **Flatpak** for all flavors.

**WhatsApp** doesn't have an official Linux version yet. Hence, I downloaded **WhatsApp Desktop** (Unofficial WhatsApp Web Desktop client) **Flatpak**, which works like a charm.

## A note regarding Signal, Telegram, and WhatsApp

**Signal** is a 100% open source encrypted method of personal communication.

**Telegram** has **closed/proprietary** parts of its code. It may soon add ads, and *only secret messages are encrypted*. In *2019* and *2020*, Telegram was *involved in scandals related to leaked user data*.

**Signal** and **Telegram** both come as official packages in **Pamac**.

**WhatsApp** encrypts messages but is owned by **Facebook/Meta**.

# Text editors

The **Linux** community has given us excellent text and source code editors. Many of them have been ported to **Windows** and **macOS** as well. Many such tools exist, but I present the most prominent ones in Manjaro's official repositories. If you are not interested in SW development, use **KWrite** for simple text editing or **Kate** for more features, and skip the rest of this section.

Before going on, we must know what **Integrated Development Environment** (**IDE**) means. IDEs combine a proper code/text editor with tools for building and debugging SW integrated inside its GUI. They are made especially for SW development. I will review the list of editors and IDEs in four groups, and at the end of the section, I will provide a few general recommendations for different use cases.

## Lightweight/simple editors

**gedit** has a simplistic look but supports syntax highlighting, word completion, spellcheck, a file browser panel, an embedded Terminal, a **Python** console, and other features. Check its preferences menu for more options. It is a **GNOME** project that started in *1998*; thus, it is mature and high quality.

Another good light editor is **KWrite**. It is a **KDE** project that started in 2002 and is the rich **Kate** editor with multiple features turned off. **KWrite** offers more text editing features and options than **gedit**. It includes configurable auto-complete, different input modes, file encoding choice, optional Vi input mode, extended menus and options, among others. It lacks an embedded **Terminal** (present in its full version, **Kate**) and is my recommendation for simple text editing.

Our third option is **Notepadqq**. It was inspired by **Notepad**++, which was designed only for Windows (and never ported to Linux due to a lack of resources). A team of developers gathered and developed a Qt-based editor that looks like Notepad++. It is not that powerful and lacks many features of the original, **KWrite**, and **gedit**. If you want to learn coding, help, or contribute to **Notepadqq**, this is an excellent opportunity, so contact the team.

## Sophisticated editors

In *Table 6.1*, we have a list of the more advanced text editors with short descriptions:

| Kate | Kate adds to the basic feature set of KWrite a file explorer, an integrated Terminal, Git connection, and project and session management. It also adds many plugins (e.g., for Clang, C++, and Go). All options are configurable. |
| --- | --- |
| **Sublime text** | A powerful text editor that is free for personal use. It supports themes, plugins, and a folder view and has a great community. |

| gVim | gVim is a GUI for the remarkable Vim command line editor that can help you quickly learn all common Vim commands. It is excellent for beginners to Vim. |
|------|------|
| **Atom** | Atom is powerful, completely configurable, and supports themes. It explicitly aims at SW developers and has been GitHub's official text editor since 2011. Microsoft owns GitHub, and in December 2022 they stopped Atom's development and support. Atom was used as the base of the development of the Electron framework and Visual Studio Code. Despite this, it is still available as a Flatpak. I left it here as it was famous for years.<br><br>The original author of the editor is planning an improved successor named Zed, developed separately from the Electron framework. You can check out the development status at `https://zed.dev/releases`. |

Table 6.1 – Overview of the most popular sophisticated text editors for Linux

## IDEs

In *Table 6.2*, we have a list of some of the most famous IDEs, with a few words on each:

| **Geany** | Geany is an award-winning lightweight IDE with an integrated parser for 50 programming languages, a compiler, a Terminal, messages (for status, compilations, find results, and others), and many view options. It also supports plugins and is recommended for developers. |
|------|------|
| **Visual Studio Code** | Visual Studio Code is a FOSS editor developed by Microsoft with hundreds of extensions, plugins, debug and development tools, Terminal integration, Git integration, etc. According to Stack Overflow, it is the most popular IDE since 2018, and I confirm it is magnificent. It is frequently referred to as VS Code. |
| **Bluefish** | Bluefish is rich in features and aimed at SW development, focusing on web languages and technologies. It is the middle ground between a powerful editor and an IDE, supporting regular expressions, scripts, and advanced search-and-replace. |
| **Eclipse** | This was one of the most potent IDEs for many years, with ports and versions for many programming languages. Due to this, it was one of the most loved IDEs by millions, including as an official company-approved tool (I have used it for commercial projects). Lately, however, the Eclipse Foundation has been focusing on other projects. As many other editors got better, the FOSS community switched to other options. |

Table 6.2 – Overview of the most popular text editors for Linux

## Terminal editors

We finish our list with a few essential Terminal editors listed in *Table 6.3*:

| **Nano** | Nano is a minimalistic Terminal-based editor, available by default on Manjaro and many other distributions. Its main shortcuts are displayed directly on its terminal interface. Nano provides syntax coloring, interactive search and replace, auto-indentation, line numbers, word completion, and other features. |
|---|---|
| **Vim** | Vim stands for Vi Improved and has been one of the most popular text and code editors in the Linux world for decades as it is powerful and based on the Vi text editor. As its commands are specific, find a beginner guide if you don't know them – the internet offers hundreds. It is suitable for code maniacs and Linux environments without graphical environments. |
| **GNU Emacs** | This old Terminal-based editor is actively developed and supported. Just like for Vim, finding an online beginner guide to work with it is recommended. |

Table 6.3 – Overview of the most popular text editors for Linux

## Recommendations

It is hard to say that a particular text editor is the best for you or me. Still, I will make some recommendations depending on the purpose:

- *For basic text editing*: **KWrite** is an excellent combination of basic features and speed. **gedit** is a good option if you need an integrated Terminal and basic **Git** integration.

- *Terminal/console editing*: **nano** is the recommended choice here thanks to its simplicity. If you want more features, choose **Vim**, but consider finding an online beginner's guide and/or use **gVim** to exercise its commands before this.

- *For development*: In this list, there are some IDEs not covered so far. The recommendations come from multiple online comparisons of the best IDEs in 2022 and 2023. From all IDEs, only VS Code is a good option in all cases:

  - **Web**: Bluefish, jEdit, Brackets, VS Code, Kate, Komodo, NetBeans, or SeaMonkey

  - **C/C++**: Kate, VS Code, Code::Blocks, Sublime, Geany, or Qt Creator

  - **Java**: Kate, VS Code, Intellij IDEA, BlueJ, or NetBeans

  - **Rust**: VS Code or Intellij Rust

  - **Ruby**: VS Code or Sublime Text

  - **D development**: VS Code, Sublime, or Kate

# Drivers, tools, and simple games

Each Manjaro flavor comes with system tools; the Linux world is literally full of them. If you open the main menu and type `monitor`, `task`, or `system` in the search box, you will find several tools available from the initial installation. The KDE Plasma default **System Monitor** is magnificent in my experience. If those are not enough for you, open **Pamac** and search, for example, for `System Monitor`.

I have selected applications that I find interesting and useful: **System Monitor** (`gnome-system-monitor`), **MATE System Monitor**, **KSysGuard**, **NVIDIA System Monitor** (if you are using an NVIDIA graphics card), **Conky**, **CopyQ**, and **CoreStats**. There are hundreds more.

Consider two things – applications designed for one flavor may not work on others. Here, Flatpak helps a lot, as it has the necessary dependencies. The second point is that if an application fails, removing it takes seconds.

You can search for description words in Pamac; for example, try `investigate`. For me, the results contained the **Universal Radio Hacker** – a tool for checking WiFi networks and protocols running in your area.

Another point – don't think that tools with *hacker* in the name do something bad. In the Linux world, **hacking** means *investigation and understanding how things work*, while **cracking** is used for illegal actions. Despite this, even tools that have *crack* in their names are often created for investigation, development, and tests.

All Manjaro flavors support **Bluetooth** by default.

To inspect connected **USB** devices, you can install **USBView**.

## Drivers on Manjaro Linux

The **Manjaro Hardware Detection** (**mhwd**) tool is run during installation. Thus, regular USB, Bluetooth, PCI devices, and other drivers are installed automatically. **mhwd** provides no choice menus, as the installation image contains the latest ones from its creation date. After this, you will get any necessary updates. The only exception is the initial LiveUSB choice for open source or proprietary graphic card drivers.

Historically, installing a device driver (or any system-level SW) happened via a Terminal on Linux. Manjaro has also improved this; all driver packages are available in **Pamac** from official or **AUR** repositories.

All vendors typically try to provide drivers in **AUR**, **Debian**, and a few other major distribution servers. From there, they spread to all child distributions. For Manjaro, there are rare occasions when specific hardware (**HW**) doesn't have a driver at least in **AUR**. Your vendor must have provided the corresponding Linux driver installation files and instructions in this case. When facing such issues, searching the internet for solutions usually helps, as they are discussed widely.

To get information on the current HW and drivers, we can use the **hwinfo** application from the Terminal. It is preinstalled on all Manjaro flavors. Open the main menu, write `Terminal`, start it, and then type `hwinfo`. When you press *Enter*, you will get a detailed listing of any HW module on your machine. An alternative for shorter output is the **lshw** tool, but it is available by default only on GNOME. For **Xfce** and **KDE**, you have to install it yourself via **Pamac** or the following Terminal command:

```
$ sudo pamac install lshw
```

A detailed troubleshooting example with one driver problem is provided in *Chapter 14*, *System Cleanup, Troubleshooting, Defragmentation, and Reinstallation*. The WiFi chip on a new *HP ProBook 440 G9* laptop didn't work by default, while on its previous version, *G8*, it worked like a charm. No update helped with the issue, as it was related to a not-yet-officially supported new **WiFi** chip. The new driver was already available in **AUR**. To understand the *Chapter 14* solution, you should understand the **Terminal** and command line (presented in *Chapter 7*), and the **pacman** and **Pamac** basics from *Chapter 8*. The troubleshooting example from *Chapter 14* also gives examples with Manjaro forum topics, which help in most such occasions.

## NVIDIA, open source, and other hardware drivers

The **Manjaro Settings Manager** is available by default on all Manjaro flavors. It contains the settings for graphics card drivers in the **Hardware | Hardware Configuration** section. Open it to access controls for switching between proprietary and open source drivers. There, you will see options based on your HW, i.e., if **mhwd** detected a graphics card with more than one set of drivers. As mentioned in the installation explanations in *Chapter 2*, we usually first try the open source drivers. If they don't work well, we try a proprietary driver, if present. Thousands of users have shared their experiences, issues, and ways to solve them on different HW/laptops/desktops in the Manjaro forum. If you have issues, search it for information; proprietary drivers may be recommended for specific graphic cards.

Last but not least, the **Manjaro Settings Manager** has in the **Hardware Configuration** the clickable checkbox **Show all devices**. When selected, we will also see in the list **memory**, **PCI**, **USB**, and other controllers. **hwinfo** and **lshw** will provide more detailed information, which many times helps.

## Screenshot tools

On **Xfce**, we have as default the *limited* **Xfce screenshot** tool, and on GNOME, the also *limited* **Gnome screenshot**. **KDE** comes with the *great* **Spectacle** tool.

A *good screenshot tool* provides a choice of the region to capture. Then, it allows you to annotate with rectangles, text, arrows, and other visual elements, acting as a basic image editor. On Manjaro, I can recommend *three great alternatives* that cover these requirements – **Ksnip**, **Spectacle** (the default **KDE** tool), and **Flameshot**. Here they are ordered from the best to the lightest.

**Ksnip** provides a separate tabbed interface for multiple screenshots, the ability to edit annotation elements that have already been added, an automatic numbering tool, icon addition, crop, rotate, scale, blur, and many other options.

**Spectacle** has a more limited set of options but still allows the editing of already added annotations, automatic numbering, shapes, and blur.

**Flameshot** doesn't allow editing of added elements but has a great interface and allows direct upload to `https://imgur.com/` or another program.

All three tools have **Undo** and **Redo** buttons. Any of them can be installed from the Official repositories, and they are all small and fast applications. *The only exception is that in June 2023,* **Ksnip** *doesn't work well on* **GNOME**. The instructions for changing the screenshot tool on each official Manjaro flavor follow.

For **Xfce**, open the main menu, type `keyboard`, go to the second tab, **Application Shortcuts**, and delete the `print` shortcut assigned to **xfce4-screenshooter-fd 1**. Click **Add**, type for command correspondingly `flameshot gui`, `spectacle -g`, or `ksnip -r`, click **OK**, and press the *Print Screen* key when asked to select a keyboard shortcut. There are two more custom print shortcuts here – one for region, *Shift+Ctrl+Print*, and one for the currently active window *Shif+Print*. I keep them as they are. The regular *Print* assignment is enough for me.

For **KDE**, open the main menu, type `shortcuts`, and open the **Shortcuts** (`Configure Keyboard Shortcuts`). In the shortcuts listing, find **Spectacle**, and when you hover over it, a trash can icon will appear to delete it. Click on it. Then, click **Add Application**, use the search field to find **Flameshot** or **Ksnip**, add it, and set the **Take Screenshot** action to **Print**. Finally, press **Apply** in the bottom right corner.

On **GNOME**, open the main menu, type `shortcuts`, and start the **Keyboard** settings application. It has an incremental search, which we leave for now. Scroll down to the **Custom Shortcuts** section and click it. Here, click the + sign to add a new one. Name it as you wish (e.g., `MyPrint`); for command enter correspondingly `flameshot gui`, or `spectacle -g`, and click the **Set Shortcut** button. Now click the *Print Screen* keyboard key and select **Replace** in the top right corner. You can use the incremental search now to check assignments, e.g., for `print`.

There is one more option – you can create *application shortcuts* for both tools on the main menu bar (both on **Xfce** and **GNOME**), as explained in *Chapter 3*. For Flameshot, you can start the application to get a tray/main menu bar icon to trigger it. So, if you add **Flameshot** to the startup apps, it will stay by default in the main bar.

## Virtual machines

I used **Oracle's VirtualBox** for years. It is available in the official repository. It works flawlessly and provides a good amount of options. The minimum required settings are presented in *Chapter 2*, in the *Installation on a virtual machine* section.

As of July 2023, **VMware Workstation Player** is not officially supported and is hard to install. More information can be found at `https://wiki.manjaro.org/index.php/VMware`.

**QEMU** is great but complex to handle. To work with it, you must download the `qemu-desktop` package and all the necessary dependencies. Even as a snap application, it is unavailable as a single centralized app but only as a set of packages. It is definitely unsuitable for beginners. More information is available here: `https://fosspost.org/use-qemu-test-operating-systems-distributions/`.

**Gnome Boxes** is another option, but it has limited settings. Although it's easy to set up, I could install Manjaro Xfce on KDE Plasma but not start it. The **Manjaro Gnome** host could potentially work well.

As a result, I always recommend using **Oracle's VirtualBox**.

## Education and Science

Every application in the **Education & Science** category is worth exploring. I've made a limited list of a few great tools mostly from it, related to mathematics, CAD SW, and science:

- **Kig** – interactive geometry
- **Tux, of Math Command** – entertaining math game for kids
- **Mathomatic** – for symbolic mathematics and quick calculations
- **Cantor** – a GUI frontend for powerful mathematics and statistics packages
- **Funkcio** – draw functions from points
- **GeoGebra** – dynamic mathematics SW with interactive graphics, algebra, and spreadsheet functionalities
- **GNU Octave** – a famous interactive programming environment for numerical computations
- **Hopsan** – a modeling and simulation tool for fluid power and mechatronic systems
- **Plot** – a simple graphics SW
- **Scilab** – a numerical computational package with GUI and a high-level numerically oriented programming language
- **Veusz** – scientific plotting package
- **OpenSCAD** – solid 3D CAD modeler
- **SolveSpace** – a free parametric 3D CAD tool
- **Makhaber** and **SciDAVis** – for scientific data analysis and visualization
- **LabPlot** – for interactive data analysis and visualization
- **Avogadro** – molecular editor
- **KCalc** and **Liri** – nice calculators; there are many like them

## For video and/or audio conversion

Some examples are **FFaudioConverter**, **MakeMKV**, the **Transmageddon** media transcoder, **SoundConverter**, **HandBrake**, **Gnome Subtitles**, and **Ciano**.

**VLC** and the **FFmpeg** libraries can also convert video or audio for you, but **FFmpeg** is only used on Terminal.

## Others

Here are a few more interesting applications:

- **Marble**: Provides virtual maps and a 3D earth globe explorer
- **Bookworm**: An e-book reader
- **Calibre**: E-book device editor
- **QGIS desktop**: FOSS geographic information system
- **Kadas Albireo**: A mapping application for non-specialized users based on QGIS
- **Valgrind**: The most famous Linux profiling tool for SW developers, with GUI frontends – **KCachegrind** and **Massif Visualizer**
- **GPXSee**: GPS log file viewer and analyzer
- **Health**: Track your fitness goals
- **Krop**: A tool to crop PDF files
- **PinApp**: Create and edit application shortcuts
- **YACReader**: Comic book reader
- **Komikku**: Read your favorite manga online or offline
- **Buoh**: Online comic strip reader

## Simple Games

An example of a simple game is **Solitaire**. Many others are well-known; check them out. We have **kigo** – an implementation of the Go game, and at least six **chess** applications. We also have **Armagetron**, **Aqueducts**, **Tetris**, **Minecraft**, **Checkers**, **Reversi**, **Backgammon**, **Solitaire**, **Mines**, **Puzzles**, **Tetris**, **ReTux**, **The Open Racing Car Simulator**, **X- Moto**, **Mahjong**, and **Sudoku**.

There are *over 600* such apps in the **Pamac Games** category of Manjaro **KDE** and *over 650* for **Xfce**! **GNOME** offers only 80, but you can download most of the natively available ones for **KDE** or **Xfce** as Snaps or Flatpaks. There are Gameboy and Nintendo emulators, space shooters, arcade games – anything you can think of.

## Advanced 2D/3D game support on Linux

Before diving in, here is a periodically updated list of free Linux games (31 in July 2023): `https://itsfoss.com/free-linux-games/`. Search the web for `best free Linux games 2023` to find more. Here, I will provide more information on Linux game development, as there are paid platforms and frameworks offering a lot in this area.

**The Steam platform** is a game distribution platform owned by **Valve Corporation**. It is supported officially on Linux and is available in the Official repositories. When installing, choosing all optional dependencies is recommended. Currently, this is one of the most up-to-date installation guides: `https://linuxnightly.com/install-steam-manjaro/`. I have installed Steam, and it worked flawlessly, starting the **Dota2** game on my machine in a few minutes (not considering the 40 GB game download time). For issues, visit `https://wiki.archlinux.org/title/Steam/Troubleshooting#Native_runtime`.

**Proton** is a compatibility layer to run Microsoft Windows games on Linux-based operating systems, developed by **Valve** in cooperation with developers from **CodeWeavers**. It is a collection of SW and libraries combined with a patched version of Wine to improve the performance and compatibility of Windows games. **Proton** is designed for integration into the **Steam** client, known as **Steam Play**. It is officially distributed through the Steam client, although third-party forks can be manually installed. Its official website is `https://www.protondb.com/`.

**Vulkan** – `https://en.wikipedia.org/wiki/Vulkan` – is a high-performance real-time 3D graphics applications API developed by **AMD** and **DICE**. It is cross-platform, which means it is supported on Linux.

**Lutris** is a FOSS game manager for Linux-based operating systems. For games that require Wine, community installer scripts are available for automatic Wine environment configuration. Lutris also provides integration and directly launches games purchased from **GOG**, **Humble Bundle**, **Steam**, and **Epic Games Store**. It is available in **Pamac** from the Official repositories.

Furthermore, regarding advanced 2D and 3D games on Linux, we have the magnificent website `https://www.gamingonlinux.com/`. If you are interested in gaming, I would say checking it out periodically for any related news is almost obligatory.

**Wine** is the solution to play many old Windows games on Linux; its community has put much effort into developing and setting up hundreds of them. To get a list of supported ones, go to `https://appdb.winehq.org/votestats.php` and filter the last 200 packages for category games. There is also a community rate; the highest-rated ones work flawlessly.

## Game controller support and further points

**DualShock 3** and **4**, **Xbox 360**, **Xbox One**, **8BitDo**, and the **Xbox One Wireless** controllers all have drivers for Linux. Check the web for articles on Linux game controller support.

> **Important note**
>
> Consider that many games have better performance on Linux than on Windows. Why? Because Linux has better resource utilization, management, and filesystem.

Follow these links to see the progress of game development in the short period between 2019 and 2020:

- *Microsoft Should be VERY Afraid - Noob's Guide to Linux Gaming*: `https://www.youtube.com/watch?v=Co6FePZoNgE`

- *Linux gaming is BETTER than Windows?*: `https://www.youtube.com/watch?v=6T_-HMkgxt0`

These links are from the **Linus Tech Tips** YouTube channel, currently with over 15 million subscribers. Thanks to its high-quality content, it is one of the most famous tech channels nowadays.

**OpenRGB**, `https://openrgb.org/`, is a great tool for controlling the RGB lighting for your gaming setup, and it is natively cross-platform, that is, it is also made for Linux.

Many games are not yet available for Linux. Despite this, considering the support of the big companies mentioned so far, and when we add **AMD** and **Nvidia**, the area is developing lightning fast.

## For advanced users

The following is a list of the game engines supported on Linux:

- The **Unity game engine**, considered one of the best

- The **Godot game engine**, which is FOSS

- **Unreal Engine** has also started to add support for Linux since 2022: `https://www.gamingonlinux.com/2022/07/unreal-engine-5-editor-quietly-gets-a-proper-linux-version/`

- The **Defold game engine** is FOSS and has official Ubuntu support; its `.deb` package can be converted for Manjaro and any other distribution

# Creating application shortcuts, and converting .deb and .rpm packages

**KDE** and **GNOME** support a two-click way to create links and shortcuts. For **KDE**, you open the main menu, find your program, right-click, then click + **Add to Panel (Widget)** or + **Add to Desktop** – and there it is. For **GNOME**, again right-click and select **Pin to Dash**.

For **Xfce**, selecting +**Add to Panel** adds an icon for the application at the end of the list of elements in the given panel. This isn't ideal, but it's easily solved by right-clicking the application icon and selecting ->**Move** to move it on the panel. The other option is to click anywhere on the panel, select the **Panel** option at the bottom of the right-click panel, select **Panel Preferences**, then go to the third menu tab, **Items**, and reorder all elements the way you want. The only problem is that you will see the application icon twice – once as a launcher and one more time if the application is running (the ancient Windows style). I find this annoying – space is precious on small screens, and it is already solved on **KDE**.

Unfortunately, there is no easy way to switch to another standard panel or dash on **Xfce**. The reason is that such panels or dash bar SW are often made explicitly for a given environment (Xfce, KDE Plasma, Gnome, Cinnamon, etc.) and so tightly related to it. There are currently *two potential external SW alternatives* for an Xfce dock menu, **Plank** and **Cairo Dock**, both available from Official repositories. However, you will have to play with their options and edit the default Manjaro menu bar preferences to make it comfortable for you. I hope that in the future, more such options will be added to the default Xfce panel.

For an **AppImage**, you may want to create a launcher, easily done on **KDE** and **Xfce** by right-clicking on the desktop. For **Xfce**, select the first option, **Create Launcher**, and fill in the top two fields as you wish. For **Command**, give the location of the executable file. For **Working Directory**, I provided the dedicated directory I created for AppImages called `Installed`. Finally, click **Create**, and you will have it. The filled fields are shown in *Figure 6.1*:

Figure 6.1 – Xfce Create Launcher window fill for AppImage

On **KDE**, right-click on the desktop and choose **Create New | Link to Application**. In **General**, name the launcher as you wish. In the **Permissions** tab, mark **Is executable**. In **Application**, for **Command** provide the executable file path, and for **Work path** provide the directory containing the AppImage. Finally click **OK**. An example is shown in *Figure 6.2*:

Figure 6.2 – KDE Plasma – Create a new link to an AppImage application

## Converting .deb and .rpm packages

Before continuing, remember that `.deb` packages are created for **Debian** and `.rpm` for **Fedora**. They are often natively supported on their respective child distributions *only as formats*. As a result, if you try installing an `.rpm` package *created for* **Fedora** on **SUSE** Linux, *or vice versa*, this typically results in *issues*. The reason is that apart from the package format, the given distributions differ in file structure, settings, and many others. Hence, the information inside the package of where to install configuration, executable, and `.so` files differs.

Arch and Manjaro use a format designated with the extension `.pkg.tar.zst`. Conversion to it is theoretically possible if the converted package or its target have no peculiarities or specifics. After all, a simple application typically comes with one or a few executables, shared libraries, and config files. If an installer puts them in the correct locations and creates an entry in the applications list of the target OS, everything *shall* work. The problem is in the word *shall* – no converter can guarantee 100% correct operation.

That's why, if you find an application available only as an `.rpm` or `.deb` package, always try to find alternatives in the following way: a **Flatpak** version, an **AppImage**, an **AUR** version, and finally, if you are using Snap – a **Snap** version.

Apart from this, to try converting `.deb` packages, you can use **debtap**. Remember – converting a `.deb` package to the `.pkg.tar.zst` package format will allow you to *attempt installing it on Manjaro* without *granting success*. Still, *it is worth trying if you have no other option*, as enough times, this might succeed. So, open Pamac, enable **AUR** in **Preferences | Third Party**, find the `debtap` package, and install it. Then, open a **Terminal** and, to update the `debtap pkgfile` database, run the following:

```
$ sudo debtap -u
```

Then, to convert `somePackage.deb`, call the following:

```
$ debtap -Q somePackage.deb
```

The conversion (depending on your machine and the package size) can take from a few tens of seconds up to tens of minutes, so be patient. If successful, you will have a similarly named file with the `.pkg.tar.zst` extension. Double-clicking it in your file manager will automatically open the Pamac GUI installer, and once it finishes, you will have the application in your main menu.

Option two is `deb2appimage`; however, it requires an additional JSON configuration file and is not trivial at all.

The name of the `.rpm` extension comes from **Red Hat Package Manager**. While there are ways to extract the rpm contents and then pack them as `.pkg.tar.zst`, avoiding this is *strongly advised*. Red Hat- and Fedora-based SW is quite different from Arch.

# How to use Windows software on Linux

This section will cover all ways of running Windows applications on Linux.

## Short history recap

In the 90s, Windows was the main widespread OS with a rich GUI experience and enough SW for it. Back then, Linux was making its first steps as a free OS and tools suite. As the years went by, it grew widely in contributors and users and initially won a lot of business users in the server market. It also became *the leading OS for scientific tools*. Free alternatives to Microsoft Office were developed. Now, there are many more Linux tools and SW than Windows ones, and they are flourishing in a great distributed ecosystem. During the 90s, when there were few GUI tools on Linux, many users and industries oriented themselves to Windows. As a result, thousands of legacy and modern Windows-based applications will never have a Linux version. Porting SW from Linux to Windows and vice versa is expensive due to the essential differences in the OSs' architectures, libraries, and design.

Amusingly, tons of free Windows tools emerged from the FOSS world of GNU/Linux. Of course, there are many tools *available exclusively for Linux*. Windows made no effort to provide a way to run them until *August 2016*, when the **Windows Subsystem for Linux** (**WSL**) was released. In *2019*, **WSL2** came with the capability of running a Linux virtual machine on Windows. It is, unfortunately, suitable only for advanced users.

To make a difference, Linux has been developing the **Wine** framework for running Windows applications *since 1993*.

## The Linux tools for Windows software

Running Windows software is easily done via **Oracle's VirtualBox**. However, running a whole virtual machine only to start one or two programs is generally a waste of resources (except when it is the only solution). The following method guarantees maximum efficiency and smoothly running applications (when the setup/support exists). We currently have in **Pamac** from official repositories **Wine** and **Winetricks**. As Flatpaks, we have the **Wine**-based **Phoenicis PlayOnLinux**, **q4wine**, and **Bottles**.

The parent project is **Wine**. Its website, `https://www.winehq.org/`, has been running since 1997. In *Table 6.4*, we have short information on each of the packages:

| Package | Description |
|---|---|
| Wine | Wine stands for Wine Is Not an Emulator – a framework for translating Windows OS, allowing a Windows executable to run on Linux. Wine has a GUI but requires user configuration for some of its packages. |
| Winetricks | As explained on their GitHub page, `https://github.com/Winetricks/winetricks`, Winetricks is an easy way to work around problems in Wine. It has a list of games/apps for which it applies the workarounds automatically. It also allows installation of missing DLLs and tweaking of various Wine settings. |
| PlayOnLinux | This is a Wine-based GUI frontend for simplifying the install/setup/configuration process for some Windows applications in Wine. Though it offers a lot of applications online, its local installation requires a 32-bit Wine version, and there are only a few locally listed applications. Considering its complicated installation, I don't recommend it. |
| q4wine | A Qt-based GUI for Wine. |
| Bottles | Another Wine-based SW with a lot easier configuration and setup of Wine-based installations. |

Table 6.4 – Overview of packages for execution of Windows applications on Linux

The installation of **Wine** depends on the age of the Windows SW you are running. If it's too old, it may require an older Wine version. All the information related to versions and support is available at `https://www.winehq.org/` in the *Application Database* section. There are over 28,000 applications (at the end of 2022), but many are old versions. **Notepad++** is one of my favorite Windows applications, and finding it on the Wine website was easy.

All the software in the database has a rating. The ratings are defined at `https://wiki.winehq.org/AppDB_Rating_Definitions`. *Table 6.5* describes them:

| Rating | Description |
|---|---|
| Platinum | Works as well as (or better than) on Windows out of the box |
| Gold | Works as well as (or better than) on Windows with workarounds |
| Silver | Works excellently for normal use, but has some problems for which there are no workarounds |
| Bronze | Works, but has severe problems in normal use |
| Garbage | Doesn't work at all |

Table 6.5 – Overview of Wine package ratings with explanations

As you can see, the *Platinum* and *Gold* statuses mean you can use the SW without a problem. *Silver* has some limitations; reading the description in the database usually explains them. The other two mean the SW is unusable and needs more work.

It is worth reading the Wine Wiki as it is good: `https://wiki.winehq.org/Main_Page`.

A search on `https://www.winehq.org/` looks not only in the Wine AppDB but also in the bug tracker. Thus, you can find help when facing an issue, or if there is no information, you can file a report. Despite this, I advise you to carefully review the Wine Bugzilla (`https://bugs.winehq.org/`) before filing a new report. Otherwise, you make the contributors work harder by duplicating issues. An example of a search for Notepad++-related bugs is shown here; there are hundreds: `https://bugs.winehq.org/buglist.cgi?quicksearch=notepad%2B%2B`. Despite this, its latest version works perfectly.

OK, now let's install **Notepad++**. You must first install the latest **Wine** and **wine-mono** (for .NET-based applications) versions, and it is strongly advised also to install **Winetricks**. It helps **Wine**, and is necessary for the **PlayOnLinux**, **q4wine**, and **Bottles** GUI frontends. If you are asked to install additional dependencies, I recommend it. Then, open the Wine AppDB and find Notepad++. There is info for each major application SW version in the first column and the Wine version it was tested with in the fourth: `https://appdb.winehq.org/objectManager.php?sClass=application&iId=2983`. I chose the latest Notepad++ version, *8.x*, tested with Wine *7.2*. My installed latest Wine version was *8.8-1*.

Open a Terminal and run the `$ winecfg &` command. The `&` (ampersand) sign will start the process in the background, freeing up the Terminal for further commands. This command will start the basic configuration script and create a `wineprefix`. It is a virtual C drive required by any Windows application. The default `wineprefix` C directory is placed in `/home/YourUser/.wine`. The `winecfg` command will also automatically open the **Wine configuration** GUI. There is nothing important you should do here now, but taking a quick look will help you know what you can control from it.

Now, download the Windows installer (in my case, the 64-bit version from `https://notepad-plus-plus.org/downloads/v8.5.4/`), right-click on it in the file manager, and select **Open with Wine Windows Program Loader**. This will start the installer. Carry out all the steps and, if you wish, also enable the **Run now** and the **Create Shortcut** options. After the installation, I had a shortcut on the desktop and could find **Notepad++** in the application launcher search. The whole process worked like a charm for me on all Manjaro flavors. I could even add application shortcuts on the corresponding panels and dash bars. The installed files were located for me in `/home/luke/.wine/drive_c/Program Files/Notepad++/`.

### q4wine

Another option we have is **q4wine**. The first time I ran it, I got the standard FOSS application notification and an explicit disclaimer that no help would be provided. On the other hand, the installer had a lot of options, and at the end gave me a link to a nice page for new users: `https://q4wine.brezblock.org.ua/documentation/en_US/05-first-steps.html?version=1.3.13`. After working with it – apart from giving access to some additional tools – I could not see any significant advantage over **Wine**. **q4wine** might be helpful occasionally for an experienced Windows and Wine user.

### Bottles

**Bottles** is our last option. It has even more options than **q4wine**. I liked it a lot. Unfortunately, after installing **Notepad++**, I could not make a desktop shortcut as some privileges were missing. When attempting to make a shortcut, I got a warning and was sent straight to the Bottles online documentation. After executing a single-line Terminal command, the privileges problem was solved, but the shortcut was missing. Despite this, the documentation is excellent, and I believe that with more digging, I would have succeeded. Here is the link: `https://docs.usebottles.com/`.

Working directly with **Wine** and **Winetricks** is the best of all tested applications. However, adding some complex or custom dependencies (including **DLLs**) might be easier with **q4wine** and **Bottles**.

### Microsoft Office

**MS Office**, unfortunately, is quite a challenge, as explained by the statuses here: `https://appdb.winehq.org/objectManager.php?sClass=application&iId=31`. To summarize – since MS Office *2013*, no later version has achieved a status higher than *Silver*. Theoretically, the *2019* version and the *365 Personal* will work at a satisfying level, but practically, I would expect problems. The reason is that Microsoft Office is an extensive, rich suite of multiple applications, depending on many libraries, OS calls, and modules. As a result, it generally doesn't work under Wine or any derivative.

### Windows software on Linux - Conclusion

Having a database with thousands of applications, the majority old and the others games, is insufficient to say we have good Windows application support.

On the other hand, if you need any old application or don't mind downgrading to an older version, it would most probably be supported. Wine supports Adobe, multimedia, and other products, and having over 28,000 applications in its database is significant.

A big advantage is that you can make a custom setup of a Windows application of your choice, but it would require advanced knowledge. **.NET** support is available in **Wine**. Many basic additional Windows libraries are provided, so it has enough resources. Any regular monolithic application shall work. You may need additional DLLs, which should be downloaded from servers (if available) or copied from Windows installations. **Q4wine** and **Bottles** can be helpful for such occasions.

# Summary

In this chapter, I have presented the rest of the regular/common software you might need. Keep in mind we didn't cover the hundreds of other available applications. As showcased, Linux offers a wealth of options across many fields.

In *Chapter 7*, we will continue with the Terminal and see how easy and effective it is to work with it. It will start with a good deal of essential explanations. Further, each presented command will have examples and explanations of the presented uses.

# All Basic Terminal Commands – Easy and with Examples

Even though modern Manjaro offers graphical SW for almost all daily tasks, the Terminal remains the primary place to carry out special tasks and advanced system management on Linux. Like almost all its users, I've never taken a course on it, nor has someone taught me how to use it. A simple explanation of approximately 30-40 commands can show a newbie how easy it is to work with, and that's how we all start. We'll separate these commands into basic and intermediate levels.

The sections in this chapter are as follows:

- The most important commands for newbies
- A bit of advanced (and still easy) commands
- Getting advanced help directly in the Terminal

## The most important commands for newbies

OK, to start, first we need to know *why*. *Why* the *Terminal* and not some fancy, easy **GUI** with buttons, menus, and a results window?

Between 1970 and 1998, computers were *not powerful, and many* basic administrative tasks were done via a **command line** (or a console) in a **Terminal**. Additionally, **GUI** SW required too many HW resources and was much harder to write. There were no rich graphics libraries, and for developers, it was a lot easier to write applications that get input from the terminal and then print the results there.

Even nowadays, many of the tasks presented in this chapter can be done in the fastest way via a terminal. Let's say we have an application that accepts 10 different commands, with 5 to 10 options each. This results in a minimum of 50 tasks that it can perform. A terminal-only application will only have one *input command processor*, directly triggering any given function. Although that's not easy, it's a lot easier than adding a rich GUI. The **User Experience/User Interface (UX/UI)** design also plays a role. Nowadays, this is a separate profession, as designing and creating interfaces requires clear and correct ways of controlling and presenting the results.

Last but not least, terminals are *rich* in options, often providing *colored output* for faster and easier comprehension. They are the fastest way to remotely connect to a system or access and control one without a graphical environment such as **Xfce** or **KDE**.

## What are Shell, BASH, and Zsh?

A **shell** (sometimes called a command-line interpreter) is a program in the OS user space that takes text commands from a terminal and interacts with the OS kernel to execute them. The first **Unix shell** was created by Ken Thompson at **Bell Labs** in *1971*. In *1979*, Steven Bourne made an improved version, named after him, called the **Bourne shell (sh)**. As **Unix** was **proprietary**, so were these two shells. **BASH** stands for **Bourne-Again Shell** and was released as GNU GPL **FOSS** SW in *1989*. It has been the *default shell* in many Linux distributions *for decades*, as it was the first one ported to Linux by Linus Torvalds. It is powerful and still accepted by many people as the primary standard for a **command-line** *input processor*.

There are other shells with *more* or *less* similar syntax to **Bash**, such as **Zsh**, **ksh**, **tsch**, and **fish**. Many of them also add more features. However, the differences between these shells are important only for advanced shell usage and scripts. For simple command execution, which we review in this chapter, they are *irrelevant*.

## What is a Terminal?

Nowadays, the Terminal is the **GUI**-based application that opens when you start the application with this name. It is independent of the shell that processes the commands. This GUI is called a **Terminal application** or a **Terminal emulator**.

Historically, terminals were single points of access to a *multi-user system*. Let's say we have one powerful, expensive CAD design *computer* to be used by multiple users. It is often called a server. Let's say we also have in the building many weak computers to access it remotely via the internal network. In the past, these less powerful computers were called **terminals**.

In *2023*, Manjaro **KDE** comes with the terminal application **Konsole**. Its *default shell is* Zsh (not **Bash**). **GNOME** comes with **GNOME Terminal** and its *default shell* is also Zsh. Xfce comes with **Xfce Terminal** and its default shell is **Bash**. To emphasize again, in terms of *basic features and the regular terminal commands presented in this chapter, there is no difference* between **Bash** and **Zsh**.

Let's start with the commands. Try them out while reading. After all, learning by doing is often the most effective method.

# For beginners

Open the main menu, type `terminal`, and press *Enter*. By default, the terminal starts in *your user home directory*. In the following examples and throughout the book, when I write a full command, I've preceded it with the dollar sign $ to designate this. Don't type this dollar sign in your terminal.

## Pure basics – pwd, cd, mkdir, rmdir, and exit

**pwd** is the *print working directory* command. When executed in a freshly opened terminal, it will return the home directory as follows:

```
$ pwd
/home/luke
```

**ls** (a lowercase L, then s) is the `list` command. We will explore it in detail later. For now, know that $ `ls -l` (lowercase L, s, space, minus, lowercase L) lists the files and subdirectories of the current directory.

**cd** is the **change directory** command. We execute `cd DirectoryName` to navigate to a directory such as `Downloads`. Then, we can list its contents with `ls -l`, as shown in *Figure 7.1*:

```
 ▥  ▞ ~   ls -l
total 1584
drwxr-xr-x 3 luke luke    4096 Aug 25 20:59 build
-rw-r--r-- 1 root root     149 Sep  7 13:17 crontab.config
-rw-r--r-- 1 luke luke  150018 Oct 15 12:33 DateReport.log
drwxr-xr-x 2 luke luke    4096 Jul 24 11:16 Desktop
drwxr-xr-x 2 luke luke    4096 Jul 24 11:15 Documents
drwxr-xrwx 2 luke luke    4096 Oct 15 12:31 Downloads
drwxr-xr-x 3 luke luke    4096 Jul 24 13:11 git
drwxr-xr-x 3 luke luke    4096 Sep  6 17:39 go
-rw-r--r-- 1 luke luke    1247 Oct 15 12:33 log
drwxr-xr-x 9 luke luke    4096 Oct 15 12:27 man_Pages
-rwxr--x-x 1 luke luke     878 Sep  6 21:02 myScript.sh
drwxr-xr-x 2 luke luke    4096 Jul 24 11:15 Pictures
-rw-r--r-- 1 luke luke   11222 Sep  1 20:08 SYSTEMCTL_COMMAND_Options.cpp
-rw-r--r-- 1 luke luke  104705 Aug 31 15:57 SYSTEMD_BOOT_SEQ1.svg
-rw-r--r-- 1 luke luke   82283 Jul 24 11:56 SYSTEMD_BOOT_SEQ.svg
-rw-r--r-- 1 luke luke 1142606 Aug 31 19:21 systemd_dump.log
-rw-r--r-- 1 luke luke   62445 Sep  1 16:05 systemd_user_start_Seq.svg
-rw-r--r-- 1 luke luke    2655 Oct 15 12:30 UserReport.log
drwxr-xr-x 2 luke luke    4096 Jul 24 11:15 Videos
 ▥  ▞ ~   cd Downloads
 ▥  ⌂ ~/Downloads   ls -l
total 16416
-rw-r--r-- 1 luke luke 16746944 Aug 23 16:57 expressvpn-3.53.0.0-1-x86_64
-rw-r--r-- 1 luke luke     9314 Oct 15 12:31 images1.jpg
-rw-r--r-- 1 luke luke     7226 Oct 15 12:31 images2.jpg
-rw-r--r-- 1 luke luke     6657 Oct 15 12:30 images.jpg
-rw-r--r-- 1 luke luke    30935 Oct 15 12:30 panda.avif
 ▥  ⌂ ~/Downloads   ▮
```

Figure 7.1 – Example of the ls and cd commands

cd .. (cd with a *space* and *two dots*) will return you to the *parent directory* (in this case, home).

' . ', the single dot character, is used as an argument in paths *to designate the current directory*. We will see it in use later.

**mkdir** is the **make directory** command. For example, with mkdir TestDir, we will create a directory called TestDir in the current one. The ls  -l results will now also contain the new directory.

**rmdir** is the **remove directory** command. To delete TestDir we type $ rmdir TestDir. The deleted directory must be empty, or you will get an error message. We also must be "outside" of it (one level up).

mkdir and rmdir provide status if the -v **option** (also called a **parameter**) is provided before the argument, like this: $ mkdir -v TestDir.

The **exit** command closes the current terminal. If we have multiple open tabs in the terminal, it will close the current tab. It has no parameters. After trying it, open a new terminal to continue with the chapter.

## Commands, options, parameters, and arguments

**Arguments** are any strings after a command's name that modify its operation. For GNU programs, arguments starting with *dashes* are **options** or **parameters**. Thus, in $ mkdir -v TestDir, TestDir is an *argument* and -v is an *option*. All **GNU** programs use a *single dash* for *short options* and *a double dash* for *long ones*. The long alternative to -v is --verbose; you can use whichever form you want. Most of the programs presented in this chapter are GNU ones and they follow this styling.

Many modern programs don't use dashes for all their parameters/options, so *don't consider this a law*. In addition, some people use the words *parameters* and *arguments* as synonyms, despite the GNU, Unix, and other definitions.

> **Important note**
>
> Strings in any Linux terminal are *case-sensitive by default*! This includes *all commands, parameters, options*, and *arguments*. Some modern terminals can correct mistakes between lowercase and uppercase, but this is rare! So, LS and ls are *two different commands*. Generally, *all commands and their options are written in lowercase*. The main exception is for some programs with many options – in this case, some parameters can be written in uppercase.

## Zooming and terminal keyboard controls/shortcuts

All the terminals – **KDE Konsole**, **Xfce**, and **GNOME** – support **zooming** with *Left_Ctrl ++* and *Left_Ctrl +-*. However, in **GNOME**, it only works with the + and – keys next to the *backspace* key, not the numeric pad + and –. **KDE's** terminal supports **zooming** with *Left_Ctrl* + mouse scroll, while on **Xfce**, this is done with *Left_Shift+Left_Ctrl+* mouse scroll. All terminals support paging up and down with Shift+PgUp/PgDn.

Most terminal applications have many GUI and functional options via onscreen buttons or right-click menus. In **Xfce**, they can be found by going to **Edit | Preferences**. In **GNOME**, you must go to **Preferences** from the three-dots menu.

It is a bit complicated for **KDE's Konsole**, as they often improve and change their menus. Buttons might not be visible by default, so right-click on the top bar space and open the **Configure Toolbars** menu. In it, select **Main Toolbar < konsole >**, enter the word `Manage` in the **Available actions** filter, and then add the **Manage Profiles…** action to the **Current actions** box using the arrow button, as shown in *Figure 7.2*:

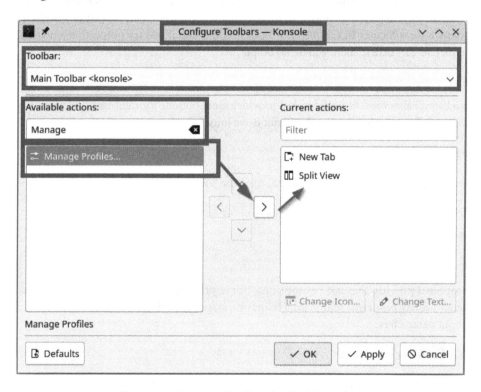

Figure 7.2 – Configure Toolbars for KDE's Konsole

Then click **Apply**. In the same way, I added the **Switch Profile** and **Edit Current Profile** icons to the **Session Toolbar**. Then, in **Manage Profiles**, first set **Breath** as the default profile (if it is not already). Now click the **+ New** button to create a new profile. Name it as you wish, and you can now edit its scrollback, color scheme, keyboard shortcuts, and so on. Finally, set the new profile as *default*.

In **Xfce**, the keyboard shortcuts are unfortunately *not listed*. According to the official Xfce documentation (`https://docs.xfce.org/apps/xfce4-terminal/start`), we have to edit a `.scm` file located in `/home/OurHomeDirectory/.config/xfce4/terminal/accels.scm`. Still, due to a current issue (as of July 2023), keyboard shortcuts are not present on Manjaro Xfce.

### echo and printenv

**echo** prints on the terminal the string it is given as an argument. If the argument is an *environment variable*, it will output its value. Here is an example:

```
$ echo Hello
Hello
```

`$ echo $USER` outputs the current user *name*, and the $ prefix refers to an *environment variable* (in my example, the output would be `luke`). Environment variables provide information about the current context; we will review them in *Chapter 8*. The `echo` *string* output is particularly useful in **shell scripts** (which we will cover in *Chapter 15*). Keep in mind that if you try outputting a non-existent variable, such as $HsdGER, there will be no output.

**printenv** lists all *environment variables*.

**env** again lists all *environment variables* but provides more options for their manipulation. We have two commands for historical reasons I will not delve into.

### Getting help

*To get short help for almost any command*, you can type one of the following:

- `$ command --help`
- `$ command -h`
- `$ command -?` (really rare)

Most programs provide the first two options. Some provide only the first or only the second. Additionally, many programs will print a helper message on calling their help if you provide the wrong input parameters.

### The rich listing command

`ls` is the list directory contents command, as we have seen. Its syntax is as follows:

```
ls [options] [argument]
```

`argument` is a path to a directory or file. It is typically used with a directory to list its contents. If a file is provided, it lists its properties. If *no argument* is provided, it lists the *current directory contents*. Its parameters are rich (almost the complete English alphabet), and they modify the listing details and formatting. Here are the most important ones:

- No parameters – this will list directories *in a row*
- `-l` – this is the *long vertically listing* option, providing file/directory *permissions* at the beginning, then user *ownership*, *size* in bytes, and the last *modification* date

- -h – when added to -l, the *size* in MB is in a *human-readable format*

- -a – includes hidden files, which in Linux *start with a dot*

An example of their combination, $ ls -lha, is shown in *Figure 7.3*:

```
[atanasr@atanas-virtualbox          ]$ ls -lha
total 327M
drwxr-xr-x  2 atanasr atanasr 4.0K Jul 10 19:00  .
drwx------ 22 atanasr atanasr 4.0K Jul 10 17:10  ..
-rw-r--r--  1 atanasr atanasr  24M Jul  6 19:06  Game-of-Fifteen-x86-64.appimage
-rw-r--r--  1 atanasr atanasr 2.6M Jul  6 20:44  gnome-mahjongg_3.38.3-2_amd64.deb
-rw-r--r--  1 atanasr atanasr 4.5M Jul  7 12:40  npp.8.5.4.Installer.x64.exe
-rw-r--r--  1 atanasr atanasr 9.7M Jul  3 18:40 'Romeo Santo - Eres Mía.mp3'
-rwxr-xr-x  1 atanasr atanasr 287M Jul  4 11:38  viber.AppImage
[atanasr@atanas-virtualbox          ]$ █
```

Figure 7.3 – ls detailed listing example

For ls, *the parameters order* after the dash is *unimportant*. Some commands are like it, while for others, having the wrong order or parameter combination may lead to errors. You will usually get at least a basic error message on such occasions. The following are a few other interesting ls options:

- -g – like –l, but *doesn't* list the *owner*

- -r – list in *reverse* order

- -R – lists subdirectories *recursively*, which means listing all subdirectories of the current one, with their contents and sub-subdirectories, their contents, etc.

- -S – sorts files (not directories) by *size*, starting with the *largest file*

- -t – sorts listing by *last modified*, with the *newest* first

ls is rarely used without the -l option, as most modifiers work with it. Here are a few nice combinations to try:

- ls -lhag – lists contents with details in *alphabetical* order, including *hidden* items, with *permissions* and *human-readable sizes*, and hides the group

- ls -lShag – lists contents with details in *reverse order of size*, including hidden items, with permissions and human-readable sizes, and hides the group

- ls -ltag – lists contents with details in *reverse* order of *most recent modification*, including hidden items, with permissions and human-readable sizes, and hides the group

To see all possible ls options execute $ls --help. Think of some combinations and try them out. The permissions are explained later in the chapter.

### Usage of the up and down arrows for the command buffer

Once you have executed *several commands*, pressing the *Up* arrow on your keyboard allows you to go through the list of *already executed expressions*. Pressing the *Down* arrow will return you to the most recent commands. This is helpful, especially for longer and more complex calls you want to execute again. Try it out.

### Long outputs, Ctrl+C, and less

When we execute $ ls  -1Ra in our home directory (listing all files, including the hidden ones, **recursively**), the result is typically *up to thousands of lines*. If the command is called by mistake, pressing *Ctrl+C* on the keyboard will send an interrupt signal (SIGINT) to the executed application and *stop it*. To try this, run $ ls  -1Rga  /. (which will start listing *all files starting from the root directory*), and then press *Ctrl+C*.

For some applications, pressing *Esc* is also an option. However, we often *need this long output*, and we can use the **less** or **more** programs to scroll through this result output in the terminal.

The less program *is better than* more. To use it, we need to *understand the concept of piping*.

### Streams, piping, and less

Each program (like ls, cd, and echo) has three **Input/Output (IO) streams**, listed in *Table 7.1*:

| IO stream ID | Name | Description |
| --- | --- | --- |
| 0 | stdin | The standard input is all arguments, parameters, and data (including files) we provide to a program when calling it. When a longer program is running, and it interactively asks you to type a choice via the keyboard, this is also stdin. |
| 1 | stdout | The standard output is all the data provided as results, such as strings, result status, listings, and output files. |
| 2 | stderr | The standard error are typically error messages or error logs in a file. |

Table 7.1 – Standard streams of programs

Any shell allows us to direct the stdout of one command to be the stdin of another. This is done via the vertical line symbol, |, named piping. In other words, **piping** is the *consecutive usage of commands* in this way so that the output of one command is the input of the next one. You can pipe tens of commands, but typically, people pipe up to several. Try it now with the long ls listings in your home directory:

```
$ ls -1Rah | less
```

This will open the text output via the less program, and there you can use the main less options:

- *PgUp* or *PgDn* (or *b* and *the spacebar*) – to navigate up and down in the list.
- *Up/down/left/right* arrows – for navigation.
- /SomeString – type a forward slash / and a string (e.g., jpg) to search forward for the string.
- ?SomeString – type a question mark and a string (e.g., jpg) to search backward for the string.
- g – goes to the start of the listing.
- G – goes to the end of the listing.
- ng – jumps to line number *n*. The default is the start of the listing.
- nG – jumps to line number *n*. The default is the end of the listing.
- n – goes to the next match (after a successful search).
- N – goes to the previous match.
- s – saves the current content (got from another program such as grep) in a file. This option is explained in the next *Important note*.
- h – displays help; press *q* to exit the help.
- q – quits less.

> **Important note**
>
> When you exit less, the *long output will disappear*! Thus, if you need to save it to a file, press *S*, enter the filename, and press *Enter*. Finally, press *Q* to *quit when ready*. You can open this file with **nano**, **Kate**, or other editors. By default, less will save the file in the *current working directory*.

During my previous run of less, I saved a log file in the home directory. To open any such text file, simply provide it as an argument like this: $ less logFile.log.

## Dumping files with cat

The **cat** command dumps (prints) character (text) files to the terminal. To do so for the previous log file, we write the following:

```
$ cat logFile.log
```

There is a tac program, which does the same, but in reverse line order. We also have the head and tail programs, which print, respectively, the first and last 10 lines of a text file.

## Clear and the terminal buffer

After so many commands, we have a long listing to scroll through in the terminal buffer. To clean it, we call `clear` and press *Enter*.

The terminal buffer length depends on the settings of your terminal. In **GNOME**, we have a short limit of only *10,000 lines* by default. **Xfce** has a limit of *999,999 lines*, and **KDE** has *the ability* for *unlimited scrollback* (saved to the `/tmp` directory). When you exit the terminal, the current buffer is *deleted*. All three terminals support tabs and have scrollback options in their preferences menus. **Konsole** has *split views* and rich settings. Though the **GNOME** Terminal is *minimalistic*, it has plenty of basic options in **Preferences | Profiles**. Check them out.

## Path usage

We will go deeper into the **Linux** directory/filesystem structure in *Chapter 9*, but here, we need to cover some basics to understand **paths**.

The `root` directory (with a path of a single slash, `/`) is where everything exists and starts. There are many system directories in it, such as `bin`, `dev`, `etc`, `lib`, `lib64`, `proc`, `root`, `sys`, and `tmp`. The `home` directory, in which we have *our user directory* (in my case, `luke`), is also there. The full path to my `Downloads` directory is as follows:

```
/home/luke/Downloads
```

In a path, each next forward slash designates another subdirectory. Only the last item in the string can be *a file* instead of a subdirectory. *Just like the commands and their parameters, all Linux/Unix-based filesystems are case-sensitive*. Thus, `Downloads` and `downloads` are two different directories!

We can work with this or similar paths *no matter where we are currently*, and we call this an **absolute path**. If we go into the `Pictures` directory to do some work but need a listing of the `Downloads` directory, we can run `ls` with a full path as shown in *Figure 7.4*. By using `pwd`, we can check where we are at the current moment.

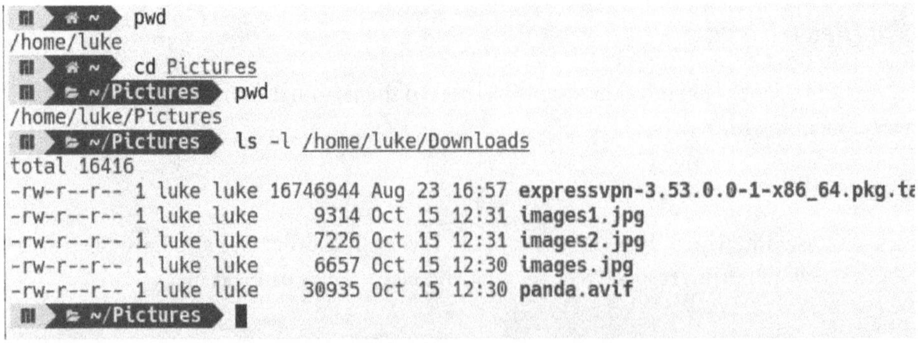

```
▥  ⌂ ~   pwd
/home/luke
▥  ⌂ ~   cd Pictures
▥  ⌂ ~/Pictures   pwd
/home/luke/Pictures
▥  ⌂ ~/Pictures   ls -l /home/luke/Downloads
total 16416
-rw-r--r-- 1 luke luke 16746944 Aug 23 16:57 expressvpn-3.53.0.0-1-x86_64.pkg.ta
-rw-r--r-- 1 luke luke     9314 Oct 15 12:31 images1.jpg
-rw-r--r-- 1 luke luke     7226 Oct 15 12:31 images2.jpg
-rw-r--r-- 1 luke luke     6657 Oct 15 12:30 images.jpg
-rw-r--r-- 1 luke luke    30935 Oct 15 12:30 panda.avif
▥  ⌂ ~/Pictures   █
```

Figure 7.4 – Example usage of an absolute path

## The tilde symbol (~)

The tilde symbol ~ is typically (and historically) used to designate an **absolute path** to a user's home directory. In a terminal command, it is substituted for me with /home/luke. Thus, if I write $ ls -l ~/Downloads, the command will be extended to $ ls -l /home/luke/Downloads. That's why, in some of the screenshots (mostly taken from Konsole on KDE or GNOME Terminal), you can see a tilde at the beginning of the line, like in *Figure 7.4*.

## Usage of the Tab key for autocomplete

Now that we know about **absolute paths**, there is an easy way to work with them in the terminal. Say we want to use the /home/luke/Downloads path. It is enough to write *the first few letters* from each directory and press *Tab* to *finish the given subdirectory name automatically*. If there are more than one option, they will be displayed. Try the following:

```
[luke@minis home]$   ls -l /h
```

Press *Tab* to get the following result:

```
[luke@minis home]$ ls -l /home/
```

Then, *add the first two letters* of your user name, in my case, lu:

```
[luke@minis home]$ ls -l /home/lu
```

Press *Tab* again and you will get this result:

```
[luke@minis home]$ ls -l /home/luke/
```

Then, add *the first two letters* of the Downloads directory:

```
[luke@minis home]$ ls -l /home/luke/Do
```

Press *Tab* again, and the results provide two options:

```
[luke@minis home]$ ls -l /home/luke/Do
Documents/ Downloads/
```

Just type w:

```
[luke@minis home]$ ls -l /home/luke/Dow
```

Press *Tab* again, and the path is *automatically filled* with Downloads:

```
[luke@minis home]$ ls -l /home/luke/Downloads/
```

Finally, press *Enter* to execute the command.

## Commands and applications

Every **command** is an **application**. Thus, hwinfo, ls, cd, echo, rmdir, mkdir, and so on are applications *installed* on your system. When one command is missing, *you should simply install it to use it*. Once installed, it can be used with your terminal's *Tab* autocomplete functionality, which is *the standard for a modern Linux terminal*. This is also valid for GUI applications such as firefox and brave. Try it by typing fire and pressing *Tab* to fill the letters fox, then press *Enter* to execute it. *Ctrl+C* will also work here from the terminal as an alternative to the GUI close button.

### tree

**tree** is a nice tool for exploring the directory structure in Linux. Executing it without arguments will list all files, directories, and subdirectories files in the current directory. Adding –L X (where X is a number) will do so X levels deep. Try this command:

```
$ tree -L 1 ~
```

The result for me is shown in *Figure 7.5*:

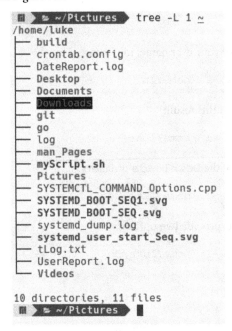

Figure 7.5 – Example of tree

You can try this as well to see the root directory structure. Check the results of $ tree -L 1 /. Try the same with -L 2 /.

### Working with relative paths

While we are in `Pictures`, say we want to list the *parent* directory. We can use *two consecutive dots* to designate this path. What does this mean? The following command, `$ ls -l ..`, is equivalent to saying to `ls`, *run one directory up*. This forces `ls` to calculate that we want to run it *one level above the current one*. For example, with `/home/luke/Pictures`, it removes the last directory and runs `ls` as if I had written `$ ls -l /home/luke`.

We can even list `home` from `Pictures`. Just type `$ ls -l ../..`; as shown, we again use the forward slash separator `/` to designate sub- or upper directories.

This is called working with **relative paths**, as we make the location relative to our current directory. From `Pictures`, we can also say *go one directory up*, then enter a subdirectory, such as `snap`, and *list its contents*:

```
$ ls -lah ../snap/
```

For many commands, using *one dot designates the current directory*. So, if we want to list the `git` subdirectory with `tree` (located in my `home` directory) with all sublevels, we use one dot to designate the current directory:

```
$ tree ./git/
```

It shows one subdirectory with its contents, as shown in *Figure 7.6*:

```
                    tree ./git
./git
└── CPP11_ThreadPool
    ├── LICENSE
    ├── License.h
    ├── main.cpp
    ├── README.md
    ├── thread_pool.cpp
    └── thread_pool.h

2 directories, 6 files
```

Figure 7.6 – Example of tree used from the current directory to list a subdirectory

We can *skip* the *single dot and first slash* ( . / ) in the last example, but I've included it to show the meaning of the single dot.

### history and session

A session is the time between one opening of a terminal application and its **exit**. This time can be from seconds up to weeks or months.

**history** is a simple command to list the history of commands you have used at least during your current terminal session. Some terminals preserve it for the complete user login session, even when opening and closing terminals. We scroll through these entries when we use the up and down arrows. After we have tried so many commands, try using the arrows first, then try out the `history` command to see them all.

### Permission settings, chmod, and chown

Now that we have listed so many files using `ls` with the long listing option `-l`, we get a curious and *essential* string at the beginning of each entry line:

```
.rw-r--r-- 1 luke 18K  Jul 11 16:15 2023-07-11_16-15.png
drwxr-xr-x 3 luke 4.0K Jul 11 15:39  git
```

This string is the file or directory **permissions**. Having a dash means no permission, while letters have a dedicated meaning. The full format is `drwxrwxrwx`.

If this is a **directory**, the first symbol is `d`; if this is a file, it is a `dot` or a `dash` (depending on the terminal).

`r`, `w`, and `x` refer to **read**, **write**, and **execute**. If *the given letter is present*, we have the given permission. We have *three groups* of `rwx` for the file/directory `owner`, its `group`, and `others`. Thus, `"- rwx rwx r--"` means the following:

- This is a **file** (not a directory), designated by the beginning dash
- *Its owner and the group have all permissions* (read, write, and execute)
- *Any other users* belonging to other groups can only read it

`r`, `w`, and `x` can be presented by an octal value: 4 for read, 2 for write, and 1 for execute. Here are all the possible combinations:

- 1 means only execute.
- 2 means only write.
- 3 means write and execute but not read (2 + 1). This is extremely rare.
- 4 means only read.
- 5 means read and execute, but not write (4 + 1).
- 6 means read and write, but not execute (4 + 2).
- 7 means all permissions (4 + 2 + 1).

The program used to modify the permissions is `chmod`, and the one to change the owner is `chown`.

**chmod** stands for **change mode**. Writing `$ chmod 760 example.txt` in the terminal will set these permissions: `"-rwx rw- ---"`.

There is another common input form:

```
$ chmod u=rwx,g=rx,o=r fileName
```

u is for user (the owner), g is for group, and o is for others.

To create a file with fileName, execute:

```
$ touch fileName
```

After each of the following examples, you can use $ ls -l fileName to check the results.

You can *update one of the permission groups* like this:

```
$ chmod u=rwx fileName
```

With g= and then nothing else, you *remove all permissions for the group*.

We can also use the + sign to add explicitly one permission. The next command will add execution permissions to the group:

```
$ chmod g+x fileName
```

With the minus sign, –, we *can remove specific permissions*, such as *disallowing* **others** to execute the file here:

```
$ chmod o-x fileName
```

With a, we can *set the permissions for all groups*, granting, for example, write permission to *everyone*:

```
$ chmod a+w fileName
```

We can also modify two permission groups at once: $ chmod ug+x fileName.

The -R option allows us to *recursively change the permissions for a directory and all files in it*. The following command will give *all permissions to the user*, only r and w permissions to the group, and *no permissions* to others for the directory and all files in it:

```
$ chmod -R 760 ./Directory_With_Files
```

**chown** is the command we use to *alter the ownership of the file*. It is used to change the user or group ownership of the file. Let's look at this listing:

```
-rw-r--r-- 1 luke luke 18K  Jul 11 16:15 img.png
drwxr-xr-x 3 luke luke 4.0K Jul 11 15:39 git
```

The first `luke` is the user, and the second `luke` is the group the user belongs to. A group with the user's name is automatically created when the user is created, so *the duplication is not an error*. *Chapter 13* has more information on this and another important command, `chgrp`, for modifying *a user's group memberships*.

### sudo

When a user is created, they can operate only with files in their home directory. They can also execute programs from some system directories, but without writing to them or changing their files. As the *first Manjaro user created during installation* is added to several groups for system management, it is of type **administrator**. Thus, we can carry out system-critical actions such as SW (de)installation, network management, and power options. Despite being an administrator, *we are asked for our password*.

As an administrator and first user, we are also added to the `sudoers` users group, which can *perform actions from the* **superuser**'s *name*. This way, we can temporarily execute a command as if we are **root**. To do this, we need to put the `sudo` command before a command and its arguments, and before the execution, we are asked for our *user password*. This allows us, for example, to read **root**-owned files located in the `/etc` directory. Let's see an example.

In the **Terminal**, if we try to read most of the files in the `/etc/` directory, we will not be allowed to. We know we have such a case when we see a `Permission denied` operation result. Then, we add `sudo` at the beginning of the command to request temporary **root** user **privileges**. **sudo** stands for **superuser do** or **substitute user do**. Let's try dumping `/etc/sudoers`:

```
$ cat /etc/sudoers
```

The result will be as follows:

```
$ cat /etc/sudoers
cat: /etc/sudoers: Permission denied
```

How can we know this in advance? When listing the `/etc/` directory contents with `ls -l`, we see that *the permissions for this file* are as follows:

```
.r--r----- 1 root root      3319 Jul 11 13:49 sudoers
```

When we add `sudo` and provide our user password, we get the following result:

```
$ sudo cat /etc/sudoers
[sudo] password for luke:
## sudoers file.
##
## This file MUST be edited with the 'visudo' command as root.
...
```

More information on **sudo**, **su**, user management, and **root** privileges will be provided in *Chapter 13*.

# A bit of advanced (and still easy) commands

We continue in this section with more frequently used commands. Though we've called these *advanced*, most of the examples we'll provide are easy. If you don't need some complex parts, such as the regular expressions for grep, just skim through them and keep the book easily accessible to return to those sections when you need them.

## Installing a missing command (application)

Almost all mentioned commands are preinstalled on all Manjaro flavors and well spread on most Linux distributions. However, installation is easy if you have installed a minimal version or don't have some of them. You can either use the **Pamac GUI** or, better yet, directly type the following in the terminal:

```
$ sudo pamac install someApplicationName
```

For the `neofetch` tool, this would be:

```
$ sudo pamac install neofetch
```

We will explore the subject of Pamac and its rich command-line options in *Chapter 8*.

There is one great replacement for `ls` called `exa`, but I chose to use `ls` in all my examples as this is the most widespread tool. `ls` is also used in tons of existing scripts and tutorials. Install `exa` with `$ sudo pamac install exa`; it has better coloring than `ls`. Try `$ exa -lah`, `$ exa -lgRah`, and `$ exa -Tlah`. You can find its other options in the help. Unfortunately, sending the results to `less` *clears the coloring*.

## grep and more piping

The next widely used command, even sometimes as a verb, is `grep`. The name stands for *Globally search based on a Regular Expression and Print it*. This has been one of the most used commands for decades, as it allows us to find any complex strings in any text file. Its first release was for Unix in *1973*.

Remember that `grep` looks for a given pattern until a new-line character named **Line Feed** (**LF**). This is the **Linux** and **macOS** style. If you provide **Windows** text files, each line will end with a **Carriage Return-Line Feed** (**CR-LF**) pattern.

The standard form of `grep` is as follows:

```
grep [OPTIONS] 'string' [FILE LIST]
```

The options are not obligatory, but `string` and `FILE LIST` are. The last can be a list of exact filenames in the current directory or paths to files in another directory. The only exception is if you use `grep` piped to the output of another command; in this case, only the search string is obligatory.

If you use the -r option for *recursive* directory search (without FILE LIST), it will look into all files *in the current directory and its subdirectories*. If you use an asterisk symbol * for FILE LIST, it will look into files only in the current directory. We will see several examples of -r and * usage later.

A *simple search string* (without *dashes*, *spaces*, and *special symbols*) can be without quotes. With the following command, we look for the word hello, *case insensitive* (-i), in the files menu.h, main.c, and titles.cpp:

```
$ grep -i hello menu.h main.c titles.cpp
```

The next example is of a recursive listing with ls for the / directory:

```
$ sudo ls -lgRah / | grep socket | less
```

This will first list all the contents of the / directory *recursively and without printing them on the* terminal. We need sudo, as most files would *not be listed* otherwise. Then, we pipe the output to grep with | and look for the string socket. We finally *pipe* to less, as the result is long.

### The options for grep

The following are the essential options for grep:

- -i – case insensitive.
- -w – whole words only.
- -v – inverts to only print out lines that don't match the searched string.
- -f – takes the strings to search for from a file.
- -m NUM (e.g., -m 100) – stops after NUM matches are found.
- -n – prints the line number of each printed matched line.
- -r, --recursive – looks for strings recursively.
- -H – prints the *filename* before the results from it. This is good if you're looking into multiple files.
- --color – prints the *matched pattern in color*.
- -o – prints *only the matched part* of the line.
- -l – prints the filename only *without the matching line*.
- -B NUM – prints NUM lines *before* the matched line.
- -A NUM – prints NUM lines *after* the matched line.
- -C NUM – prints NUM lines of output context (*before* and *after* the matched line.
- -s – suppresses error messages (about non-existent or unreadable files).

The next options are for **regular expressions**, referred to in the Linux world with the abbreviations **regex** and **regexp**. We will go through a few examples with explanations for them later in the section:

- `-E, --extended-regexp` – *extended* regular expression

- `-F, --fixed-strings` – *strings*

- `-G, --basic-regexp` – *basic* regular expression

- `-P, --perl-regexp` – Perl regular expression

- `-e, --regexp=PATTERNS` – using `PATTERNS` for matching

There are *even more options*; you can look on the web or the original full `grep` manual at `https://www.gnu.org/software/grep/manual/grep.html`.

However, let's start with some simple practical examples.

*Example 1*: Look recursively for `socket` *in any file* (depicted by `*`) in the `etc` directory. *Color* the output. As we look through many files, *their full paths will be listed*, along with the *line numbers*. This example shows we don't need to **pipe** `ls` in front:

```
$ sudo grep --color -n -w socket -d recurse /etc/*
```

*Figure 7.7* presents part of the results:

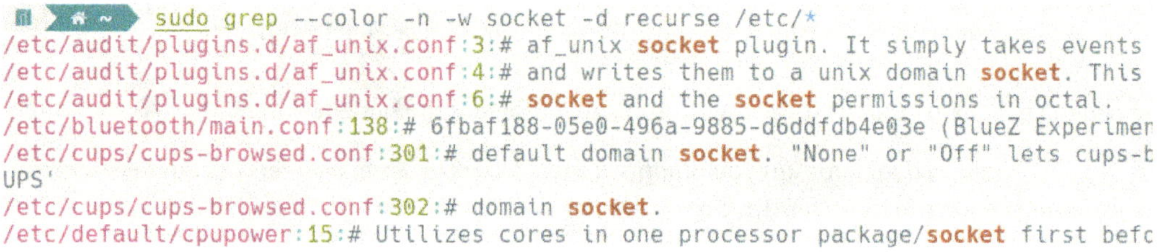

```
#  ~    sudo grep --color -n -w socket -d recurse /etc/*
/etc/audit/plugins.d/af_unix.conf:3:# af_unix socket plugin. It simply takes events
/etc/audit/plugins.d/af_unix.conf:4:# and writes them to a unix domain socket. This
/etc/audit/plugins.d/af_unix.conf:6:# socket and the socket permissions in octal.
/etc/bluetooth/main.conf:138:# 6fbaf188-05e0-496a-9885-d6ddfdb4e03e (BlueZ Experimer
/etc/cups/cups-browsed.conf:301:# default domain socket. "None" or "Off" lets cups-b
UPS'
/etc/cups/cups-browsed.conf:302:# domain socket.
/etc/default/cpupower:15:# Utilizes cores in one processor package/socket first befc
```

Figure 7.7 – Complex grep results with coloring and line numbers

*Example 2*: Look *only in the current* directory for all files containing `socket`:

```
$ sudo grep socket *
```

*Example 2.1*: Look *recursively* in the *current* directory for all files containing `socket`:

```
$ sudo grep -r socket *
```

*Example 2.2*: Look *only in the current* directory for all files containing `socket` and color the output:

```
$ sudo grep --color socket *
```

*Example 2.3*: Look *only in the current* directory for all files containing socket and color the output, but suppress any extra messages for directories or unreadable files:

```
$ sudo grep -s --color socket *
```

*Example 3*: When using the *wildcard character*, which is a dot '.', grep looks for no specific symbol in its place. Thus, the following command will look for any string that starts with 19, is followed by any two symbols, continues with 168, and has six more random symbols. This gives us any *IP address* that starts with 19x.168 in the files contained in the etc directory:

```
$ sudo grep -w '19..168......' --color -n -d recurse /etc/*
```

*Example 4*: The next example shows how we can look for combinations of strings – in this case, starting with Ma and ending with n *or* p:

```
$ sudo grep -w 'Ma[n,p]' --color -n -d recurse /etc/*
```

The result for me is as shown in *Figure 7.8*:

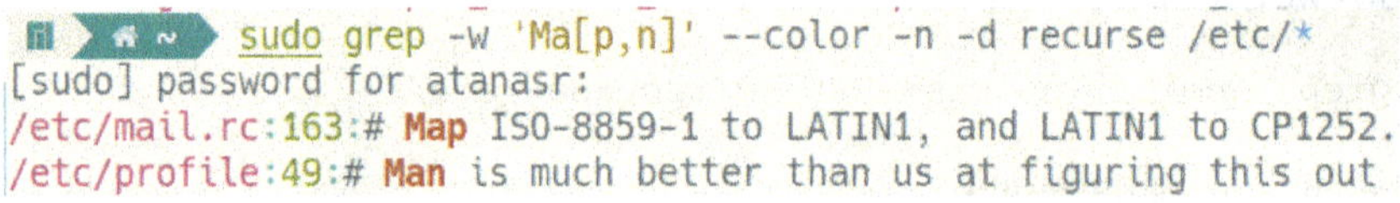

```
  ⚡ ⌂ ~   sudo grep -w 'Ma[p,n]' --color -n -d recurse /etc/*
[sudo] password for atanasr:
/etc/mail.rc:163:# Map ISO-8859-1 to LATIN1, and LATIN1 to CP1252.
/etc/profile:49:# Man is much better than us at figuring this out
```

Figure 7.8 – Matching two words with only a different last letter with grep

The comma in square brackets is *optional*. Omitting it provides the same result.

*Example 5*: We can also look for *any combination* of two letters given in two sets of square brackets. The following command looks for dp, dn, mp, or mn:

```
$ sudo grep -w '[dm][pn]' --color -n -d recurse /etc/*
```

### regex and regexp

Regular expressions are powerful, as they can extend the options of the searched strings. **grep** calls them **regexp,** while the **find** command uses **regex.** The following examples help us understand the basics.

*Example 1*: This command uses a regular expression to look for either 44 or blue (case insensitive):

```
$ sudo grep -E -w '44|blue' -i --color -n -d recurse /etc/*
```

This one does the same:

```
$ sudo grep -E -w '44|blue' -i --color -n -r /etc/
```

These two variants show the equivalence of -d recurse and the short option -r.

*Example 2*: Look for a line beginning or ending with a certain string with ^ or $.

^http matches all lines that start with the http string:

```
$ sudo grep -E -w "^http" -i --color -n -r /etc/
```

conf$ matches all lines that end with the conf string:

```
$ sudo grep -E -w "conf$" -i --color -n -r /etc/
```

*Example 3*: . is used as a **wildcard character** – that is, to *match any character*.

^ . $ will match all lines with *a random single character*. If we want to display the two lines after each search result, we need to add -A 2:

```
sudo grep -E -w "^.$" -i --color -A 2 -n -r /etc
```

### Looking for special characters and using square brackets

Use the backslash, also called an **escape sequence**, to look for special characters. \ * will match the lines that contain the * character. However, to match $, you may need to add two backslashes. Thus, to look for $*, try \ \$\ *. All the special characters are:

```
.  ^  $  [  \  *  +  ?  {  }  |  (  )
```

Square brackets ( [ ] ) are used for *character choices or ranges*:

- [a-e] – this matches all lines that contain any of these letters: a, b, c, d, or e
- [^aeiou] – ^ applies inversion, so this looks for all lines that do not contain a vowel
- ^[0-9] – this will match all lines that start with a digit

Check these two links for examples: https://linuxhint.com/find_strings_regular_expressions/ and https://linuxhint.com/use-special-characters-in-the-grep-command/.

## touch, rm, mv, cp, reboot, and shutdown

**touch** *updates* the *timestamp* of a file to the *current time* or *creates* it if it doesn't exist. Suppose we have an application that will load a conf file located somewhere in /etc/ during its initialization, *but only if it has a timestamp from today*. We can close the application, use touch on the given *conf file*, and then start the application again.

**rm** removes a file. -f is used to *force* the action and doesn't ask us for confirmation. -i forces rm to *ask explicitly for every file to be removed*. -d *removes empty directories*. -v stands for **verbose**, so rm prints detailed terminal log output while rm works. We usually use it like this: $ rm someFile.

**mv** *moves* or *renames* a file. The standard form is mv [OPTION]... [-T] SOURCE DEST. -f is *force*, which will not ask us if the operation results in overwriting an existing file. -n is used to explicitly not overwrite files in the destination directory. -v is for **verbose** output. -u is used to move the file only if the SOURCE file is newer than the DESTINATION file. Here are a few simple examples:

```
$ touch aFile.txt                # Create a file.
$ mkdir ~/Documents/NewDir/      # Create a target directory.
$ mv -f -v aFile.txt ~/Documents/NewDir/ # move or rename
# The result is as follows:
renamed 'testFile.txt' -> '/home/luke/Documents/NewDir/testFile.txt'
```

Why does it say *it renamed it*? Because mv will always first attempt to *rename* the location in the filesystem by calling the rename system command. It will not *move the file* if not necessary, as moving is slower and only required in certain cases, such as when changing the drive.

mv can also be used to *rename a directory* like this:

```
$ mv ~/Documents/SomeDir ~/Documents/NewDirName/
```

**cp** is a tool to copy files or directories. Here are a few examples of its use.

Copy a file to a new one with another name. With the first command, we copy to the current directory, and with the second, one directory up:

```
$ cp originalFile.mp4 newFile.mp4
$ cp originalFile.mp4 ../newFile.mp4
```

Copy several files to someDirectory:

```
$ cp file1.mp4 file2.mp4 file3.mp4 ~/someDirectory
```

Copy a whole directory recursively to a new one, creating the directory if necessary:

```
$ cp -R originalDirectory ~/newDirectory
```

By default, cp will ask you whether there is already a file on the target path to be overridden.

**reboot** and **shutdown** are used to *reset* or *shut down* your system. Use --help for the provided options. Note that reboot *acts immediately*, while shutdown works with a *one-minute delay*.

## nano, vi, and vim

**nano** is a great, simple *terminal text editor*. It shows its generic controls at the bottom of the terminal screen, and most of them begin with the ^ sign and a capital letter, which means you should press *Ctrl* and the *letter*. The *Meta key* sequences are marked with M-, but on **Xfce** and **GNOME**, they work with *Meta+Alt+Key*. On **KDE's Konsole**, *Meta* combinations are with *Alt+Shift+Key*. *Table 7.2* shows some of the main controls:

| Key | Action | Key | Action | Key | Action |
|---|---|---|---|---|---|
| Ctrl+S | Save | Ctrl+G | Help | Ctrl+A | Home |
| Ctrl+K | Cut | Ctrl+U | Paste | Ctrl+E | End |
| Meta+Alt+6<br>Alt+Shift+6 | Copy | Ctrl+Arrows | Navigate | Ctrl+X | Exit |
| Meta+Alt+U<br>Alt+Shift+U | Undo | Meta+Alt+Q<br>Alt+Shift+Q | Find next | Ctrl+W | Search Forward |
| Meta+Alt+E<br>Alt+Shift+E | Redo | Meta+Alt+W<br>Alt+Shift+W | Find prev. | Ctrl+Q | Search Backward |

Table 7.2 – nano editor main controls

nano has a lot more capabilities. Check its help with *Ctrl+G*. To open a file with nano, type $ `nano someFile.log`. The file extension after the dot is irrelevant if it is a character/text file (and not a binary or executable). If the file doesn't exist, it will be created in the current directory. If a full path is provided, it will open or create the file in the given location.

**vi** is a historically renowned terminal text editor that is a bit hard to start with but extremely powerful once learned. **vim** is based on vi but adds autocomplete, spell check, comparison, Unicode support, regular expressions, scripting languages, plugins, and syntax highlighting. vim stands for **vi improved**. Installing and learning vim is recommended if you need to do a lot of text, source, or script editing via the terminal. vi is installed by default on Manjaro, while vim must be installed manually via $ `sudo pamac install vim`, or via Pamac. As mentioned in *Chapter 6*, finding a how-to or beginner guide on these is *strongly advised*. Starting with the GUI-based gvim may also help.

## find

The find command is a utility for *searching* **recursively** *through the filesystem*. It can be used to find files and directories and perform subsequent operations on them. It supports searching by file, folder, name, creation date, modification date, owner, permissions, and so on. Here's an example of its usage:

```
$ find OPTIONS PATH expression
```

PATH defines the top-level directory where `find` begins searching. If *omitted*, it will assume we want to *start from the current directory*. The expression contains *specific search options* and potentially a partial or complete filename string.

To continue with the explanation, we need to understand symbolic links. These are shortcuts to files located in another directory. Links are explained in more detail in *Chapter 9*. By default, `find` will *not follow these shortcuts*. To change this, replace `OPTIONS` with `-L`. I will skip the other special options here as we concentrate on the basics in this section.

The most used option is `-name`, followed by a single string, which will look for it in file names. Using `*` replaces any string, so the following command will look for any file ending with `png`, not following symbolic links:

```
$ find -name "*png"
```

As mentioned, we can *pipe* `| less` at the end if there are too many results.

The next example will look for any file ending with `conf` in `/opt/` and its subdirectories:

```
$ find /opt/ -name "*conf"
```

Let's try a long search (*Ctrl+C* will stop it):

```
$ sudo find -L / -name "*.sh"
```

This will find *all files* ending with `.sh` (shell scripts) on our system. As we have added `-L`, it will also follow symbolic links. If we have links for one directory in another, and link to the first one in the second, following them will result in jumping from one to the other forever. `find` will detect this and print the warning `File system loop detected`, but will break this loop and continue its execution. Despite its detection, this should typically be avoided. If it happens, you might need to remove `-L` or change the expression.

Some users think `find` is slow, and new faster alternatives exist, which is valid for many specific use cases. However, the previous command *without* `-L` *finished in 9 seconds* on *a slow old PC*, which suggests that `find` is fast enough for any regular search.

The following are explanations for the most frequently used `find` expression options with examples:

- `-iname` – can replace `-name` for a *case-insensitive* search.
- `-regex` – for searching based on *regular expressions*. The following command will look for all `txt` and `sh` files:

  ```
  $ find / -regex ".*\.txt" || ".*\.sh"
  ```

- -mmin – find files *modified* within the last *X* minutes:

  ```
  $ find ~/ -mmin -50
  ```

- -cmin – find files whose metadata *changed* within the last *X* minutes. This can be, for example, a permissions change. We will learn about this in *Chapter 9*:

  ```
  $ find ~/ -cmin -50
  ```

- -mtime – find files *modified* within the last *X* days:

  ```
  $ sudo find / -mtime 3
  ```

- -atime – find files *accessed* within the last *X* days:

  ```
  $ sudo find / -atime 10
  ```

- -type – find a specific **POSIX/Unix** file type. Use b for *block* special files, c for *character* special files, d for *directory*, l for *symbolic link*, p for *FIFO*, f for *regular* file, and s for *socket* (more details will be provided in *Chapter 9*). The following command will look for all *subdirectories* in the *current directory*:

  ```
  $ find . -type d
  ```

- -empty – find *empty* directories in the *current one*:

  ```
  $ find . -empty
  ```

  We can also find empty files (of size 0 B), in this case, in the /tmp directory:

  ```
  $ find /tmp -type f -empty
  ```

- -perm – find files with *specific permissions*, for example, ones that are writable by either their owner or their group:

  ```
  $ find . -perm /220
  $ find . -perm /u+w,g+w
  $ find . -perm /u=w,g=w
  ```

  Here is how to find *read-only files* for their owner:

  ```
  $ find . -perm /u=r
  ```

  We can also find files *executable* by anyone:

  ```
  $ find . -perm /a=x
  ```

- -user – find files related to a *specific user* (replace luke with the user you want to look for):

  ```
  $ find . -user luke
  ```

- -size – find files of a *specific size*. The + prefix indicates bigger than, - is for smaller than, and *no prefix* is for an *exact size*. Use c for *bytes*, M for *megabytes*, and G for *gigabytes*. Remember that the size is rounded up or down and *matched to the unit*. Therefore, -size -1M will match files that are 0 M (0 megabytes) in size, while -size -1048576c will look for files *smaller than* 1 megabyte. The following will look in the current directory for files bigger than 10 megabytes:

  ```
  $ find . -size +10M
  ```

  A combination like the following is possible for files bigger than 50 MB and smaller than 100 MB:

  ```
  $ sudo find / -size +50M -size -100M
  ```

- -group – this finds all the files belonging to a specific group:

  ```
  $ find . -group developer
  ```

- -and or -a – this combines *multiple conditions*, although both -and and -a can be omitted (as seen in the size example):

  ```
  $ find . -size +50k -and -name "*.png"
  ```

- ! – this inverts the expression directly after it. For example, the following conversion of the upper command will find files for which +50k is false (that is, smaller than 50 KB) and that end with .png:

  ```
  $ find . ! -size +50k -name "*.png"
  ```

- The next one will look for sizes less than 50k, but also names *not ending with* .png and *user* and *group* having write permissions:

  ```
  $ find . ! -size +50k -and ! -name "*.png" -and -perm /u+w,g+w"
  ```

- -o – this is an OR condition, so when *the first succeeds*, the *second is not evaluated at all*, which is a nice speed optimization:

  ```
  $ find . -size +50k -o -name "*.png"
  ```

## Advanced find – execute a command for each result

This section will show how to execute a command for each find result. We can either use -exec or, better, pipe to xargs.

-exec executes a command for each file found. After this option, you must type a command with its arguments, provide the proper place of the find result string with { }, and terminate the command to execute with \;. In the next example, we will copy each .png file in the current directory to the one called Images. { } represents the filename:

```
$ find . -type f -name "*.png" -exec cp -t Images '{}' \;
```

Unfortunately, the `find` official manual says `-exec` can be *unsafe for some operations*. The same manual strongly recommends using the `xargs` program *piped* to the `find` output. The upper line shall then be rewritten like this:

```
$ find . -name '*.png' -type f | xargs -r /bin/cp -t Images
```

The reason not to write some special designator for each result found is that `xargs` will have the *file path copied* at the end of the command for each result.

Here are a few other examples:

- In all `.c` files, find those that contain `buffer` and *only print the filenames*:

```
$ sudo find / -type f -name "*.c" | xargs grep -l buffer
```

- Find all `.log` files in the `/var/` directory and delete them:

```
$ sudo find /var -type f -name "*.log" | xargs rm
```

- Find all `.c` files that contain `buffer` and then pipe again to `ls` to *provide detailed information for each file*:

```
$ sudo find / -type f -name "*.c" | xargs grep -l buffer | xargs
ls -ld
```

The last example is not necessarily the best but shows how we can pipe even more commands (in this case, pipe `xargs` twice) if needed. Here are two pages with many more `xargs` examples: `https://linuxhint.com/xargs-find-linux/` and `https://iq.opengenus.org/xargs-command-in-linux/`.

## Getting system information

The following simple commands provide useful system information, which is of great help when asking for help, checking your system characteristics, and so on:

- `neofetch` – a great tool that provides general info for your system without arguments. You may need to install it.

- `free` – used to check the amount of free and used **RAM** and **swap** space.

- `id` – used to print information for the user and group IDs, and others.

- `lsblk` – lists information about *block devices* such as **HDD partitions** and other *storage devices*.

- `lsb_release -a` – provides information about your Linux distribution and release.

- `timedatectl` – lists information about your current time zone.

- `w` – prints who is logged in on the system.

- `df` – provides information about the disk space on all mounted *filesystems*.

- `ip addr show` – lists all *network interfaces* with **IPv4/IPv6** and **MAC addresses**. More on this topic in *Chapters 10* and *11*.

- `duf` – lists *all HDD devices* and their partitions with their sizes, percentage of used space, mount points, and filesystem. You must install it.

## Getting HW information

Often, when you ask for help regarding *your HW or issues with drivers not working*, the **Manjaro forum** members will ask you for the output of some of the following commands. Remember that when posting, you must enclose the **Terminal** results in `[code] [/code]`:

- `$ inxi -v -Fx` – provides a detailed but tight listing of your *HW configuration*. Its output is almost always requested for help with HW issues.

- `lshw` and `hwinfo` – provide detailed information about your *HW* with *node IDs*, but `hwinfo`'s output is several times bigger and more detailed. Without `grep`, `hwinfo`'s output is a bit overwhelming.

- `lscpu` – provides detailed information about your **CPU**.

- `lsusb` – provides detailed information about your **USB** devices.

- `lspci` – provides detailed information about your **PCI** devices (e.g., the *graphic card*, *network controller*, **IDE**, **USB**, and other **PCI**-connected controllers).

## Piping and redirection

**Redirection** changes where a program's `stdin`, `stdout`, and `stderr` streams go. A reminder: on the **Terminal**, `stdin` is the data *going into a program* (via *user commands and input files*), `stdout` is the printed result, and `stderr` is the *error output*, usually also printed on the **Terminal**. These are the *inputs* for an operation, the *output* of an operation, and, eventually, some *error output*, identified with 0, 1, and 2, respectively.

Redirections are processed *in the order they appear*, from *left to right*. We have three special symbols for this:

- `>` – the **greater-than** symbol is used to *send the output of a command to a file* and leave *no output on the* terminal, *overriding* the previous file content

- `>>` – the double **greater-than** symbol is used to *send the output of a command to a file*, but the original file content is kept, and *the new one is added*

- `<` – the **less-than** symbol *provides input to a command* from a file

The following are two simple examples:

- `$ grep -r bash /etc >results.txt` – this will put all the regular output in the file, but the **Permission Denied** error messages will be printed on the **Terminal** and not in our file

- `$ grep -r bash /etc >results1.txt 2>err2.txt` – this is like the previous example, but the **Permission Denied** error messages will go to a separate file, `err2.txt`

The **ampersand** (&) sign means *the address of*. Thus, `2>&1` means that `stderr` shall go to the address of `stdout`. In addition, note that *the order of redirection is essential*. Here are two more examples to illustrate this:

- `$ ls > dirlist.txt 2>&1` – directs `stdout` (1) to the `dirlist.txt` file and then `stderr` (2) to where `stdout` already points (i.e., the `txt` file).

- `$ ls 2>&1 > dirlist.txt` – directs `stderr` to where `stdout` currently points (i.e., the Terminal) and then says `stdout` shall go to `dirlist.txt`. So, `stderr` remains in the Terminal, and only `stdout` goes to the file. This second example is rarely used, as it is the same as writing `$ ls > dirlist.txt`, which will direct only `stdout` to the file. It is often used to explain the operators' ordering differences.

The > redirect operator can also be used to create files. Type `$ > MyNewFile.txt`, press *Enter*, then type *whatever you want* in the terminal. When ready, press *Ctrl+C* and your file will be saved.

Another form of the previous tactic is to use `cat` to enter a string directly into a text file, like this:

```
$ cat "https://some.address.I.Want.To.Write.com/" > MyNewTextFile.txt
```

Now, let's look at the other two operators, < and >>.

< is used to *direct the contents of a file as input to a program* (or command), like so:

```
$ wc -l < barry.txt > myoutput.txt
```

**wc** is a **word count** program and `barry.txt` is the *input* file. The `-l` option means *count the lines from this input*. `myoutput.txt` is where to print the result.

Let's say we have a text file, `results.txt`, with some lines, a few of which include `matches`, and we want to search for `matches` with `grep`:

```
$ grep matches <results.txt
```

We can easily *direct the results* to another file:

```
$ grep matches <results.txt >results1.txt
```

If you try to use the same file as *input* and *output*, the shell will complain. If we execute this command many times, `results1.txt` will always have *the same content* – each time completely *overridden*.

If we want to *append* contents to `results1.txt`, we need to use `>>`:

```
$ grep matches <results.txt >>results1.txt
```

## Investigating and killing running processes

**top** is a command that shows us *all the running processes on your system*. To get help, you can use `-h`, but it is brief. `top` is frequently used *without options*. It shows the **Process Identifier** (**PID**) in the first column and the user in the second. There is information for the amount of memory and CPU used, the execution time, and which command started each process. Many of the processes are from **root**, and you will also see processes from your **user** if you have started SW *in your current session*. Also, if your system is idle (not doing any work), most processes will show 0% usage and 0 for the time worked. Any SW written for Linux or any modern OS sleeps when no work is done (it doesn't use the CPU). If you pipe it with `grep` for a specific user, it will show only processes from this user. Press *Q* to stop its execution.

**kill** is a great command. With the **PID** from `top`, you can *terminate* any running or stuck program. Of course, any unsaved changes will be lost. It is usually used only for unwanted processes, as this is an abnormal termination for any Linux application. Be careful what you `kill`. Also, remember that many basic processes might be impossible to kill without `sudo`, and even with `sudo`, some essential processes may be hard to kill.

**pidof** is a simple tool to get you the **PID** of a given application *if it is currently executed*. If the application has been terminated, `pidof` returns nothing.

There are a few more commands to get the list of current processes. You can check out `htop` (which is better than `top`), `ps`, `pstree`, and `pgrep`. More information can be found at `https://www.cyberciti.biz/faq/show-all-running-processes-in-linux/` and `https://phoenixnap.com/kb/list-processes-linux`.

We review the topic of processes and their investigation in detail in *Chapter 13*.

## Executing an application in the foreground and background

Each program we start in the Terminal is executed in the **foreground**. This means it blocks the user input while it finishes. Let's take **VLC** Player as an example in the case of a GUI program. If we type `vlc` and press *Enter*, it will start the VLC Player's GUI and *block* the **Terminal**; the Terminal's standard output will be redirected to the input of `vlc`. However, sometimes, we want to start a given process in the **background**, leaving the Terminal free for other actions. In the case of **GNOME**, install it or try with the GUI applications `gnome-calendar` or `gnome-calculator`; use the *Tab* autocomplete starting with gnome. I haven't used firefox as an example on purpose, as it has special behavior in this case, though you can also start it in the background without problems.

A process started in the foreground is a child process to the started Terminal session. To start an application *in the background*, add the **ampersand** symbol (&) at the end of the call like this: $ vlc&. Additional space before the **ampersand** is ignored. This call also provides us with the **PID** of the process.

Remember that when you start regular applications such as firefox and vlc from the Terminal, any standard error or status reporting will clutter it. Thus, this is a standard way to start applications *without a GUI*. In addition, many GUI applications have arguments to turn off the error output when starting them from a Terminal. How can we change this? If we want these messages to be saved, we can use the **redirection** command from earlier and add an **ampersand** at the end like this:

```
$ firefox > firefoxLog.txt 2>&1 &
```

This will keep all messages in the given file.

There is one more solution – to send all such messages to the /dev/null device. This character device file (we will learn more about device files in *Chapter 9*) deletes any string sent to it. Due to this, it is often referred to as a **bit bucket** or **black hole**. We can do it with the following:

```
$ firefox > /dev/null 2>&1 &
```

Skipping the 2>&1 part will leave error messages in the Terminal.

It is interesting now that by using fg and bg without arguments, you can move this process to the **foreground** or **background**. While in the foreground, pressing *Ctrl+X* will kill the process. If, instead, we press *Cltr+Z* – this will suspend the process in the background, and executing bg will resume it, again in the background. There are more options for fg and bg, but they differ depending on the Terminal and shell, so I will not explain them.

No matter where stdout and stderr go, the applications started by the current terminal session can be listed with the jobs command. Start the following programs several times in the background: on Xfce, galculator and gparted, on KDE, kcalc and partitionmanager, and on GNOME, gnome-calculator. Now, execute jobs, and you will get a numbered list of all applications running in the background, *along with the commands used to start them*. Running ps without arguments will list the current session processes (including those running in the **background**) with their system **PID**, which you can use as an argument to kill.

It is essential that, in most cases, if we *close the terminal with background running applications*, all its child processes *will also be terminated*. firefox can be an exception – in some cases it is automatically completely detached from the session a few seconds after being started in the background (on Xfce with BASH). This behavior may change based on the terminal and shell settings, so don't count on it.

People typically use background applications to execute Terminal-only applications and to avoid the necessity of clicking a lot with the **mouse**.

Read more about background processes here: `https://technology.amis.nl/tech/` `linux-background-process-and-redirecting-the-standard-input-output-` `and-error-stream/`. To learn more about `/dev/null`, check out this article: `https://www.` `geeksforgeeks.org/what-is-dev-null-in-linux/`.

# Getting advanced help directly in the Terminal

Any of the presented commands provide a detailed manual, which is essential for any advanced knowledge and usage. We look here into the two programs to access it.

## man

We have seen how the `-h` and `--help` options can be used for basic help on the terminal. However, the detailed manual for each program, often called *man pages*, is downloaded and installed with it. To open it, we use man like this: `$ man find`. Man pages often include examples and may contain conceptual explanations of the application design.

Basic *navigation* in the man pages is done with the *PgUp/PgDn* keys. Pressing *H* will open the help, and *Q* will close it. Besides the *Up* and *Down arrows*, *e/y* or *j/k* can scroll *up/down* by one line. *D/U* scrolls *up/down* by *half a screen*. To search *forward* (case sensitive), use the *forward slash* (/). With the question mark, ?, you can search *backward*. Use n/N for the *next/previous* match.

Enter q if you want to quit.

You can also pipe a man page to `less` or `grep`. Let's say we are concerned with any mentions of the word `directory` and also want three lines before and after the mention. We would write the following:

```
$ man find | grep -i directory -n --color -C 3
```

Last but not least, man allows exporting a whole page to an HTML file. For this, use the `--html` parameter, which accepts a browser command after it:

```
$ man --html=firefox man
```

I have also tested this with `brave`, which I installed via `$ pamac install brave-browser`. No matter the browser, executing the man command as shown will start the browser in a child process and block the terminal. If you press *Ctrl+C*, this will *unblock* the terminal but *also kill the child browser process*. So, it is best to *start the browser in advance*. In this case, once the page is loaded and you press *Ctrl+C*, the HTML page will remain in the browser, and it will not be closed.

Starting the man html output in a *background* process is possible but *not smart*. This will leave a dormant, unused process, which might not take up serious resources but is pointless. If you open 30 man pages in the background, you will have 30 dead, pointless processes. So, again, *opening the browser in advance is the best idea.*

You can also save the web page in a directory of your choice to have it ready the next time you need it. There is only one potential issue when using the HMTL export option. Sometimes tables are not fully exported; thus, for the rare manuals with tables – keep them open in an additional terminal.

## info

**info** is an alternative to man and the *official GNU project documentation utility*. It is easily navigated with the *arrows* and *Page* keys, has better navigation than man, and has a short help via *Shift+H*. It also has better formatting and, in particular for find, *more information, including some history*. Quitting info is done via q.

Each man or info page usually has a SEE ALSO section referring to other manuals. While in man these references are only text, in info they are *links*, which you can activate via the keyboard and go directly to the referred manuals. Many basic tools, such as ls, mv, rm, cp, pwd, and mkdir, are parts of a big package, in this case, the *GNU Core Utils*. If you open, for example, the mkdir info page, you can navigate to any other *GNU Core Utils manual*. In addition, references inside the info page text are also active links. All tools from this package are listed here: https://en.wikipedia.org/wiki/List_of_GNU_Core_Utilities_commands. More information and alternatives for many of them are provided by the Arch community here: https://wiki.archlinux.org/title/core_utilities.

For a lot of commands, info and man provide the same results. However, often, there are cases when *they are not the same*. Thus, if you're digging into them, always check both. If only one of them is available for some program, this is your way to find the manual you need.

Both info and man provide options, but mostly, you don't need those. Of course, you can look at --help, read their man/info pages, and search the web.

Though info is better, I use man more often with the **HTML** output to navigate nicely in a browser and outside the Terminal.

To close this section, *there are hundreds of excellent online guides and discussions*. Sometimes, they also *solve a specific issue* you need help with. Use the search engines, as Linux has, after all, the largest online tech community in the world.

# Summary

We have now mastered *all the basics* of working with the **Terminal**. You have seen over *50 commands* and how to combine them with **piping**. Imagine writing **GUI** frontends for all these tasks – this would be a lot of work. All presented tools are periodically updated, always providing more and more information and features.

*For the rest of the book*, most of the tasks we will perform will be done via the **Terminal**. In the next chapter, we continue with the *Package management* and all the information on dependencies. We will explore **pamac** and **pacman** in detail. We will then switch to Environment variables, which allow us to modify how some applications work. We will finish with a short introduction to Licenses, which are the basis of Free and Open source software.

If you find the content so far valuable, I will be eternally grateful to you for a short book review on the platform you purchased it from. With this, you will help people interested in learning more about Manjaro and Linux.

# Part 3:
# Intermediate Topics for Daily Usage

In this part, we will take a more severe look at Manjaro's and Linux's design from several points of view. The hundreds of thousands of free software applications for Linux frequently depend on each other to achieve modularity and share existing resources. *Chapter 8* thus starts with a dependencies explanation, providing a tool to investigate them. We then move on to package managers, the primary tool for managing all installed SW modules and their dependencies. The next stop is Environment Variables, often used for application behavior control, followed by a brief note on open source licenses.

As everything on Linux is a file, *Chapter 9* continues with an extensive primer on filesystem basics, structure, and types. It also explains the Windows NTFS usage and external storage automount, an essential feature for any modern OS. *Chapter 10* extends the topic with storage management, describing its shrinking, extension, and mounting. Linux is famous for its security, and our first step into it is hard disk partition encryption. The next intermediate daily topic is backups, for which several tools are reviewed. The best one (rsync) is presented in detail.

The part continues with two chapters dedicated to networking. *Chapter 11* explains the fundamentals and emphasizes local network usage and scanning with examples. It then explains file sharing in detail, with the two main technologies for it. At the end of the chapter, it reviews the usage of SSH – the main secured remote access channel, adding SSHFS for accessing the whole filesystem on a target machine.

*Chapter 12* dives into the basics of the internet and security by explaining ports, protocols, and package transfers. It further details how those are used in the main types of network attacks. The chapter continues by reviewing the two most used free firewalls, covering their relation to the Linux kernel and explaining the **ufw** full setup in detail. The last part of the chapter details VPNs (the best way to stay anonymous online), explains how to evaluate a VPN provider, and presents the setup of the current best one.

This part has the following chapters:

- *Chapter 8, Package Management, Dependencies, Environment Variables, and Licenses*

- *Chapter 9, Filesystem Basics, Structure and Types, NTFS, Automount, and RAID*

- *Chapter 10, Storage, Mounting, Encryption, and Backups*

- *Chapter 11, Network Fundamentals, File Sharing, and SSH*

- *Chapter 12, Internet, Network Security, Firewalls, and VPNs*

# 8

# Package Management, Dependencies, Environment Variables, and Licenses

In this chapter, we will examine Manjaro further by diving into dependencies and **packages**. The **pacman** package manager is the default for **Arch Linux** and is thus supported and actively used in Manjaro. Manjaro adds the **pamac** pacman frontend, which provides a **GUI**, more options, and, most importantly, **Flatpak**, **Snap**, and **AUR** support. Furthermore, we will explain the basics of **environment variables**, often used as modification points for installed applications. We will finalize the chapter by briefly examining **open source licenses**, as they are the base of the **Free and Open Source Software** (**FOSS**).

The topics we'll cover in this chapter are as follows:

- Dependencies
- Pacman, pamac, Octopi, and package management
- Environment variables
- Licenses

## Dependencies

Applications in Linux depend on multiple modules to run. These libraries and executables are called **dependencies**. In Windows, they are `.dll` libraries (**dynamic link libraries**) and `.exe` files. In Linux, the `.so` (**shared object**) file format is used for dynamic libraries, and the executables are controlled via file permissions, so they don't have a dedicated extension. Read more about permissions in *Chapters 7*, *9*, and *13*.

As a metaphor, imagine an application as a car. For the car to run, it needs the required dependencies of roads, charging stations, tires, and maintenance, while optional dependencies could be a carwash, a park assistance feature, expensive tires, a garage, and so on. On Windows and macOS, such dependencies are installed during the initial installation/setup procedures. It is the same for Linux, but your package manager explicitly informs you when such are required and will be installed. This is because Linux is based on the fundamental values of **transparency**, **modularity**, **reusability**, and full user control.

Each distribution family has a different **package manager**; there are around 20 in total. In Arch Linux, this is the **command-line interface (CLI)** application **pacman**, supported on Manjaro by default. In addition, Manjaro's team developed **pamac**, which is a **pacman** frontend providing the **Add/Remove Software GUI**. As we will see, **pamac** also provides a rich **CLI**.

We discussed the rare (but possible) dependency collisions in *Chapter 5*. The Manjaro Official Repositories software (SW) rarely has issues (as it is tested before being released to the *Stable branch*), and as mentioned in *Chapter 5*, it is usually faster. We also understood how such rare problems are solved with the Flatpak, AppImage, and Snap application containers. Their advantages are that we have all dependencies packed in them, they have an additional isolation layer for security, and often provide applications unavailable in Manjaro's Official Repositories. Their disadvantages, however, are extra usage of HDD storage, RAM, and CPU resources.

The .so shared libraries were made to prevent duplication of the same functionality and to provide modularity. Without the modularity and reusability of the Linux dependencies, the containers would not even exist, as they also contain .so files.

The information in this chapter is related mainly to Official Repositories SW and AUR packages. The package managers download the information for each application and its dependencies from the servers. They also handle all installations, uninstallations, and a **database (DB)** of the currently installed packages. Thus, they are a key Linux component and contribute to its modularity. In the next section, we will explore all the powerful features of **pacman** and **pamac**. In the end, we will briefly examine the alternative **Octopi** pacman frontend.

This is the moment to say when all these options will be helpful to us as users. Here are a few use cases:

- To analyze our installed packages and their relations
- To identify collisions and sometimes to solve issues
- To clean up our system
- To learn about a given application, find libraries, their sources, and official websites for more information

Before we go on, apart from **pamac** or **pacman**, to get the list of dependencies for a package, we can use the magnificent tool **ldd**, which will list *all shared libraries dependencies* of an application. As an example, let's use **bash**. With $ where bash, we will get the location of the binary, while $ whereis bash will additionally provide its include and library paths. It is located in /usr/

bin/bash. Call $ `ldd /usr/bin/bash`, to list the `.so` files bash needs. The first is always `linux-vdso.so.1`, which is *not an actual* `.so` file but a *virtual library connecting the application with* the Linux **kernel**. The result of `ldd` for bash is as follows:

```
linux-vdso.so.1 (…
libreadline.so.8 => /usr/lib/libreadline.so.8 (…
libdl.so.2 => /usr/lib/libdl.so.2 (…
libc.so.6 => /usr/lib/libc.so.6 (…
libncursesw.so.6 => /usr/lib/libncursesw.so.6 (…
/lib64/ld-linux-x86-64.so.2 => /usr/lib64/ld-linux-x86-64.so.2 (…
```

# Pacman, pamac, Octopi, and package management

We will start with **pacman** and then switch to the improvements in **pamac**. It is important to note that **Flatpaks** and **Snaps** cannot be installed *with any of these tools* in the terminal.

## Pacman

At `https://wiki.archlinux.org/`, you can track the history of **pacman** back to *October 2005*, though it was created before this. It is stable, with great functionality, follows the Arch paradigm of providing only a CLI interface, and is based on the `libalpm` library. **Pacman** holds a local **DB** of all the available packages in `/var/lib/pacman/` and downloads the packages in `/var/cache/pacman/pkg/`. It requires root privileges for commands changing local installations and settings but not for query and information commands. Keep in mind that some applications, libraries, or packages belong to groups, and when we use *the group name* for installation, *all the group's packages* will be installed.

Let's examine the short help with $ `pacman -h`. It will list *only its general options*. Having angle brackets <…> means the parameter or argument is *required*. Having square brackets around a parameter or argument [...] means it is *optional*:

```
pacman {-h --help}
pacman {-V --version}
pacman {-D --database} <options> <package(s)>
pacman {-F --files}    [options] [file(s)]
pacman {-Q --query}    [options] [package(s)]
pacman {-R --remove}   [options] <package(s)>
pacman {-S --sync}     [options] [package(s)]
pacman {-T --deptest}  [options] [package(s)]
pacman {-U --upgrade}  [options] <file(s)>
```

To perform actions with pacman, you will often need more single-letter options. Here are the most common ways it is used:

- `pacman -S [package]` – *install* a package, e.g., `pacman -S duf`.

- `pacman -Rs [package]` – *uninstall* a package, e.g., `pacman -Rs duf`. The small s is optional and triggers the *removal of all dependencies* if other packages don't require them or if not explicitly installed by the user.

- `pacman -Sy` – *refresh* the local database with the latest from the servers. Using `-Syy` will force an explicit refresh of all DBs, even if they are marked as *up to date*.

- `pacman -Su` – *upgrade* all out-of-date packages.

- `pacman -Syu` – refresh the local DB and then upgrade the system. You can also combine `-Syyu` to force the DB refresh, which is *strongly recommended* if you have any updates issues.

- `pacman -Suu` – upgrade with downgrade ability. Usually, uu is used with yy in the form `-Syyuu` to switch from/to the Unstable, Testing, or Stable branch. Many packages will have lower versions when switching to Stable from the others. `-Syyuu` tells pamac to replace your local DB with the one from the server and then change all SW to the version from the changed DB, even if this means downgrading.

- `pacman -Ss <regexp>` – search for a package (*case insensitive*) with a regular expression, *both in the name and description*. An example is `$ pacman -Ss 'python.api'` (the dot represents any symbol; for more information, see *Chapter 7*, in the part on advanced commands, the section on *regex and regexp*). Filtering the results with `grep` can help, as it can highlight and has additional search options.

- `pacman -Qs [package]` – search locally installed packages (case insensitive) for names or descriptions matching a regular expression pattern. Try it with `$ pacman -Qs bash`. The results will include a few other tools mentioning `bash` in their description.

- `pacman -Qi [package]` – provides detailed information about a package, including the architecture, website (URL), dependencies, other packages requiring it, when it was built, and so on.

- `pacman -Si [package]` – provides detailed information about a package from the repositories, including for packages not installed on our system.

- `pacman -Q` – list all installed packages with their versions.

- `pacman -Q [package]` – gives the version of an installed package.

- `pacman -Ql [package]` – lists all paths of files owned by a package. The output is not optimized as it duplicates each sub-path, but the information is valuable.

- `pacman -Sg` – list all the groups that can be installed. Some popular ones are `base-devel`, `linux-tools`, `kde-utilities`, and `qt`. List the group packages using the command with a group name, e.g., `$ pacman -Sg qt5`.

- `pacman -U local.file.pkg.tar.gz` – install a package from a local archive, e.g., downloaded from a custom location. This command can also be used with a URL of a package on a specific server like this: `pacman -U https://mirror.alpix.eu/manjaro/stable/community/x86_64/glances-3.3.0-1-any.pkg.tar.zst`.

Here are a few more helpful package tools installed by default in Manjaro:

- `pactree [package]` – gives the complete **dependency tree** for a given package. Try it with `duf`, `exa`, or `vlc`.

- `pacsearch [package]` – searches for a package locally and remotely, printing in the result the repository, package name and version, and its short description.

- `pacman-conf -l` – lists all the configured repositories.

- `paclist [repository]` – lists all packages installed from a given repository. For example, try `$ paclist core`.

For `pacman` (apart from `man` and `info`), you can also check this Manjaro wiki: `https://wiki.manjaro.org/index.php/Pacman_Overview`.

## pamac

We covered **pamac**'s **GUI** basics in *Chapter 5*. In it, you can click for detailed information on any package (both installed and uninstalled). It contains a short human-language description, the *dependencies* (including the optional ones), and the *list of installation files*.

In this section, we will dive into its CLI options. Compared to pacman, **pamac** more or less duplicates its options but in a *human-readable format and with improvements*. Its man page describes it as "*a libalpm(3) frontend with AUR support*".

Before going further, from the additional tools mentioned at the end of the previous section, **pactree** remains unique. I strongly recommend using it to untangle dependencies.

The general form of **pamac** is pamac <action> [options] [targets]. Again, the *angle brackets* are a *required* argument (action), while the *square brackets* are *optional*. Here is its short help:

```
$ pamac -h
Available actions:
pamac --version
pamac --help, -h        [action]
pamac search           [options]  <package(s)>
pamac list             [options]  <package(s)>
pamac info             [options]  <package(s)>
pamac install          [options]  <package(s)>
pamac reinstall        [options]  <package(s)>
pamac remove           [options]  [package(s)]
pamac checkupdates     [options]
pamac update,upgrade   [options]
pamac clone            [options]  <package(s)>
pamac build            [options]  [package(s)]
pamac clean            [options]
```

While **pacman** requires sudo for executing system-critical tasks, **pamac** doesn't. It will instead *prompt you for a password* only if you *perform an action requiring* sudo privileges. There is only one exception, which we will mention later. This way, pamac improves security, especially for unchecked **AUR** packages, as it will build them as a regular user and then ask for the password only for the installation. You can run it with sudo in front, but it is discouraged for AUR packages. Another nice pamac feature is that its first argument (the primary command) is available with the *Tab* key autocomplete on KDE's **Konsole** and **GNOME** terminals. Now, let's proceed with options, examples, and detailed explanations.

### search, list, and info

Here's a list of the various options:

- pamac search [options] <package> – search *case insensitive* (both for name and description text), for example, $ pamac search exa. Sometimes, you might need to filter with grep, as it provides a maximum number of results. The -i option searches only in the installed packages, -r searches only in repositories, and -a is for *adding AUR results*.

- pamac search -f filenamePart – search for packages that own a file with the given string in its name; for example, $ pamac search -f Breath.prof will return results from Konsole (the KDE terminal package), as it contains files named Breath.profile. Keep in mind that *this search is case sensitive*.

- `pamac list -i` – list all installed packages, with versions, repository, source, and size, *in a better format* than pacman's. As this is the default option, omitting the `-i` suffix does the same.

- `pamac list -e` – list explicitly installed packages with versions, repository source, and size. Explicitly installed packages are those not installed as dependencies of other packages.

- `pamac list -g` – list all available groups from the currently set repositories.

- `pamac list -g [group-name]` – list all packages belonging to a given group.

- `pamac list -o` – list installed orphaned dependencies packages, which are no longer required as dependencies, with versions, repository source, and size.

- `pamac list -r [repository]` – list all packages from a given repository, with installation status. The current default repositories are `community`, `core`, `extra`, and `multilib`. Skipping the repository name will list the available repositories set on your system.

- `pamac list -f <package>` – list all files installed for the package in *a better format* than `pacman`.

- `pamac info <package>` – list the detailed info for the package; adding `-a` before the package name is *necessary for* **AUR** *packages*. This command will provide the installation status of the dependencies. It provides information even if the package is not installed.

## Install and remove

The `install` function installs packages from repositories, paths, or URLs and has the following format:

```
pamac install [options] <package(s),group(s)>
```

As mentioned, if you install a package from official repositories, you don't need to add `sudo` in front; you will be *prompted for the password*. Here comes the *exception* – installing a manually downloaded `.pkg.tar.zst` format package requires `sudo`:

```
$ sudo pamac install /path_to_file/XXXXX.pkg.tar.zst
```

This form also works with a URL. However, keep in mind that *if the package requires dependencies*, pamac *will download them*. Also, when the file is downloaded locally, double-clicking it in your **GUI** *file manager* (e.g., **Thunar** or **Dolphin**) will open the **pamac GUI** *automatically*.

The important options of the `install` command are listed in *Table 8.1*:

| | |
|---|---|
| `--ignore` | Ignore a package upgrade. Multiple packages can be specified by separating them with a comma. This is helpful when you work with a group of packages and don't want to upgrade some of them. |
| `-w` | Download all packages, but do not install/upgrade anything. |
| `-d or --dry-run` | Only print what will be done, but do not run the transaction. It's good to see what would be installed in advance. |
| `--upgrade` | Check for updates. |
| `--no-upgrade` | Do not check for updates. |
| `--no-confirm` | Bypass any and all confirmation messages. |

Table 8.1 – pamac install command options

pamac also provides `reinstall` and `checkupdates` commands. `reinstall` is rarely necessary, but if an application or package has issues or you need it in its initial state, you can easily do it. With `checkupdates`, you can check for updates without modifying your local copy of the databases. Both provide additional options – check `$ man pamac` for this.

A note about optional *dependencies*: often, packages will ask you to provide your choice of optional dependencies in numerical form. We will try this with `glances` – a nice system monitor for the Terminal (and with an optional web interface). Run `$ pamac install glances`; after the authentication step, you will be asked for them. You can use commas to list some and specify a range with a dash, as shown in *Figure 9.1*:

```
==== AUTHENTICATION COMPLETE ====
Synchronizing package databases...
Refreshing AUR...

Choose optional dependencies for glances:
1:  hddtemp: HDD temperature monitoring support
2:  python-bottle: web server support
3:  python-docker: for the Docker monitoring support
4:  python-matplotlib: for graphical/chart support
5:  python-netifaces: for the IP plugin
6:  python-zeroconf: for the autodiscover mode
7:  python-pystache: templating engine
8:  python-prometheus_client: for the Prometheus export module

Enter a selection (default=none): 1-4,6,7
```

Figure 9.1 – Choosing optional dependencies from the Terminal

The `remove` command provides a few interesting options:

- `pamac remove -u <package(s)>` – remove package(s) only if any other packages do not require them. A space-separated list is accepted for more than one package.

- `pamac remove -c <package(s)>` – remove all target packages and packages that depend on one or more target packages. *Be extremely careful with this option!* Again, a space-separated list is accepted for more than one.

- `pamac remove -o [package(s)]` – Remove *orphaned dependencies* not required by other packages. If this option is used without a package name, *it will remove all orphans*. It is excellent for freeing space.

- `pamac remove -n <package(s)>` – uninstall package(s), but don't remove orphaned dependencies.

- `pamac remove -d [options] [package(s)]` – a dry run that will only show what will be done.

- `pamac remove --no-confirm <package(s)>` – skip all confirmation messages.

I would always first use `-c`, `-o`, and `-n` with `-d` to see in a *dry run* what would be done.

## upgrade/update

These two commands are *the same*. Using either of them *without options* will *check for available updates*. The most important additional options are listed in *Table 8.2*:

| | |
|---|---|
| `--force-refresh` | Force the refresh of the databases. |
| `--enable-downgrade`<br><br>`--disable-downgrade` | Enable or disable package downgrades. |
| `-w` | Download all packages, but do not install/upgrade anything. |
| `-d` or `--dry-run` | Only print what will be done, but do not run the transaction. |
| `--ignore` | Ignore a package upgrade. Multiple packages can be specified by separating them with commas. |
| `--overwrite` | Overwrite conflicting files. Multiple patterns can be specified by separating them with commas. |
| `--no-confirm` | Skip any confirmation messages. |
| `-a` and `--aur`<br><br>`--no-aur` | Upgrade packages installed from AUR.<br><br>Do not upgrade packages installed from AUR. |
| `-devel`<br><br>`--no-devel` | Upgrade development packages (use with `--aur`).<br><br>Do not upgrade development packages. |
| `--builddir <dir>` | Specify the build directory (use with `--aur`). If no directory is given, the one specified in the `/etc/pamac.conf` file is used. |

Table 8.2 – pamac update/upgrade command options

Of all the options, the most important is the first one. If you have *any update issues*, usually when you haven't updated for months, executing this solves them all:

```
$ sudo pamac update --force-update
```

Regarding `--overwrite` and the `downgrade` options, I would use them after checking in advance with `-d` what will be done. If you need to downgrade some packages, check this Manjaro wiki: `https://wiki.manjaro.org/index.php/Downgrading_packages`. It will require setting an additional environmental variable, which we will look into later in the chapter.

We will look at the last five **AUR**-related options in the last part of this section.

### Number of locally kept package versions

When downloaded, packages are kept after a successful installation. This allows *fast reinstallation* or *downgrade* in case of failure. For the last few years, up to July 2023, the default number of previous versions kept is *three*. Thus, if I download a new version of the **Brave** browser every two months in the next year, I will always have the latest three.

The number of package versions to be kept is saved in the `/etc/pamac.conf` file, which you can edit with any text editor. Look for the `KeepNumPackages = 3` configuration. You can also open the pamac GUI and change the number in **Preferences** | **General** | **Cache**. The pamac GUI also has a **Clean** button there.

If you have less space or test many or bigger applications, they may consume a considerable amount of space in pacman's cache. A great tool to check this (or visually inspect your partitions' HDD space consumption) is the official repositories' **Filelight GUI** application. Check with it the `/var/cache/pacman/pkg/` path. You can also check your root partition's available and used space with the `duf` command.

There are two situations when keeping more than one package version is worthwhile. The first is if a current version has bugs, and the second is if you want to compare functionality between versions or test newer ones. As Manjaro is a **rolling release** distribution when you make periodic updates, a new application version may have bugs or problematic changes. In this case, if the older versions are already downloaded, it will be easy for you to switch back to an older version. Of course, such problems are rare, and from my experience, after a few years of extensive Manjaro usage, this *never happened to me*. This feature is *thus primarily valuable to developers and people testing new versions* before releasing applications to the **Stable** branch.

### Downgrading

Any **branch** (*Stable*, *Testing*, and *Unstable*) always keeps only one application version. If you don't have the old versions downloaded on your PC, you cannot get them from the official repositories. Keeping many old versions of all packages would increase the workload of the servers immensely, which would result in the deterioration of the provided service. After four years of extensive usage of Manjaro, I can say that the *Manjaro servers' response and speed are great*, and I would not be happy if the next update download was several times slower.

*There is*, of course, *a way to downgrade packages* described on this Manjaro page: `https://wiki.manjaro.org/index.php/Downgrading_packages`. It also contains information on where to get old packages. Check it if you need to downgrade.

## Clean

Let's now discuss the `clean` command, which can help us free up **HDD** space:

- `pamac clean -d` – *dry* run a generic clean to understand whether something should be cleaned and report the number of kept versions.

- `pamac clean -u` – remove all *not installed* (i.e., only downloaded) packages.

- `pamac clean -k [number]` – change the number of *kept* versions and apply directly. It will also clean the uninstalled packages if you run it with `-k 0`, but it will not change the value of `KeepNumPackages`; it is only temporary. This command is good if you want an explicit space cleanup at the current moment; with the subsequent updates in the following months, `KeepNumPackages` versions of each updated package will be accumulated.

- `pamac clean -v` – when cleaning, print all the names of cleaned files (*verbose*).

- `pamac clean --no-confirm` – this will skip any confirmation messages.

## The AUR options clone, build, and clean

Before continuing, we need a short introduction to the **Arch User Repository** (**AUR**) packages. They don't contain actual application source code but instead contain *configurations* containing *server locations of packages' source code* (the *input* for building an application). They also contain some build instructions. These configurations are written in two files named PKGBUILD and .SRCINFO, put together in a .tar.gz archive.

On Linux distributions, tools such as **yay**, **paru**, and **aurutils** are used for **AUR** packages. All of them can download an AUR package, its sources, and its dependencies from servers worldwide. Then, they build them based on the configuration, clean the temporary build and source files, and *install it on your system*. **Pamac** is excellent, providing both GUI and CLI interfaces for this task.

**Cloning** means *downloading the sources* from a server (usually a **Git** repository). **Building** is the process of compiling them from source code to executable form. **Cloning AUR** files just downloads the *configuration* files.

In software development, **cleaning** is the process of the removal of all intermediate files. Thus, this `clean` command for AUR builds differs from the previous section's `clean` command.

The **AUR** packages' dependencies can be standard ones (often pre-installed by default in Manjaro and other distributions) such as `glibc`, `zlib`, and `libpng`, but also other **AUR** packages. In the second case, **pamac** will download and build them to complete the initial **AUR** package build process. AUR packages are cloned and built by default in `/var/tmp` (defined in `/etc/pamac.conf`). This directory is also used for any additional downloads.

To build **AUR** packages with `pamac`, you will *need* the core `base-devel` package. Whether you need it depends on your Manjaro flavor, configuration, and whether it was installed as a dependency from another package. To avoid any potential issues, execute:

```
$ pamac install base-devel
```

If it is installed/up to date, you will get a warning message, and nothing will be done.

> **Important note**
>
> Sometimes, the current state of an AUR package may *generate build errors and fail*. In this case, you can't do anything but wait for fixes for the package or its Git repository. This task may be challenging even for developers and requires advanced knowledge. Searching for an alternative AUR package is an option.

Let's take **Clementine** as an example, a famous audio player application supporting multiple formats and streaming services. Such software is complex and has several *dependencies*. It is available in the official repositories but also exists in **AUR**.

We first search for it with the **AUR** `-a` parameter, as shown in *Figure 8.2*:

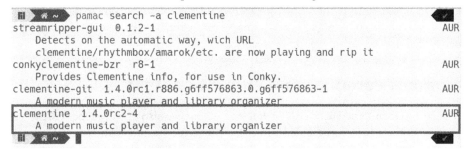

Figure 8.2 – Searching for an AUR package via the terminal

Performing all actions (clone, build, and clean) at once is done by running:

```
$ pamac build clementine
```

You can add the `-d` option, short for `--dry-run`. The next command only downloads the **AUR** configurations file package. The `-r` option adds the necessary dependencies:

```
$ pamac clone -r clementine
```

Digging into `/var/` directories is unpleasant, so to download and then build in a directory of our choice, we use the following two variations:

```
$ pamac clone -r --builddir ~/luke/build clementine
$ pamac build -k --builddir ~/luke/build clementine
```

The `-k` (or `--keep`) flag keeps the built packages in the cache after installation; otherwise, they are deleted. As in this case, we also specify a directory, they will be kept in it.

The last two separate options are helpful to SW developers. The rest of the users don't need them. Unfortunately, the sources' full copy is deleted after the build, but we have the Git repositories' **URL**s in the downloaded `PKGBUILD` and `.SRCINFO` files.

You can also use the regular `install` command; however, if the package is in the official repositories – it will be downloaded from there! If not, then pamac typically warns you it will download and build an **AUR** package.

**AUR** contains a lot of packages. Many times, some of them are in a current development phase, while others are stable versions. From the previous screenshot, in *July 2023*, the `clementine` package was built successfully, but `clementine-git` failed to compile.

## Octopi

**Octopi** is an alternative to the pamac GUI. It offers a not-so-modern but excellent systematic interface. With one click, the user can go to package groups, a file, or information for a package. The search is helpful, as the results are presented in a table. However, the file lists don't work for many of the packages. You can install it from the Official Repositories. I will not provide more information here, as I don't see it offering any other significant advantages over **pamac** or **pacman**, and its development goes slowly. Still, it may be suitable for more systematic package searches.

# Environment variables and current context

We mentioned **Environment Variables** in *Chapter 7*, where we also explored the basic shell and terminal usage. Environment variables are simple key-value pairs separated by an equal sign. With the `env` or `printenv` commands, we can see all the variables valid in the *current shell session*. Here are some from my `printenv` output:

```
PWD=/home/luke
HOME=/home/luke
LANG=en_US.UTF-8
LANGUAGE=
LOGNAME=luke
USERNAME=luke
MAIL=/var/spool/mail/luke
XDG_SESSION_DESKTOP=gnome
```

We can access these values in the shell with the dollar sign $ like this: `$ echo $LOGNAME`. In this case, the shell will replace $LOGNAME with `luke` when printing. Each *program or script started in the shell can read them*, and some can add new ones. Thus, the Environment Variables are *part of the current context* in which programs are executed. Remember that Environment Variables are *case sensitive*, like any terminal *command* and its *parameters*.

All the variables we see when we use `printenv` are set at different points in time from different applications. Some are set from the **root session** that starts the **OS**; others during the *current user login*. Next, we have variables set from the *current user's graphical Desktop Environment* and others from the *current shell* (**bash**, **zsh**, **csh**, etc.). Each following *child process inherits the variables from its parent* and adds new ones.

Many of them are standard, like the ones listed at the beginning of this subsection. They are often used to control the way specific packages, applications, and daemons behave, including to enable some of their features. In this way, they also inform the system, other applications, and the user of specific settings.

A nice example is the $XDG_SESSION_DESKTOP variable from the `freedesktop` package. It returns the current environment (**KDE**, **XFCE**, **GNOME**, **LXDE**, **LXQt**, etc.). Its supported values are listed at `https://specifications.freedesktop.org/menu-spec/latest/apb.html`. Say I write an application supporting **KDE** Plasma, **Xfce**, and **GNOME** environments. Reading this variable can help me trigger code related to environment-specific graphic libraries.

### Environment variables syntax

The commonly accepted format uses *capital letters* and *underscores* on the left side and *a string* on the right side of an *equals sign*. The name can also contain small letters and digits, but beginning with them is intentionally avoided. The variable's *name cannot contain special symbols*.

We must enclose our string values on the right in single quotes to add spaces or special characters (e.g., !, @, #, $, %, ^, &, *, etc.). *Double quotes* are also accepted but have specific rules for some special symbols. They depend on the shell interpreter; read the ones for **BASH** at `https://www.gnu.org/software/bash/manual/html_node/Double-Quotes.html`.

*Any special symbols shall only be used in the value string for specific purposes.* This is because most shells use special symbols for special functionalities. Theoretically, environment variables can be strings of thousands of symbols, but *making them as short as possible is strongly advised*. The strings they hold shall also be not too long. You have probably noticed that no matter the value, the single quotes are not printed by env and `printenv`.

Here is a simple example of a string in quotes:

```
MY_ENV_VAR='Some string with spaces and numbers 123'
```

Testing your environment variables via `echo` and checking the results of your actions with a test run is obligatory; this is also a general rule of thumb in SW engineering. A nice addition is that the environment variables are also added to the **Tab** autocomplete of most shells when you type them with a dollar sign in front.

### Making a temporary environment variable

To create a temporary variable only for the *currently open Terminal session* (i.e., the *current shell environment*), just write `VARIABLE=value`. As it is *temporary*, closing the current session will *delete* the variable. If you open *another shell* after making the variable (and before closing the given session), it will not be available in the new one.

To make this *temporary variable available to all programs started in the current shell session*, we have to use the `export` command like this:

```
VARIABLE=value
export VARIABLE
```

If we now call `$ printenv VARIABLE`, we will get the string `value`; any program started in the *current session* will have access to it. Temporary variables are used mainly in shell scripts (scripts are reviewed in *Chapter 15*), but we examine them here because understanding the mechanism is essential. This mechanism says that the child processes of a parent inherit its environmental variables.

### Adding a permanent variable

There are several files where you can put expressions such as `VARIABLE=value` to have the variables accessible from *any started shell*. They depend on your distribution, flavor, and setup. Here, we will explore two typical locations on **Manjaro** – via the `/etc/environment` file and the *current shell configuration file*.

The `/etc/environment` file is read from the `pam_env` package during system loading. As a result, its child processes inherit its environment variables. `pam_env` was started *over 20 years ago* and is part of the **Pluggable Authentication Module (PAM)** package. `$ man pam_env` provides more information about it, together with this article: `https://likegeeks.com/linux-pam-easy-guide/`.

We can edit the `environment` file with any text editor, like **Kate** or `nano`. `nano` needs `sudo` in front because `/etc` and its contents are root-owned:

```
$ sudo nano /etc/environment
```

Now add a new line with your definition VAR=2223 at the bottom of the file, and for nano, press *Ctrl+S* to *save* and *Ctrl+X* to *exit*. Upon **reboot**, this variable will be accessible in any shell started *by any user*, including root.

For me, the best example is setting the default terminal editor to be different from **vi** or **vim**. Just add the following line to /etc/environment:

```
EDITOR=/usr/bin/nano
```

To make a variable *available only for your user shell environment*, add the corresponding variable definition in your shell environment configuration file. We can add the definition at the end of the ~/.bashrc file for **Xfce**. It will be in the ~/.zshrc file for **GNOME** or **KDE**. There are other possible files, but we will not explore them here. Open the corresponding file with a text editor, add the line, and save the file. Open a new terminal shell session, and you will have your variable in the tab autocomplete when you print it with $ echo $VAR.

This **Stack Exchange** post answers explore other possibilities for adding variables: https://unix.stackexchange.com/questions/117467/how-to-permanently-set-environmental-variables. Keep in mind that some of the offered solutions differ between distributions.

## Licenses

Many people know we have **GPL**, **MIT**, **Apache**, and other **licenses**, but sometimes it is hard to understand when to use each and what it implies. Here, I will try to explain the very basics.

The idea of open source is to provide the public with free SW to use and learn from. Some licenses, however, explicitly want to force people to make any changes *available* so that other people can use and learn further from them. Other licenses state explicitly that *any derivative can be used in any way*. They all state that there is *no guarantee nor liability* regarding the SW, what it does, and whether it has defects.

Furthermore, all those licenses started being used for *documents*, digital *artwork*, web *technologies*, and others. One of the best sites to help you choose a license for your work is https://choosealicense.com/. This site is also helpful: https://fossa.com/blog/how-choose-right-open-source-license/.

### GPL and all its versions

The **GNU General Public License (GPL)** is the first widely used **FOSS** license. It states that the given SW is free to use, study, share, and modify, but *any derivative should also be open source and provided free to the public under GPL*. This last part is called **copyleft** – *the same obligations remain in any copies or derivatives* of the SW. Using the SW in commercial products is allowed as long as it is stated and any changes are stated and shared. If there are no changes in commercial applications, GPL requires that the license and the unchanged GPL source should also be shared with the end customer. For strictly private use, users may do whatever they want.

The **GPL** has three versions.

*Version 2* added more clarifications for libraries and other software components. An important one is that users *may combine and redistribute any version of the SW with any SW/libraries/packages licensed under different licenses.*

**Digital Rights Management** (**DRM**) is a technology that detects whether a given user of some digital SW or data is allowed to use it and prohibits the not-allowed people or systems from using it. For example, if I don't enter my Spotify account credentials into my smart TV, some implementation of DRM will not allow me to access its library and listen to any music in its libraries.

**Tivoization** is designing **hardware** that incorporates FOSS GPL SW but contains restrictions for anyone to run modified versions of the original FOSS GPL SW on it. It comes from the company **TiVo**, which used the Linux kernel for their product for profit but used DRM techniques in their hardware.

**GPL** *Version 3* added that SW licensed under it is not allowed to be part of systems with **DRM** and **Tivoization**-like techniques, along with a few other restrictive clauses. *Linus Torvalds* rejected using *Version 3* for the **Linux kernel**, saying it had become *too restrictive* and would harm the usage and development of the kernel. This version is rarely used. *The most widely used is Version 2.*

The **GNU Lesser General Public License** (**LGPL**) states explicitly that any SW under it can be *combined with any other SW*, including for profit. Still, if you modify it, you are *obliged to release the code publicly* and again *under* **LGPL**. It also has three versions (initial, 2.1, and 3.0); the most widely used is *version 2.1.*

**Permissive licenses** explicitly state that you *can do whatever you want with the SW*. Such is the family of **BSD** licenses, **MIT**, and **Apache** licenses.

In general, in the last 5-10 years, the most widely used licenses for FOSS SW have been **MIT**, followed by **Apache**, and some versions of **GPL**. More information on what can be used for commercial work can be found in this article:

```
https://triplehelix-consulting.com/open-free-and-limited-source-
licenses-in-the-world-of-software-development/
```

You can do anything you want unless you integrate FOSS SW into commercial products. You have no guarantee of the quality, nor does the creator have liability for any resulting issues. *Considering that most worldwide servers rely partially or entirely on FOSS SW, it is obviously of the highest quality.*

# Summary

In the first part of this chapter, we covered dependencies and package management in detail, focusing on **pacman** and **pamac**. We then looked at the basics of Environment Variables, as they are essential for many applications. We finished by explaining the essence of FOSS licenses, as they define whether or not we can use specific SW commercially and when we must disclose our changes to the public.

In *Chapter 9*, we will continue with the fundamental aspects of the filesystem, beginning with an overview of its basics and the different file types. We will cover topics such as links, attributes, and inodes in depth. Next, we will examine the history and types of filesystems and conclude by examining the Linux filesystem structure, focusing on Manjaro. We will finally investigate external storage volumes, go deeper into the NTFS filesystem usage and automount, and close with a short section on self-healing and RAID.

# 9

# Filesystem Basics, Structure, and Types, NTFS, Automount, and RAID

This chapter will explore **Linux** further by diving deeper into its **filesystems** (**FSs**). As we have already seen, it has features to control files and directories access and permissions (presented in *Chapter 7*). Devices, drivers, and communication means are also presented in the **FS** *as files*, so the access control features also transfer to them.

We will start here with all file type details, as they are independent of the **FS**. Then, we will explore the **FS** types and continue with the FS structure implemented in **Manjaro**. Our next stop will be **external volumes**, **NTFS** support, and **automount**. We will close the chapter by briefly reviewing snapshots and **RAID**.

The sections in this chapter are as follows:

- Linux FS basics
- Linux FS types
- Linux FS structure
- External volumes, NTFS, and automount
- FS snapshots and RAID

## Linux filesystems basics

This section will first look at Linux/Unix file types. Further, we will go deeper into the subject by reviewing where and how the different drive partitions are presented in the FS. We will then explore the inodes, links, and the advanced topic of extended attributes.

## Linux/Unix file types

The possible file types date back to **UNIX**'s *1970s* and *1980s* developments. In *1988*, they were added to the **Portable Operating System Interface (POSIX) IEEE** standard. As a result, we have *standard files*, *directories*, and five *special file types* – **symbolic link**, **socket**, **block device**, **character device**, and **FIFO** (with the alternative name **named pipe**). Let's look at each of them.

When listing with details with `exa -lahg` or `ls -lah`, these types are designated for each result at the beginning of the line. The first sign just before the **permissions** string is a single letter with the following meanings:

- `.rwxr-x---` or `-rwxr-x---`: A file
- `drwxr-x---`: A directory
- `lrwxr-x---`: A link
- `srwxr-x---`: A socket
- `brwxr-x---`: A block device
- `crwxr-x---`: A character device
- `prwxr-x---` or `|rwxr-x---`: A FIFO (a named pipe)

### Hard and symbolic/soft links

A **path** is *any file or directory location in the FS hierarchy* represented like this – `/home/luke/Documents/SomeFile.xxx`.

Every data file in the Linux FS has a first **hard link**, which relates its *name* and *path* with the *physical location* of its content on the **hard disk drive (HDD)** or **solid-state drive (SSD)**. A **soft/symbolic/symlink** is another small file in whichever FS location containing the path to the **hard link** of the first file. In the Windows world, a **symlink** file is called a **shortcut**. When listing a directory's contents with `exa -l` or `ls -l`, such a file would begin with `l` in the detailed output:

```
lrwxrwxrwx myNewSymLink -> /home/luke/Pictures/Image.png
```

Creating a symbolic link is done with the `ln` command and the `-s` parameter like this:

```
$ ln -s <OriginalPath_Of_File> <the_New_Sym_Link_Name>
```

Here's an example:

```
$ ln -s ~/Pictures/Image.png myNewSymLink
```

The last command will create the symlink in the *current directory*, but the *last argument* may also be a **path** in the FS hierarchy, like this: `~/MyDir/the_New_Sym_Link_Name`.

A directory will have in the **FS** an entry relating its name only to its location (path) in the FS hierarchy. It is *not allowed to have a* **hard link** but *can have* a **symlink** created in the same way as shown previously.

Each file can have additional hard links and unlimited soft links. We will learn more about this further in the *Inodes and links* section.

### FIFO (a named pipe)

**FIFO** (**First-In-First-Out**), or a **named pipe**, is one of the simple forms of **Inter-Process Communication** (**IPC**). In Unix, it is represented as a file and allows sending information from one process/application to another. It is a *unidirectional* mechanism, which means information goes only in one direction. It is designated with a p for **pipe** in the beginning string of ls -l, or a vertical line | from exa -l. Named pipes typically have one sender and one receiver, as having more results in corruption and hard-to-manage behavior. Any data sent over a **pipe** is in the form of a simple byte stream.

**Pipes** are typically created by the applications that need them, but we can also create one with the mkfifo or mknod command. mknod typically creates a **device file**, so it is not recommended to be used for this, although it provides a parameter for a pipe. Execute $ mkfifo myPipe, then check the result with ls or exa.

Now, we can call in the terminal $ exa -lah > myPipe, and in a second terminal, $ cat < myPipe. After the first call, the first terminal will stay *blocked* until *a receiver takes the data* sent to the **pipe**. After the second call, we will get in the second terminal the exa text output (with removed coloring), and the first terminal will be unblocked by exiting the successful send operation.

As a **pipe** is a file, chmod can modify its user, group, and other **permissions**. To delete the pipe, call $ rm myPipe.

Last but not least, the *commands piping* in a shell (**bash, zsh, csh**, etc.) is based on **pipes**; hence, the name is the same.

### Socket

A **Unix Domain Socket** (**UDS**, or simply a **socket**) is a *bidirectional* form of **IPC** with more features than **pipes**. Apart from a simple byte stream, the data sent over them can be discrete messages. *Bidirectional* means that both connecting processes can be a *sender* and a *receiver*. In addition, UDS is full-duplex, meaning that when connecting two points, they can send and receive data simultaneously. **Sockets** can be created only via a programming interface from an application (unlike **pipes** creation with mkfifo).

As this subject is a developer's topic, I will not go further but only provide a few examples. Sockets are used from the **X11** server (the X Window graphical framework), the **Docker** daemon (for connections on the local machine), **syslog** for system logs, **MySQL** database servers for communication with local clients, and so on.

If you have heard the term **TCP/IP sockets**, you may wonder whether they are related to **UDS**. The answer is *no*. There are some similarities in their functionality, but **UDS** is a form of **IPC**, while **TCP/IP sockets** are used for bidirectional data transfer over the **internet**. Networking and the internet are reviewed in *Chapters 11* and *12*.

### Character devices

**Character devices** can be **USB** keyboards and mice, game controllers, joysticks, sound cards, serial ports, local and remote terminal interfaces, and others. They work by sending and receiving short messages (of only a few bytes). They often require fast communication with a guaranteed message sequence, usually specified as synchronous communication.

A classic example of such a device is the **keyboard**. Each keystroke on the keyboard goes to the terminal. Any *delay* is unacceptable, and the character *sequence* shall be kept, as even a single missed character leads to errors. Another similar example is the **mouse**, which constantly sends position updates to your PC. An interesting example is a **sound card** – any *missed* or *delayed* packet would result in sound corruption during playback.

**Character** and **block** devices are located in the `/dev/` directory. List *only the character* ones with the command `$ exa -lah /dev/ | grep crw --color`.

The `mknod` command was used to create character devices, block devices, and named **pipes**. Nowadays, this is automatically handled by the **udev** SW, which is integrated into **systemd**. For more information, check the `udev` and `udevadm` man pages. We will explore **systemd** (but not **udev**) in detail in *Chapter 13*. As the `mknod` **command** is rarely used for such purposes nowadays, I will not cover it here. You can read a bit more information here: `http://www.infotinks.com/mknod-mdev-udev-udevadm/`.

An interesting article on this topic *relevant for* **developers** is provided at `https://linux-kernel-labs.github.io/refs/heads/master/labs/device_drivers.html`.

### Block devices

The most common examples of **block devices** are storage drives, such as **HDDs**, **SSDs**, external USB sticks, and DVDs. They usually communicate by sending entire data blocks, have a specific hierarchical data structure, and support random access.

Like **character** devices, the **block** ones will also suffer if a block is corrupted. However, additional checksums and other error-recovery techniques are often applied to them. Thus, if part of the data is corrupted or lost, it will be requested again.

Unlike **character** devices, **block** devices are mostly *not time-critical*. Everybody likes copying at 150 MB/sec, but if the USB/HDD speed is slower and copies at 15 MB/sec, this would typically not result in issues.

To list *only the* **block devices**, execute:

```
$ exa -lah /dev/ | grep brw --color
```

Any **storage drives** are typically named `sda` or `sdb`, followed by *digits* for **partitions** on the device. We will see how to work with them in detail in *Chapter 10*.

An interesting article on this topic *relevant for* **developers** is provided at `https://linux-kernel-labs.github.io/refs/heads/master/labs/block_device_drivers.html`.

## Drives and partitions

We have some partitions on our machine, whether on an HDD or SSD. The main one on our primary drive usually contains the **root** partition on which Manjaro is installed. In *Chapter 10*, we will dive into external drive mounting and unmounting. Here, I will provide only a few basic points.

All storage drives connected to our system are located in the `/dev/` directory, usually under the names `sda`, `sdb`, or `sdX`, where `X` is some letter from the Latin alphabet. If we have several partitions on one HDD, they will have an additional number after the primary designator. Thus, the partitions for my PC's `/dev/sdb` drive are `sdb1`, `sdb2`, and `sdb3`.

To see the partitions of the different drives in GUI, use **GParted** or **KDE Partition Manager** – write `partition` in the main menu and run the one you have.

To list the mass storage devices attached to our system, we can use the pre-installed `df` command. However, I prefer to install `duf` as it has a better-formatted output. Run `duf` without arguments and look at the first table titled `local devices`. To get more details, use `$ sudo fdisk -l`.

To inspect the *used space* and which directories are the biggest, use the **GUI** app **Filelight** (you must install it on **Xfce** and **GNOME**).

## Inodes and links

Our next stop is one of the main building blocks of the **ext**, **XFS**, **btrfs**, and other Unix/Linux-supported **FS**s – an **inode**. It is an abbreviation of **index node** and contains information about the file type (a **file**, **directory**, **socket**, **pipe**, **character**, or **block** device), size, permissions, ownership, access or modification time, counter of hard links, and its *actual data location* on an HDD. It doesn't contain the FS hierarchy **path**!

A modern conventional **HDD** consists of *metallic plates* and resembles a vinyl record. Locations on it are related to the plates' rotation, radius, and sector, read by the arm, just like the needle of a record player. However, the HDD plate's units, which represent the digital bits of information, are the size of *nanometers* (thousands of times smaller than the thickness of a human hair). On an **SSD** or a **USB** stick, these bits are located on **NAND flash** cells, again on the *nanometer* scale.

Since separating billions of bits is almost impossible, the storage devices are separated into **sectors**. On one of my current systems, with a *150-GB* HDD, each sector is *512 bytes*, and the HDD has *314,568,765 sectors* in total. The **inode** contains the **ID** of one of these sectors *as the beginning of your file* and how many sectors it spans through.

Each **inode** has *one unique identifier* in the form of a number. $ exa -li or $ ls -li will list the details of the current directory files, with the beginning **inode's unique ID** in the first column. The **inode** is like a personal ID card with your name. It is not you, but it identifies you and has a *unique number and **metadata*** – the date issued, address, in which country/state it was issued, etc. **Metadata** means *data about the actual data*.

On most Unix/Linux-supported FSs, the **filename, path** in the FS hierarchy, and a pointer to a corresponding **inode** are located in a **kernel**-managed table called **dentry** (*directory entry*).

Any file can have *multiple* **hard links**, like having *multiple names*. A new **dentry** record is written for each new **hard link**, and then the **inode** hard links counter is incremented with 1. When we have *several hard links for the same file*, the **FS** will delete its **inode** and the actual data only *after all of them are deleted*. As hard links are related to the **FS**, they *cannot be created for files from another partition or drive*. Finally, if we edit a file from one location, as the other hard links point to the same data, they will access its modified version after the changes.

In contrast, **soft links** are like a note stating that you live in Illinois, USA, at address XXX. However, this has nothing to do with your **ID**; *it is only a path*. And as on **Linux** we mount drives – if the **path** is located on another not mounted drive, having the path may be pointless after unmounting. We will learn more about mounting in *Chapter 10*.

Why is the **inode** important? Because it is **unique** and holds a file's metadata and actual location. In addition, say I want to move a 40 GB UHD movie from the AmazonDownload directory to the Movies directory *on the same drive partition*. The aforementioned **FSs** would *only modify the FS metadata information* when performing this task. If we move the file from a directory on the /dev/sda1 drive to another on the /dev/sdb2 drive, the *40 GB* movie will be *copied* to the second partition and *removed* from the first one. In addition, the old drive's **inode** will be deleted, while a new one will be created on the second drive.

The number of **hard links** is listed with $ exa -lH or $ ls -l. The count is the *first column* after the permissions string for both commands. For exa, if we add h like this $ exa -lhH, we will also have *headers* of the columns with data. A **hard link** is created via the ln command, as shown at the beginning of the chapter. To make it a **symlink**, use the additional parameter -s.

Here's an interesting fact – when we get a new **HDD**, partition it, and format it with a **FS**, a small percentage of the partition *is reserved for inodes, dentry, and metadata information*. We will be able to use the rest for storage. $ duf -inodes will give us the percentage of used inodes, as they are limited. However, on a regularly used modern file system, we will run out of space long before running out of space for metadata or available **inodes**.

## Standard and extended attributes

A file's **standard** and **extended** attributes are part of its metadata. All of them are listed in *Table 9.1*:

| Type | The Unix file type described at the section's beginning – file, directory, socket, named pipe, and block or character device. |
|------|------|
| **Permissions** | As we saw in *Chapter 7*, we have *three permissions groups* – for the owner, group, and other users, written in the **rwx** format. Replacing each letter with **a dash** means that permission *is not provided.* |
| **User ownership** | If we create files or directories with our user (in my case, luke), we are the owner. The **root** is the owner if we create them with sudo. |
| **Group ownership** | As you already know, a file also has group ownership. By default, when a user creates a file, it belongs to this user's group. If it is created by root or with sudo, it will be in the root group. To list the file's group in a second column, use $ ls -l or $ exa -lg. |
| **Timestamps** | There are three timestamps – the times of *creation, latest modification,* and *latest access.* ls doesn't offer column headers, so use $ exa -lhm for modified. Change the m with u for accessed and a capital U for created timestamp. |
| **Size** | This is the size in **bytes**, which exa reports in better form than ls. |
| **Inode** | When we add i to the exa and ls *options string*, we get the unique **inode** number in the first column. |
| **Flags** | The file flags are listed with the lsattr command. When used without parameters, it lists the flags for all files in the current directory. When used with a path to a file, it lists them for the file. Here are some of them:<br><br>• i: Immutable file (which cannot be changed)<br>• a: File can only be appended<br>• A: File time not updated<br>• c: File is compressed<br>• e: File uses extents to map blocks on the HDD<br>• E: File is encrypted<br>• s: File securely deleted<br>• S: File updates are synchronous<br>• u: File can be undeleted<br><br>There are even more flags, and some vary in support depending on the FS (**ext4, btrfs, zfs, xfs**, etc.). |

| Hard links | The number of **hard links** to a file; adding a capital **H** to the **ls** and exa options shows their count. |
|---|---|
| **Extended attributes** | In the **ext4** FS, **extended attributes** (**xattrs**) are key-value pairs that can be associated with a file or directory. They are stored in a separate inode and can be used to store more metadata about a file or directory that does not fit into the standard FS data structures.

Here are a few examples of extended attributes:

• `user.mime_type`: This can be used to store the **MIME** file type, which can be used by applications to determine how to handle the file

• `user.xdg.origin.url`: This can be used to store the **URL** from which the file was downloaded

• `security.selinux`: This is created to store the **SELinux** security context of a file or directory

It is also possible to create custom extended attributes. The naming convention followed by these custom attributes is usually `user.namespace`. |

Table 9.1 – Listing of all file attributes

**Extended attributes** are not typically used in the **Linux** kernel, most distributions, or for everyday applications. The reason is that their size and ways to use them depend on the **FS**. Hence, a **universal kernel** such as **Linux** cannot depend on them.

Conversely, extended attribute support is provided in most kernel-supported **FSs** and for any programs that use them. In particular, **extended attributes** have been supported for the **ext** FS family since **ext2**.

An excellent example of extended attribute usage is Security-Enhanced **Linux** (**SELinux**). This Linux kernel **security module** provides a mechanism to support **access control security policies**, including mandatory access controls. **SELinux** is a set of **kernel** modifications and user-space tools, designed to add extra security to Linux systems by enforcing rules on processes and files. We will not explore them further in this book.

## Linux-supported FS types

In this section, we will review the most popular Linux-supported FSs to know what they are, should you meet any of their abbreviations. We will also check several essential features with explanations in the last two subsections.

## Ext4 and a bit of history

The **MINIX** *monolithic microkernel* inspired Linus Torvalds to start writing the Linux kernel. Initially, he used the **MINIX FS**, which had limitations. To overcome them, the extended (**ext**) **FS** was made. It was the first FS *designed explicitly for* **Linux** and was based on traditional **UNIX FS** principles.

We can hardly find information on **ext**, as it was introduced in *1992* and replaced by **ext2** less than a year later, in *January 1993*. **ext2** was the first long-term **Linux** FS. It solved size and name limitation issues and was inspired by the BSD Unix **fast FS**.

In *1998*, **ext3** development began. It introduced **journaling**, which allowed tracking of changes and latest performed actions. This added the capability for *data recovery after abnormal shutdowns* and reduced the possibility of file corruption. It had good backward compatibility with **ext2**.

In *June 2006*, one of the **ext3** contributors announced the development of **ext4**, officially introduced in the kernel in *October 2008*. It has backward compatibility with **ext2** and **ext3** and introduced *online defragmentation*, *delayed allocation*, *extents*, *unlimited subdirectories*, *checksums* for the **journal**, faster FS checks, allocation, and timestamp improvements. Thanks to all these features, it is *still the default* **FS** on many distributions.

## XFS and ZFS

**XFS** is a **Silicon Graphics Inc.** FS started in *1993* and has been officially supported in the **Linux kernel** since *2001*. It is the default FS in **RedHat Enterprise Linux** and has similar features to **ext4**, along with some improvements such as **self-healing** and **RAID** array support. It has only one disadvantage – resizing an existing **XFS** partition is hard and slow. However, it supports **B-tree indexing** for scalability of free space management, a large number of concurrent operations, extensive runtime metadata consistency checking, diagnostic utilities, scalable and fast repair utilities, and many others. While it is excellent, it explicitly targets *large-scale systems*, *super-computers*, and *clusters* and, compared to alternatives, outperforms only there.

**ZFS** by the **OpenZFS** project is a good alternative to **ext4**. It is in active development and is not yet massively supported due to license incompatibilities between its **CDDL** permissive license and the **GPL** of the Linux kernel. It has several advanced features compared to **ext4** and has a **Linux** support and integration package. Adding it to Linux is done manually (and typically by advanced users). **Ubuntu** and only a few other distributions support it natively, and there are many claims that *the license issues are not problematic*. Despite this, it doesn't have wide support and usage due to the supposed licensing issues.

## btrfs

**btrfs**, or **B-tree** FS (called by some **BeTteR** FS), was designed by **Oracle** and released with the Linux kernel *2.6.29* in *2009*. It is definitely good and is currently the default **FS** in **KDE Partition Manager**. Its features include online defragmentation, online block device addition and removal, **RAID** support, configurable per file or volume compression, file cloning, checksums, and the ability to handle swap files and swap partitions.

Compared to **ext4**, **btrfs** is not a journaling FS but a **copy-on-write** FS with *logging*. It *updates* **metadata/inodes** *only after the data writing has finished, reducing potential data corruption significantly*. It manages big and small files better than **ext4** and natively supports snapshots (which is excellent for incremental backups). Its **RAID** support is configurable at different metadata and user data levels. It detects **SSDs** and conventional **HDDs**, applying specific settings for the disk drive type.

Some performance evaluations state that for some scenarios, **ext4** is better, while for others, **btrfs** is better. The majority of the Linux community expects that, at some point, **btrfs** will replace **ext4** as the default FS on most distributions.

At the *end of 2022*, most distributions continue using **ext4** as the default FS, as it is old, mature, tested, and sound. **Btrfs** is newer, some features are pretty recent, and the Linux community doesn't switch to new technologies before they are sure it is the right decision. In other words, we are a bit conservative. In addition, many **servers** and **technologies** rely on specific aspects of current or recent technologies, so making changes risks damage to **infrastructure** and **customer products** priced at billions. Changes shall never be made without considering all possible implications.

Conversely, there is no better candidate for replacing **ext4** than **btrfs**. Unfortunately, some of its features are considered too new and not completely bug-free. In addition, a few more are still planned but not yet implemented, such as in-band data deduplication, online FS check, more **RAID** features, encryption, and a persistent read and write cache.

One more positive in favor of **btrfs** is that it is already used in multiple commercial cases, most notably by **Synology NAS**, **Facebook** servers, **Oracle** Linux, **SUSE** Linux Enterprise Server, and **Netgear**. Practically millions of users already benefit from its improvements over **ext4**.

The Manjaro current installer, **Calamares**, uses **ext4** by default but also offers automatic installations on **btrfs**, **xfs**, and **f2fs** (covered in the following section).

## Other FSs used rarely

With minor differences in **KDE Partition Manager** and **GParted**, they also offer support for the **FSs** listed here.

**FAT**, which stands for File Allocation Table, was created in *1977* for **floppy disks**. It was later adapted for **HDDs**. Multiple companies contributed to it, and it has three major versions – **FAT12**, **FAT16**, and **FAT32**. The latter is still widely used on small USB flash drives for bootloaders and small or legacy systems.

**LVM2 PV** is *not a* **FS** but a *Logical Volume Manager*. **PV** stands for **Physical Volume**. One of the primary uses it was created for was to manage more than four partitions on an **MBR**-formatted disk. Making one of them an **LVM** partition allows more sub-partitions to be created. **LVM2 PV** is still used for this today. In addition, it helps with volume management, including for large disk farms (supporting hot swapping), making snapshots, and encrypting multiple partitions with one password. Unless you need some of its advanced features, nowadays, you don't need it for a **GPT**-partitioned HDD, as **GPT** handles up to 128 partitions.

**JFS** (Journaling File System) is a 64-bit FS created by **IBM** that was actively supported for **Linux** in the first decade of the 21st century. No particular uses or news has come up lately, but it is possibly still used somewhere.

**Flash-Friendly File System** (F2FS) was developed for the Linux kernel in 2012 by **Samsung Electronics**. It is designed explicitly for flash memory. **Motorola**, **Google**, **Samsung**, **Huawei**, and others use it. It supports up to 16 TB flash devices; the maximum single file size is 3.94 TB. It is actively used and supported by **Arch Linux**, **Gentoo**, and **Debian**.

**EXFAT** (Extensible FAT) was created by **Microsoft** for USB flash devices and SD cards in *2006*. It is always used when **Microsoft's NTFS** is inappropriate, but **FAT32** is *insufficient* (with a maximum volume of 4 GB). In *2019*, Microsoft opened the specification.

**HFS** and **HFS+** are **Apple's** proprietary **FS**s and are supported by **GParted**. We can format a volume with them.

**NTFS** is **Microsoft's** proprietary FS. Yes, we can format a partition in this format on Linux. We'll talk more about this in the *External storage, NTFS, and automount* section.

**NILFS** (New Implementation of a Log-structured File System) was developed for the **Linux kernel** and released in *2005* under the **GNU GPL**. Despite being supported, it is not clear whether it is actively used.

**REISER** was the first journaling FS introduced in the **Linux kernel** in *2001*. **Reiser4** was its successor, but support might soon be ending.

**UDF** (Universal Disk Format) is an **ISO/IEC** open and vendor-neutral FS introduced in *1995* and is most widely used on **DVDs**. It is supported in **GParted**.

## Journaling, CoW, B-tree, and checksums

This section will briefly summarize a few important terms related to **FS**s. For more information, you will have to do additional research.

**Journaling** in a **FS** is a **circular log** that tracks all currently started **FS** operations. If *ongoing operations are interrupted* due to power loss, system errors, or unexpected system restart, the **journal** is checked, and if possible, the relevant parts are recovered. This helps a lot, as before the **journaling** introduction, system crashes and restarts often resulted in severe **FS** *corruption*, data loss, and even OS reinstallation.

**Journaling** *is done before a write or change operation.* There are many ways to implement it. It was first introduced in *1990* by **IBM** and later improved in many FSs. Since it writes a small amount of additional data, it slows the file operation a little, tangible only in cases of large amounts of data and servers with heavy workloads.

**Copy-on-write** (**CoW**) is a technique of first writing all changes to an HDD and then updating the corresponding **metadata** and **inodes** only after a successful write. Usually, the metadata and inode updates are *small* and *fast*, seriously decreasing the potentiality of a **FS** *failure*. If an event prevents the final update of the metadata, we will have simply lost the latest file copy. Consequently, preserving the data correctness with **CoW** is as good as with **journaling** but without the additional log-write overhead.

**CoW** typically writes to *free sectors* when moving data, *not overwriting* older files. Some implementations often preserve the initial data until the new data and its **metadata** have been updated, allowing better data recovery. In addition, when copying, **CoW** only increases the **hard link** count. Only when the second copy is changed is an actual copy made, hence the name **Copy-on-Write**.

A **checksum** is *a unique value* calculated from the mathematical representation of the data in a file, sector, metadata, or **inode**, allowing to prove its correctness. Some types of checksums allow partial correction of single-bit failures. It requires additional resources, but modern CPU, HDD, and memory architectures have dedicated HW for this task and do it fast. **ext4** has checksums on the **journal** and the **metadata**, while **btrfs** and **zfs** add *block-by-block data checksums* for improved file data recovery.

**B-tree** is a *self-balancing tree data structure* with a very good search time. It is practically a special binary tree with improvements and applications in **FSs**. It is used in **btrfs** and **xfs**. To avoid confusion, this is only one of the tens of the binary tree types, and *many of them are used in other* **FSs**.

## Healing and self-healing

**Self-healing** is the ability of a FS to *detect data corruption* and *recover* automatically when or after it is detected. From a purely technical perspective, *this is hard*. Such corruptions occur typically due to one of the following: **RAM** failure (extremely rare), *power failure during writing* (more frequent), SW failure (which is also rare but possible), and HDD **firmware/hardware** *failure* (more frequent).

From the earlier listed **FS**s, **self-healing** is implemented in **ZFS**, **XFS**, and **btrfs**. Multiple user reports state that, for now, **ZFS** and **XFS** are *better* at this than **btrfs**.

**Self-healing FS** *in combination with* **RAID** (reviewed later in the chapter) is present only in **ZFS**.

While different forms of **self-healing** may help in many scenarios, preventing the possibility of having such issues is also important. Our **RAM**, **HDD**s, hardware, and software are significantly improved compared to those we had decades ago. Still, there are three more things we can do by ourselves to reduce the chance of errors significantly, and here they are:

- **UPS** (Uninterruptible Power Supply): This is *the single solution to avoid data corruption due to power loss*, frequently used for servers and workstations at the workplace. I have used it in my office and with three customers at different workplaces.

- **Laptop**: A laptop doesn't need a **UPS** as it notifies the user when the battery charge reduces dangerously. In addition, you usually have a power scheme *to put the laptop to sleep/hibernate*, preventing any potential FS failure due to power loss.

- **ECC RAM**: This is a special, expensive type of **RAM** with an additional *Error Correction Code* (**ECC**), preventing *rare single-cell failures*. This memory is approximately 80% more expensive than regular RAM and is widely available for workstations, regular PCs, and servers. Although rare, it is also available for laptops and mini-PCs.

## Linux directory structure

Now that we know **file types** and the most famous FSs on **Linux**, let's explore its **FS directory structure**. Linux has a stable structure, inherited initially from Unix and developed further. You will never again be confused once you know all the *main directories* and their **usage**. Before continuing, I recommend visually inspecting your directories with the **Filelight GUI** application. It provides an excellent **HDD** space usage analysis.

If you haven't installed **tree**, just run $ `pamac install tree`. This command-line tool lists the directory **structure** up to a chosen level; find out more in *Chapter 7*.

It is important to note that there is a **FS Hierarchy Standard** (**FHS**), initially created for any **Unix**-like systems. *Version 3.0* was issued in *2015* by the **Linux Foundation**. It is not a strict obligation but more of a recommendation. In addition, different distribution families have had different approaches to some of its parts over the years, so don't expect **Manjaro**'s **hierarchy** to be the same as on other distributions. The standard is posted here: `https://refspecs.linuxfoundation.org/ FHS_3.0/fhs/index.html`.

For Manjaro, the list of directories in the root directory / are as follows:

```
$ tree -d -L 1 /
/
├── bin -> usr/bin
├── boot
├── dev
├── etc
├── home
├── lib -> usr/lib
├── lib64 -> usr/lib
├── lost+found
├── mnt
├── opt
├── proc
├── root
├── run
├── sbin -> usr/bin
├── srv
├── sys
├── tmp
├── usr
└── var
```

In **bold** and with an arrow - >, we have symlinks to subdirectories *that point to* usr ones. We will discuss them after we review each directory's standard concepts:

- /bin: This directory contains all **bin**aries, including executables such as ls, exa, grep, cat, sysctl, pamac, chmod, etc.

- /boot: The directory of the OS boot-related binaries, *including the Linux kernel itself*. It also contains the efi sub-directory with the **efi bootloader** and is typically located on /dev/sda1.

- /dev: This is where all device files reside. When listing with $ exa -lah /dev/, you will see no regular files or binaries but *mostly character and block device files*.

- /etc: This directory holds *text configuration files*. Examples include fstab, time zone, shell, terminal, user accounts and passwords, Xorg, and the graphical environment. The fastest way to inspect the configurations is by dumping them on the terminal with $ cat /etc/filename, sometimes requiring sudo.

- /home: The directory containing each user's home directory. When we open the terminal, we always start on the /home/YourUserName path. It contains the regular Desktop, Documents, Downloads, and Pictures directories, and hidden files with user configurations such as .bashrc and .zshrc.

- /lib and /lib64: Many SW applications depend on *shared object libraries* located here. Examples are libvlc.so, libturbojpeg.so, libsystemd.so, libQtxxx.so, libjson.so, libjpeg.so, libcamera.so, and thousands of others, preinstalled by the Manjaro team or installed by additional applications. lib64 has the suffix 64, as we have been working with 32-bit libraries for years, and initially, 64-bit versions were rare, separated in their own directory. Nowadays, many libraries have both versions, but we mainly use 64-bit software. Since many legacy applications count on one or the other directory, we keep them both.

- /mnt: Empty by default, this directory is to *temporarily mount* FSs, HDD partitions, network locations, and other storage.

- /opt: This directory was initially intended to install optional third-party SW with a **symlink** to it in the /bin directory. I have in my opt an **onlyoffice** installation. Additional SW such as exa, tree, glances, and others are from the category "system software" (not optional), so they are typically installed in /bin. Thus, newly installed software will rarely be in /opt.

- /proc: The directory for keeping information on all currently running processes. If we run $ ps -e, $ htop, or $ glances, we will get their list, identified by their **Process Identifier** (**PID**), explained in detail in *Chapter 13*. As on Manjaro, the boot and system manager is **systemd** – it has **PID** 1. In proc, we have a directory named after the **PID** for each process, holding pointer files to its temporary data. Thus, it is a directory used by the Linux kernel, which we can use to investigate the processes. Read more about it here: https://tldp.org/LDP/Linux-Filesystem-Hierarchy/html/proc.html.

- /root: this is the **root**'s home directory; thus, we need to use sudo to list its contents with ls or exa. It contains the same directories as the standard user /home directory, such as Desktop, Documents, and Downloads. It also has the root .bashrc and other user-specific hidden configuration files. All of them are available to us when we log in to the system as the root user.

- /run: This holds temporary directories, files, sockets, and other objects for currently running processes. It is relatively new; /proc existed before /run was accepted as an additional location.

- /sbin: Originally, this was for *system binaries* and shall contain the init directory, and binaries only run by the **root** user. In the case of Manjaro, it is a link to /usr/bin.

- /srv: This abbreviation is for *service data*, containing data for the *system's provided services*. A typical example is if you run an **HTTP** server that hosts/provides a website on your machine, its data will be stored there. Currently, for Manjaro, it contains two directories, one for HTTP and one for FTP protocols. More on networking and protocols will be revealed in *Chapters 11* and *12*.

- /sys: On **Arch**, Manjaro, and a few other distributions, this directory has a mounted *pseudo-FS* **sysfs**. It provides an *interface to kernel data structures* holding information related to HW and the OS. On Manjaro, it holds information for *block devices*, *busses* (such as USB, HID, PCI, and CPU), *devices*, firmware, the kernel, and so on.

- /tmp: The directory for *temporary files*. These are files that will never be required after reboot. Examples are temporary data from your browser and temporary results from running software. On Arch and most Linux distributions, a pseudo-FS called **tmpfs** is used in it, which is flushed upon boot.

- /usr: This is usually the location of all *directories* for executables, libraries, binaries, and installation files related to and installed by the current user. In the distant past, the /usr/bin directory was on a separate HDD, which was expensive, small, and contained all critical binaries for running the system. When placed there, the most frequently executed binaries were fast to reach and execute. Back then, many other infrequently used directories were on a slower disk. Nowadays, disks are big and fast, and such separation is unnecessary. As a result, **Debian** and **Arch Linux** merged the usr directories with the system directories, and **Debian** called the operation **UsrMerge**. Currently, many **Arch** and **Debian** derivatives use the same approach. This is why bin links to /usr/bin, lib links to /usr/lib, lib64 links to /usr/lib, and sbin links to /usr/bin.

- Var: This last directory holds any constantly changing variable data files such as **logs**, temporary data for games, mail, libraries, and caches. It is not cleaned automatically like /tmp, and inspecting it is often the best way to check what happened on a given system during the latest runs. As we saw in *Chapter 8*, here are the pacman and pamac caches. It also contains the journalctl logs, which we will explore in *Chapter 13*.

## External storage, NTFS, and automount

This section will explore topics related to external storage.

### External storage

External storage is any disk drive connected to our PC via **USB**, **SATA**, or **NVMe**, mounted *temporarily* in our system. An example can be an **HDD** from a friend's computer provided to us for data transfer.

The typical **hardware** interfaces for large storage devices are **SATA**, **NVMe**, and the already *obsolete* **IDE** and **SCSI**. Here is a short video explaining **SATA** and **NVMe** as physical interfaces: https://www.youtube.com/watch?v=s-2VrxgI49Q. The following two links show how to **exchange/mount/unmount** an **SSD** and a few other parts: https://www.youtube.com/watch?v=Sn8mFE_N2xc and https://www.youtube.com/watch?v=pHykQX455P0. Many people (and sometimes official vendors) have *video guides* on how to connect/disconnect such parts and upgrade your machine. While I do it often by myself, on certain occasions, I prefer to leave my laptop with *hardware technicians* at a local company I buy HW from. If you are not experienced but have skills with a screwdriver, watch videos and do it. If not, *use the services of hardware shops* – usually, there are people to help you. However, *always remember that whether you update your SSD, RAM, fan, or any other part, you must be sure your machine supports it. If you are unsure, the hardware shops are your best option.*

External hard drives formatted with any of the reviewed Unix/Linux **FS**s and **NTFS** are recognized by Manjaro and can be worked with flawlessly, and we can format any drives in all those **FS**s. The only exception from the earlier listed **FS** alternatives is **ZFS**.

## NTFS

We have already listed all popular **FS**s; you probably remember that **NTFS** is the **MS Windows** one. It has several versions and is supported by Linux.

Before we go on, if anyone wonders why we care about NTFS support in **Linux**, home and commercial systems often must *transfer data* from/to Windows-formatted drives. Microsoft is notorious for how long they *refused to implement the ability to format/read/write* **ext4** and any other Unix/Linux **FS** partitions on Windows. They also refused to open-source their **NTFS** specifications until *the end of 2021*. All Linux specifications are, of course, free and open by default.

### A short Linux and NTFS history

**NTFS** stands for **New Technology File System**. Its first version was 1.0 in *1993*, then 1.1 in *1995*, and 1.2 in *1996*. In *2000*, it got updated to *3.0*, and in *October 2001* to *3.1*. For historical reasons, the last two releases are referred to as *5.0* and *5.1*.

The oldest Linux NTFS driver (integrated initially in the Linux kernel around *2001*) was limited and *provided only read support.*

In *2006*, **Linux** got some write functionality with the first version of the **FOSS NTFS-3G** driver, which continues to be *developed and supported*. This is essential because it was the only way to read/write to/from NTFS partitions for 15 years, but it is unfortunately based on the **FS in USerspacE (FUSE)**. Some people claim that it is *too slow due to this*, but the original creators from **Tuxera** explain that their purpose was always stability and quality. In addition, several benchmarks prove that when used on an average CPU, this driver performs equally to a *kernel space driver*. Its **performance** *decreases* proportionally to the **CPU** power on micro or weak systems.

In *2021*, **Paragon Software**, a proprietary company *working with* **Microsoft**, started developing an official **FOSS** GPL-licensed **NTFS** driver for Linux and its kernel integration. After effort from them, the *Kernel Community*, and *Linus Torvalds* himself, the driver was integrated into kernel release *5.15* from *October 31, 2021*, under the name **ntfs3**. **Paragon** (represented by their CEO) promised to maintain it, do bug fixes, and support it. Despite some glitches in the process, *the driver has been maintained since then*. The corresponding **GitHub** repository has commits: `https://github.com/Paragon-Software-Group/linux-ntfs3`. However, the official documentation for the kernel **NTFS** support comes from **Tuxera** and is thus related to the **NTFS-3G**. You can read it here: `https://docs.kernel.org/filesystems/ntfs.html`. Regarding the **Paragon** driver, this Arch Linux page is a nice start: `https://wiki.archlinux.org/title/NTFS`.

Compared to the **FOSS NTFS-3G** driver, the new one provides approximately *20% speed improvement* and *no issues for micro systems* to work with **NTFS**.

### What do we have on the current Manjaro releases?

We have both drivers integrated and working. If we write `ntfs` and press the *Tab* key, we will see *multiple options for commands*, however, all from the `ntfs-3g` module (e.g., `ntfscluster`, `ntfsrecover`, `ntfsundelete`, and `ntfsinfo`). This means that when working with any of these tools, we use the **NTFS-3G** driver.

The **Paragon ntfs3** driver has been officially integrated into the **kernel** since version *5.15*, from *October 31, 2021*. This kernel was available in Manjaro's official releases from *December 23, 2021*. In *July 2023*, I worked with Manjaro *22.1.3 Talos* (with Kernel *6.1*), and the default activated driver was still the **Tuxera FUSE**-based **NTFS-3G**. Manjaro has a command line and a GUI interface to switch the kernel (reviewed in *Chapter 16*). I switched to a newer one (*6.4.6-1*), and after inspection, the OS worked with the new, better kernel-space driver after the first restart. After the second restart and more experiments with all official Manjaro flavors, they all fell back to the **NTFS-3G** driver each time. Connecting, copying, transferring, and using files directly on an external NTFS drive worked flawlessly on all official flavors with both drivers. Only one failed from over 30 connection attempts, and after reconnecting, the problem disappeared.

After checking tens of posts, news, and pages on the internet, it turned out that despite the new driver continuing to get fixes and being maintained, it is not the default on Manjaro. To change this, you have to edit **udisks** files. More information is here, although this driver didn't work for me: `http://storaged.org/doc/udisks2-api/latest/mount_options.html`. I also tried a manual `mount` option, which I couldn't activate in *July 2023*.

This support is important, as for any of us exchanging files with Windows via NTFS partitions, we will work faster and better.

*In conclusion*, **Paragon** will do more updates, but when the driver will be used by default depends on multiple factors, including the **udisks** project, **Arch**, and the Manjaro developers (if a custom configuration is required). As of *July 2023*, the **Tuxera** driver will remain in use, as the **Paragon** one doesn't provide any tools yet.

In *Chapter 16*, we will learn more about the **kernel**, its available versions, and how to switch them. This is important, as Manjaro leaves this to the user's control.

*Chapter 13* will review the `journalctl` and `dmesg` commands to inspect the system logs. This is essential to know what happens on our system and, in this case, to see which driver worked in the last few moments.

At this point, however, the easiest way to see which driver is used is as follows. First, double-click your NTFS partitions in your file manager (Dolphin, Thunar, or other). Then execute on the terminal `$ duf` or `$ df -T`. If the drive type is reported as `fuseblk`, it uses the **Tuxera** driver. If it is `ntfs3`, then it uses the new **Paragon** kernel driver.

## Auto and manual mount

This is a standard feature on **Windows** and **macOS** and is equally standard on **Manjaro Linux**. On all three official flavors, *when you connect* a **USB** drive, you have a standard **daemon** that will recognize the drive and mount it without any need to run a terminal command. As explored in the GUI overview sections in *Chapter 3*, each official flavor has settings for what will happen when an *external device or removable media is connected*. Automount will work equally great for **NTFS** and any Linux FS.

For internally connected **HDDs/SSDs** – I currently write using **Manjaro KDE** on a **dual-boot** Linux/Windows PC with two HDDs, one formatted with NTFS and installed with Windows *11*. I see my second HDD in the **Dolphin** file manager, and *double-clicking automatically mounts it*. This is also valid for the **Xfce** and **Gnome** flavors.

We will explore how to mount devices manually in *Chapter 10*. Here, I will mention that with $ `lsblk -fs`, we list all block devices with their identifiers, FS, UUID, and mount points when/if mounted. I always prefer `duf` for its nicer terminal output.

# FS snapshots and RAID

This short section presents each of the given topics briefly. The snapshots are related to *Chapter 10*, and the **RAID** subsection is here only if you need to learn about it.

## Snapshots

A whole HDD or a single FS partition **snapshot** is a fully usable copy of it, including its metadata, **dentry** table, **inodes** information, and *files data*, made at a single point in time. No changes shall be made to the given HDD or single partition when making the snapshot. If the original data corrupts, a user can recover the whole HDD or partition from the snapshot backup media. This restores the given partition exactly as it was when the snapshot was taken.

A snapshot will not hold versions of the files in it, unlike a version-control system. To highlight the difference, a version-control system keeps newer and older versions of files and provides the user with tools to access current or older versions. The most famous source code version-control system nowadays is **git**. While a snapshot doesn't support versioning, if we make a snapshot of a partition where a version-control database resides, we will have a backup from the moment the snapshot was taken.

Copying a whole partition is relatively slow but optimized. Instead of copying file by file, we directly copy whole data blocks, regardless of their contents. It's like if we work with whole large boxes with hundreds of products instead of opening each box and extracting each product individually. Still, the time required to make a snapshot is proportional to the size of the partition we make a snapshot of.

Snapshots are usually kept in a separate secondary HDD on your machine, a personal **NAS**, external or cloud storage.

Only btrfs and ZFS have integrated snapshot support from the earlier reviewed **FSs**. This means making snapshots with them is optimized and efficient.

We will explore the related topic of **backups** in *Chapter 10*.

## RAID

**RAID** stands for **Redundant Array of Independent Disks** (some people use the term *Inexpensive* for *I*). It was created to combine more than one HDD to create a single logical big volume with *higher data redundancy* and/or *increased performance*.

There are two primary RAID configurations. The first is *mirroring data entirely* on two identical HDDs (identical meaning precisely the same model and size). If one of the drives fails, the data has a backup copy on the other drive, implementing **redundancy**. The second option is **stripping** and combining drives *to work in parallel* so that when a data packet is written, *parts of it go simultaneously to the different HDDs*, increasing the *read/write speed*.

A combination is also possible – for example, say four identical disks are combined as two parallel drives, mirrored to the other two drives. We get both increased *speed* and *redundancy*.

There are also configurations for *combinations of dissimilar disks*. They are mostly theoretical and usually avoided, as the increased performance or redundancy is harder to achieve. There are exceptions, which I will not cover here.

**RAID** can be implemented with **SW** and **HW** methods. Using **SW** is based on additional drivers for data transmission and distribution to **HDDs** and partitions. The **HW** means are often more effective and based on dedicated chips mounted on the motherboard of a **PC**, **server**, or PCI expansion cards. When the motherboard supports **RAID**, this is always covered in its manual. All modern **NAS** servers support at least several **RAID** configurations.

Over 10 years ago, I used mirroring (RAID1) for one of my older machines, as regular HDDs were unreliable back then. This saved my data when one of the drives failed. Modern disks are *much more reliable* (especially the more expensive ones with additional *data checksums*, improved *mechanics*, and better *hardware*). However, RAID continues to be used because we can always benefit from increased **performance** and **redundancy** for servers, NAS, and personal storage. I use **RAID** on my **Synology NAS** server to ensure my data is safe in case one of the drives fails.

Be warned – working with **RAID** is generally *considered an advanced task*. In addition, investing in a **NAS** server *will provide you with both* **RAID** and *data backup*.

There are six main configurations of **RAID** and combinations of them. **RAID configurations** are called **levels**. The most common ones are presented in *Table 9.2*:

| Level | Description |
|---|---|
| **RAID0** | **Stripping** data on two parallel disks to increase speed. |
| **RAID1** | **Mirroring** data on two disks for redundancy. |
| **RAID2 and 3** | A rarely used type of **stripping** that uses *more disks* for data, with dedicated additional disks for checksums. |
| **RAID4** | **Mirroring** data on *more than two disks*, with an additional disk for parity/error corrections. This system provides *faster* and a *higher volume* of read capacity, but unless the FS is aware of **RAID4**, *writes are slow*. |
| **RAID5** | A *block-level* **stripping** technique where *parity* and *error correction* blocks are distributed on all drives; it's rarely used and requires at least three HDDs. |
| **RAID6** | The same as **RAID5**, but with one more additional disk for secondary *parity/error correction* blocks, again distributed on all HDDs from the set. |
| **RAID0+1** | Also known as **RAID01**, this takes *a pair of* **stripped** *drives* and **mirrors** them to another *pair*; it requires four HDDs and provides both increased speed and redundancy. |
| **RAID1+0** | Also known as **RAID10**, this takes two pairs of **mirrored** HDDs and **strips** the files between them, achieving a better speed of the redundant information. |
| **RAIDXX** | Other combinations – many are mostly theoretical and/or applicable only to specific large clusters/workstations with more than four HDDs. |
| **Linux MD RAID 10** | This is a software **RAID** subsystem provided by the Linux kernel. It is called **md** and supports the creation of classical **RAID10** arrays, regular **mirroring**, and various other combinations. |

Table 9.2 – RAID levels (configurations)

The **mdadm** SW package was created on Linux to build, manage, and monitor **Linux md Software RAID** arrays. It is preinstalled on **Manjaro Live** and is a terminal tool. This **Arch** article provides guidance and more information on **mdadm**: `https://wiki.archlinux.org/title/RAID`.

It is essential to note the so-called **fake RAID** – a term explained in the **Arch** wiki here: `https://wiki.archlinux.org/title/Install_Arch_Linux_with_Fake_RAID`. It designates a situation where producers implement a *low-level software RAID driver* instead of integrating an actual RAID controller (usually listed as a PCI device by `$ lspci`). This allows them to advertise their product *as if it has* **RAID**, but **SW RAID** is usually less effective than an actual hardware RAID controller. That's why machines and motherboards with real **RAID** controllers are typically expensive.

Modern **HDDs** are much better in *speed* and *capacity* than they were a decade or more ago. Unless you have a *dedicated* **HW RAID** *chip* and you explicitly want or must handle it, I suggest you don't bother with the topic for regular usage. I would always recommend backing up all the critical data on a **NAS**, some cloud, or at least on a simple external USB HDD. Also, if you want to invest in real data backup, buy at least a small **NAS** server (which will always support RAID with easy GUI settings) and a pair of expensive **NAS HDDs**. Then, you will have both **RAID** protection and a real backup solution.

Why are personal **NAS** servers highly recommended? All of them provide some form of *automatic backup* – you just set up the client software *to back up a directory* so that everything changed is copied automatically to the **NAS** server. As I use **Synology**, I have checked for Linux-based automatic data backup client software, and they don't provide any on their website other than an Ubuntu **deb** package. But in the **Pamac search**, I have client applications for **Synology** and **Qnap**. In addition, they all provide a web interface and, most importantly, **rsync** capabilities (excellent for *backup*).

Finally, for some motherboards, it might be necessary to *disable* **RAID** *from the* **BIOS** *to see drives' sizes correctly and work with their total capacity*. If you know you have **RAID** and don't use it, even if you have already installed your **OS** and set up your machine, check that you can correctly see your **HDD** sizes with `df` or `duf`.

## Summary

In this chapter, we covered the basics of FSs by looking into file types and how they relate to FSs via inodes and file attributes. We further checked all widely used **Linux FSs** and dived into the FS *hierarchy* implemented in Manjaro. We then covered the basics of *external volumes*, **NTFS** support, and *automount*. Our final stop was a short section on snapshots with FSs and a basic **RAID** overview.

In the next chapter, we will continue our journey with storage, its mounting, partition management, and encryption. We will then investigate backups, which are essential for anyone working on long-term projects and backing up data.

# 10

# Storage, Mounting, Encryption, and Backups

This chapter will teach us how to mount, inspect, partition, and manage storage. We use external storage to keep or transfer big amounts of data, but we will also review it from the perspective of backups. Here, we will also see how to encrypt partitions, which will be our first encounter with the topic of security. After that, we will dive into backups, as any long-term project requires its intermediate stages to be saved. We should also back up our personal data, documents, photos, and so on. The chapter finishes by comparing available backup tools and how to use the best one, integrated by default in Manjaro.

The sections in this chapter are:

- Storage management, partitions, and mounting
- Partition creation and encryption via a terminal
- Backups, tools, rsync, and recovery

## Storage management, partitions, and mounting

In this section, we will dive into storage management by exploring its basics, formatting, and partition creation.

### Storage

Storage can be *internal* (like an **HDD** or **SSD** on your machine), *external* (such as an **HDD**, **SSD**, or USB stick), on a *network* (a personal **NAS**), or the *cloud* (like Google Drive, Dropbox, or Amazon Drive). We will briefly look at each of them.

**Internal storage** is fast to access, cheap, and easy to use for backups. Its disadvantage is that it is not protected from theft, fire, or machine failure and is bound to the machine it is mounted on. In addition, it cannot be accessed via the internet (except with dedicated SW), and if it is also the HDD where your OS is installed, any Filesystem (FS) failure may lead to data loss. Thus, when using it for backups, it is recommended to use a separate HDD/SSD on a PC/laptop.

**External storage** is the cheapest form of secondary backup, not connected to your machine. While it is usually portable, it may sometimes be inconvenient for frequent backups of large amounts of data. Still, it is an excellent choice when you infrequently need to back up personal data such as photos, videos, and projects. It is also suitable for transferring large amounts of data between two PCs or laptops.

**Personal NAS** is a more expensive solution, designed especially to operate 24/7, providing both **RAID** array for redundancy or speed (discussed in *Chapter 9*) and network access. If it is locally connected, it is great for quickly backing up large amounts of data. Accessing it from the internet may be slower, depending on the speed and bandwidth of the connection. It also requires some special setup.

**Cloud storage** is great as you have worldwide access to it, but it usually has a regular subscription fee, potentially a complex setup, and requires an internet connection. It may become inconvenient to periodically update/upload large amounts of information, as it would require an expensive, high-bandwidth internet connection.

To make **backups**, you must have a storage location for them with *enough free space*. The cheapest solution is to buy an *external* **HDD/SSD/USB** stick or, if you have a free secondary slot for HDD or SSD, to buy one for your machine. They all start from 10-20 USD and rise in price with capacity and quality. Choosing a good brand is essential, or after several years of usage, you may suffer HW failure and data loss. There are hundreds of reviews on the internet comparing external storage quality. I have several external HDDs and USB sticks priced between 20 and 80 USD.

If your primary **HDD/SSD** is *big* (roughly over *200 GB*) and you don't want to use external storage, later in the chapter, we will see how to *shrink* the /root or /home partition to make space for an additional *backup partition*. Remember that shrinking is a simple but potentially risky operation, as any error (though hardly possible) would result in losing your current data.

The expensive option is to invest in a **personal NAS**. We will later review multiple SW tools for backups, but I must mention that **rsync** is the best for personal storage. As of early 2023, **rsync** support is already provided by the **NAS** producers *Synology, QNAP, ASUSTOR, WD My Cloud, TerraMaster, Drobo, ioSafe, Buffalo LinkStation*, and *Seagate Personal Cloud*. A good basic **NAS** will have *at least two HDD bays and* offer a **RAID** array for redundancy or speed. As of mid-2023, such **Synology** devices start from 360 USD without **HDD**s. A pair of 4 TB HDDs start from 200 USD. Thus, the total minimal price for a good NAS solution starts at roughly 560 USD.

I recommend a personal NAS for backup, as it offers **rsync** and **RAID** support, and worldwide access to your data. In addition, these devices are robust and of high quality, developed and designed especially for backup, data storage, and 24/7 operation. They usually provide client application SW on multiple machines and OSs.

For cloud storage, we can use the **rclone** application, which works with over 50 standard cloud storage services: https://rclone.org/. It has *free online configuration examples* for each provider it works with (including **Google** Drive, **Alibaba** Cloud, **Amazon** Drive, and **Dropbox**). However, if you make a system backup, you must be careful of what is supported by **rclone** *and the given cloud storage on the server side*. For example, **rclone** *doesn't preserve the extended attributes of Linux filesystems*. Working with **rclone** and **cloud storage** is generally considered an advanced task.

# Creating a new partition

In this section, we assume that you have already purchased or have at your disposal the additional storage. Before going on, it is essential to know what a partition table is and how to change it. *Chapter 2* explains some basics of the topic, and here I repeat some and add additional definitions.

A **partition table** is the *primary record on a hard drive*, defining *its partitions*. Without a **partition table**, you can't have partitions or use the given **HDD/SSD/USB** *storage*.

Any older partitions are deleted if you *write a new* **partition table** on a drive. When you *change the partitions*, this table is *modified*.

Typical partition tables related to PC and laptop OSs are the older **MBR** (named **msdos** in Linux partition tools) and the new **GPT**. MBR allows only up to 4 partitions of 2 TB of space. **GPT** can handle up to 128 partitions and millions of TB.

After a **partition table** is *written* on a storage drive, we have *unallocated space* on which we can create partitions of sizes of our choice. The Manjaro installer Calamares non-manual storage setup will use all the available space and put the /root and /home directories in one partition. Any *manual* storage setup installation option allows us to set up custom partitions, as described in *Chapter 2*.

If we already have several partitions on a drive, and one has a lot of *free space*, e.g., *100 GB*, of which only *20 GB* is in use, we can **shrink** this partition to leave **unallocated space**. *A partition* should *never be left to operate without some free space*. The typical minimum limit is *20%*; if we want to write data to it after shrinking, the free space should be more. Thus, in the previous example, we can shrink the partition from *100* to *50 GB*, leaving us with *50 GB of unallocated space*.

If we have more partitions, and one, which we don't use, we can delete it, again to leave us some unallocated space. If you choose this option, be sure that either the given partition is empty or the data is unnecessary to anyone!

With unallocated space left, we can create a new partition. We can also extend an existing one located next to the unallocated space on the storage. Both are easily visible on GUI tools such as GParted and KDE Partition Manager.

A partition we want to **shrink**, **extend**, or **delete** must *not be in use*; otherwise, the partitioning tool will *reject the operation*. In Linux/Unix, when a partition is *in use*, it is **mounted**. To change it, we have to **unmount** it.

**Formatting** a partition means *writing a new* **FS** table on it. This effectively *deletes old information on the partition (if present)*. We can format it with **ext4**, **btrfs**, or any other FS. More **FS**s are presented in *Chapter 9*.

On Linux, *each connected storage device* is located at the path /dev/sdX, where X is a small letter from the Latin alphabet in order, starting from a. Each partition on a drive starts from number 1. Thus, if a drive has three partitions and is listed as sda, we will have in /dev/ the sda1, sda2, and sda3 device files (a.k.a. device nodes). You can read more about the Linux and Manjaro FS hierarchy in *Chapter 9*.

## Storage formatting

If you do a **system backup** on *any form of personal storage* (separate HDD, USB memory stick, or external HDD), it is *recommended* that it is formatted with the same FS as the partition/data you will back up to it. If your system is *formatted* with **ext4**, the backup should also be with **ext4**; if it is formatted with **btrfs**, back up to a **btrfs**-formatted drive.

For personal **NAS**, the manufacturer has usually solved this, and the user may not have a FS *choice*. As most brands support **rsync**, they have instructions and potential limitations described in their manuals. In the case of **Synology**, they use a special combination of **btrfs**, **ext4**, and Linux **RAID** to achieve the best data retention.

If you back up only regular data files, such as images, music, and books, then it doesn't matter; your backup may be formatted with *any* FS.

If you back up to a cloud storage provider, check the provider-supported **FS**s.

## Creating a partition

Open your main menu, type `partition`, and start **GParted** on Xfce or **KDE Partition Manager** on KDE. On Manjaro GNOME, **GParted** was previously installed by default, but they have now switched to GNOME **Disks**, which I will not present. The user interface of the previous two applications is better, and I recommend installing either. There is a short subsection on **GParted** and **partitioning** in *Chapter 2*, under the *The installation itself* section; I recommend checking it if you haven't already. For more information on **KDE Partition Manager**, check out `https://docs.kde.org/trunk5/en/partitionmanager/partitionmanager/partitionmanager.pdf`. It is important to note that if you want to encrypt the partition, you can only do it with the **KDE tool**; **GParted** doesn't offer this option. One more hint before we go on: everything you can do via GUIs of any flavor, you can also do via the terminal in a unified way. Keep reading to learn both ways.

### With GParted

After connecting the drive and opening GParted from the **View** menu, activate **Device Information**. Select the correct device `/dev/sdX`, based on the size and brand, from the top right corner box. If it is *a brand-new device*, from the **Device** menu, select **Create Partition Table** and choose type **gpt**, then click **Apply**. Remember that changing the **partition table** means effectively wiping out any old information, so for old storage devices, do this only if you want to wipe them; otherwise, skip the previous step.

Now, right-click on the `unallocated` space and select **New**. By default, all the available space will be selected, and you can change it either with a number in the **New Size** cell or via the slider at the window top. Choose the FS, write a **Label** and the same **Partition name**, then click **Add**. Finally, click the *apply checkmark symbol* and that's all.

### With KDE Partition Manager

After connecting the drive and opening KDE Partition Manager, locate it on the left in the **Devices** *listing*. If it is *a brand-new device*, from the **Device** *menu*, select **New Partition Table**, choose type **GPT**, and then click the **Create…** button. Remember that changing the **partition table** means effectively wiping out any old information, so for old storage devices, do this only if you want to wipe them; otherwise, skip the previous step.

Now, right-click on the `unallocated` space and select **New**. By default, all the available space will be selected, and you can change it either with a number in the **Size** cell or via the slider at the window top. Choose the FS, write a **Label**, and then click **Add**. Now, click the apply checkmark symbol and that's all.

For **KDE Partition Manager**, marking **Permissions** for **Everyone** during creation is helpful for the future. It is recommended if you don't explicitly need to keep the partition accessible to **root only**. Of course, you can change this later from a terminal.

To encrypt in **KDE Partition Manager**, select **Encrypt with LUKS** and enter a password of your choice. You will find out more about encryption in the *Partition creation and encryption via a terminal* section.

### Via the terminal

Partition creation via the **terminal** is explained with a detailed example in the *Partition creation and encryption via a terminal* section. I will not include it here not to duplicate instructions. Check it out – it is easy, and the text explains what to do if you want only to create a partition without encrypting it. It is in *Steps 1 to 3.2*.

## Mounting

This section will briefly describe the basics of mounting and ownership of the mounted partitions. To start, we have two main commands: `mount` and `umount`. As you already know, each drive and partition has a file describing it in the `/dev/` directory. They also have the **FS type** property (**ext4**, **btrfs**, etc) and a *mount point* when mounted. With `df` or `duf`, we see the **mounted** partitions with their FSs and *mount points* in the FS hierarchy.

Usually, after a partition creation, it is automatically mounted in restricted **Read-Only (RO)** mode. In this case, right-clicking in your file manager will gray out the **Create Folder/Create Document** options. In this case, **Xfce**'s **Thunar** file manager even depicts a locked padlock on the contained files icons.

Now that we have a partition, we will explore how to:

- Mount an *external storage* device with **Read/Write (RW)** permissions
- Change the directory owner
- Automatically mount partitions permanently connected to our system

Before we start, in the following examples, I will use a *50 GB* partition called `data_BKP_Linux`. To **unmount** a partition, call `df` and find your partition based on the *name*, *size*, and **FS**, as shown in *Figure 10.1*:

```
🔲 🔳 ~ df
Filesystem      Size  Used  Avail  Use%  Mounted on
dev             3.9G     0   3.9G    0%  /dev
run             3.9G  1.1M   3.9G    1%  /run
/dev/sda2        86G   12G    71G   14%  /
tmpfs           3.9G     0   3.9G    0%  /dev/shm
tmpfs           3.9G  4.0K   3.9G    1%  /tmp
/dev/sda1       300M  288K   300M    1%  /boot/efi
tmpfs           794M   72K   793M    1%  /run/user/1000
/dev/sdb1        49G  2.1M    47G    1%  /run/media/luke/data_BKP_Linux
```

Figure 10.1 – df report for a mounted partition

Now I know it is `/dev/sdb1`, so I execute:

```
$ umount -v /dev/sdb1
```

The `-v` parameter provides a verbose execution status.

### Mounting a partition or a USB HDD with write permissions

We first *list all storage volumes connected to our machine* with `$ lsblk -f` and recognize our volume based on the *FS* and *label*. In addition, when a partition is not mounted, its `FSAVAIL` and `FSUSE%` values are not reported, as shown in *Figure 10.2*:

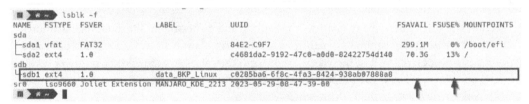

```
🔲 🔳 ~ lsblk -f
NAME   FSTYPE  FSVER            LABEL            UUID                                   FSAVAIL FSUSE% MOUNTPOINTS
sda
├─sda1 vfat    FAT32                             84E2-C9F7                               299.1M     0% /boot/efi
└─sda2 ext4    1.0                               c4681da2-9192-47c0-a0d0-82422754d140    70.3G     13% /
sdb
└─sdb1 ext4    1.0              data_BKP_Linux   c0285ba6-6f8c-4fa3-8424-938ab07888a8
sr0    iso9660 Joliet Extension MANJARO_KDE_2213 2023-05-29-08-47-39-00
🔲 🔳 ~ ▌
```

Figure 10.2 – Using lsblk to define which /dev/ device to mount

*Mounting* with the correct permissions is best to be done in *a dedicated directory in which we will mount*. The automatically mounted partitions typically go to `/run/media/ourUser/PartitionName`, but when doing it manually, we can select any target directory. In the terminal, create a directory in `/home/yourUser` with `$ mkdir BKP`. By default, `mkdir` creates a directory with **rwx r-x r-x** permissions for the **current user** who called it. *Group users* and *others* can't write. To check this, run `$ ls -l` or `$ exa -l` in the directory.

Then, we have a mount call with `sudo`:

```
$ sudo mount -w -v /dev/sdb1 /home/luke/BKP
```

`-w` is the option to assign RW permissions. In your case, you must replace my username, `luke`, with your user directory name. `-v` is used to *print verbose status*.

After this operation, you will often have the directory ownership of the target directory (`/home/luke/BKP`) changed from *your user* to **root**. In this case, execute the call in the next subsection.

> **Warning**
>
> As **Linux** offers super-flexibility, you *can mount a device to a folder that has already been mounted on*. Thus, if I have `/dev/sdb2` mounted to `/home/luke/BKP`, mounting `/dev/sdb3` at the same location will be *accepted and executed*. When you copy a file, it will go to both partitions! Always check your mounts with `$ lsblk -f` and `duf`, especially when working with multiple partitions!

**Linux** also supports *mounting one folder to several paths simultaneously*. When **unmounting** with a **device file** (such as `/dev/sdc1`), each time you call `umount`, it will *remove only one mount point*. To **unmount** everything, you must call all the mount paths, separated by spaces, like this: `$ umount /path1 ~/path2/ ...`.

I emphasize again – always check your mounts and unmounts with `duf` and `lsblk`, especially when you're working with multiple paths or partitions!

As **USB** is a *hot plug/unplug* interface, you can detach it without unmounting but don't expect to have it mounted on the same path the next time you connect it.

### Changing the directory owner

Changing the **owner** is necessary if a directory was created by another user (e.g., **root**), and we need to be the owner so that **user permissions** will apply to us. As *mounted partitions are directories*, we can apply ownership to them. To do this, execute:

```
$ sudo chown -R -v $USER:$USER /home/luke/BKP
```

`-R` will change *all the subdirectories recursively*. `-v` is for a verbose status.

Check the results with `ls -lah` or `exa -lah`.

### Mounting permanently via /etc/fstab

If you have *a partition* that is part of an **HDD** *attached to your system permanently*, then *you probably want it mounted by default at boot time*. This happens with `root` and home partitions when the kernel **boots** (starts) the **OS** from the **boot directory**. It reads **entries** described in the `/etc/fstab` file and mounts them accordingly. **fstab** stands for **filesystem table**.

A **partition** (or a volume) is *identified* by a **Universally Unique Identifier** (**UUID**), also called a **Globally Unique Identifier** (**GUID**), which is a *128-bit label* (32 hex digits). This label is reported by $ lsblk -f, present in *Figure 10.2* in the fifth column.

A partition may also be identified by its **label** (e.g., JohnDataBKP) or **device file/device node**, such as /dev/sda1, /dev/sdc3, or /dev/sdc2.

In the **fstab** file, we work most frequently with **UUIDs**, as they are *unique*, while *device nodes and labels are not*.

$ lsblk -f is the command to list the **UUIDs** for us. When we created the partition via the partition manager, it got a **UUID**. For **fstab**, simply running $ cat /etc/fstab will show us the existing entries for boot, root, and home (the last one only if located on a separate partition). If there are more drives, they will follow.

With **KDE Partition Manager**, you can automatically edit the **fstab** file by right-clicking on the volume and selecting **Edit Mount Point**. For **Identify by**, select UUID. Then, after clicking the **Select…** button, *choose the location to mount to*. From all the options, only select **No update of file access times**, to put less pressure on your hard drive. This will only set the noatime flag, which is good to ensure less wear on your HDD (whether **SSD** or regular). **Dump Frequency** and **Pass Number** should both be left as **0**. *Figure 10.3* shows the menu:

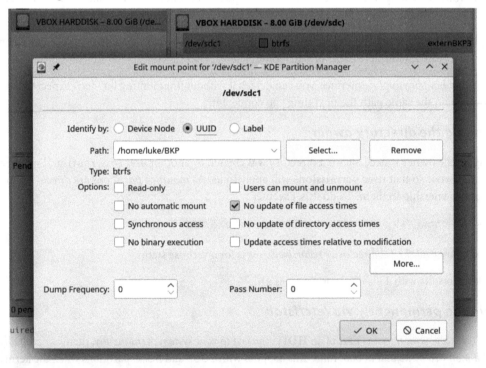

Figure 10.3 – Editing the fstab mount point via KDE Partition Manager

We use the `fstab` file mainly for permanently connected drives and partitions, as a missing drive may interrupt the OS boot process before the login GUI is shown on the screen. The mount directives in `fstab` offer options, one of which is `nofail`. *If you add it, the boot process will not be interrupted* if the given partition is missing. To do this in the menu shown in *Figure 10.3*, press the **More** button and type `nofail` in the pop-up window.

Clicking **OK** will warn you that *the changes will be written* to the **fstab** file, which is our purpose. After this, running `$ cat /etc/fstab` will *show an additional line for the partition*, identifiable by the **UUID**, FS, and its mount point. After `reboot`, the partition will always be mounted to the given path.

We can also edit `fstab` directly via the terminal. Do it with the **nano** editor run with `sudo` from a *second* **Terminal** via `$ sudo nano /etc/fstab`. This way, it is easy to copy the **UUID** from the first terminal (by *selecting and right-clicking* or pressing *Ctrl+Shift+C*). If you are unfamiliar with the options, check out the short description in *Chapter 7* and *read the short help* available with *Ctrl+G*. Then, add a line for your partition in the following form (it is all on one line in the file):

```
UUID=62bef846-0483-4088-9b4e-1886bb4984e5 /home/luke/BKP ext4
noatime,nofail 0 0
```

**Replace** the **UUID** *with yours* from `lsblk` and *the directory path* with the prepared empty directory in your home folder. Now press *Ctrl+s* to save and *Ctrl+x* to exit.

That's all – *the next time you start your machine*, this volume will be *automatically mounted at boot*. You can read more about all the possible arguments for the `fstab` format here: `https://wiki. archlinux.org/title/fstab`.

## Shrinking/resizing root or home partitions to free space

Any modern laptop/PC has **HDDs** or **SSDs** from 120 GB to 1 TB. I recommend *never buying a machine* with *less than 256 GB of storage*. Upgrading the SSD/HDD is cheap nowadays if you have an old one with size, e.g., 64 GB.

When we have a big HDD (in my experience, with a size over 200 GB), a backup can be done on a separate partition on this same HDD. The automatic Calamares installation doesn't make a separate `home` partition; we will have to shrink `root` in this case. We can shrink `home` if we did a manual installation with a separate big `home` partition.

Partitions are ordered sequentially on an HDD and cannot be split into parts. Shrinking can always be done. In contrast, *extending* one requires *the free space we will extend with* to be *just before* or *just after* it.

When doing this, I always prefer working with **GParted** or **KDE Partition Manager**, as their **GUIs** are handy. They also allow to move partitions (again, if there is free space before or after them).

The easiest way to change the `root` or `home` partition is via a **Manjaro Live USB stick**. I recommend the **Xfce** or **KDE Plasma** flavors, as they offer **GParted** and **KDE Partition Manager**, respectively.

If you don't remember how to burn a Manjaro ISO, refer to the *Installing on a USB stick* section in *Chapter 2*. Taking the latest ISO is recommended, as it will contain bug fixes and newer SW.

Be warned – although *completely safe* (it has never failed for me), this is *always a risky operation*! The problem is that *if you corrupt your root partition*, your PC will not be able to **boot**, and most of the time, as you manipulate the partition table of your **HDD**, you will explicitly need to *format* and *completely reinstall* it afterward. Before doing so, consider buying an external storage device (they are cheap nowadays – starting from 30 USD for a 120 GB device). Copy any sensitive *personal data* onto it as a backup, then shrink `root` or `home`.

Now, boot the **Manjaro live USB stick**, just like for installation. When ready, **restart** your machine and enter the **BIOS setup** described in *Chapter 2*, in the *BIOS/UEFI setup for installation on a PC* section. Once you have loaded Manjaro live, open the partition manager, select the root partition, right-click on it, and select **Resize/Move**. **Resize** it to the desired amount, *always leaving enough free space for the OS to continue working*. `root` will need *at least 20 GB* of free space; `home` will depend on your needs, but I would always leave *at least 20 GB*. Any partition should *never be filled more than 80% to operate normally*, so calculate *carefully*, considering *future additions* like data, installations, and updates.

For any partition, if it has 20 GB of data and holds the home directory, I would always leave a minimum total size of double the amount plus 25% free space. This means we'd need at least *20+20+10 GB* of size *after* **shrinking**.

Finally, click **Apply**. I will not provide further details, as **GParted** and **KDE Partition Manager** were covered at the beginning of this chapter and in *Chapter 2*.

After **reboot**, remove the USB stick when the screen goes black, and you are ready. Now, your system should load again normally. Follow the instructions in the previous subsection, *Creating a new partition*.

## Partition creation and encryption via a terminal

**Encryption** is any technique that scrambles data based on a *predictive pattern* using an **encryption key**. The resulting data is *unreadable without a special decryption algorithm and a corresponding* **decryption key** *(also called a* **passphrase***)*, calculated *precisely* for the encryption key.

The keys are mathematical numbers. Encryption is used not only for partitions but also for different **internet** and **network protocols** (such as **SSH** and **HTTPS**), separate files (such as a **7zip** or other archive files, which cannot be opened without the *key/password*), and others. We will learn more about the internet protocols in *Chapter 12*.

If you have selected the additional **encryption** *option* during installation, you have already seen that your Manjaro installation will not load until *the passphrase for its drive partitions is provided*. I haven't advised you to select it, as Linux is already secure enough. Despite this, when we talk about sensitive information such as *highly confidential projects, financial data*, or *company documents*, we might need **encryption** to ensure that if the given drive or PC is stolen, *third parties cannot read its contents*.

It is essential to note that you can back up *from or to* an **encrypted partition** without problems, as *once it is opened and mounted*, it works *like any other regular storage partition*.

## LUKS

Linux offers the **Linux Unified Key Setup (LUKS)** based on the Linux kernel subsystem, **dm-crypt**. **LUKS** is FOSS SW, widely regarded as a robust and secure encryption system many Linux distributions use. It started in 2004 and has been subjected to extensive *testing and evaluation*. It is known for providing attack-resistant *encryption capabilities* based on **AES 256-bit** encryption. **LUKS** works together with the `cryptsetup` command-line utility, which has some partial compatibility with volumes encrypted with some tools other than **LUKS**. These are **loop-AES**, **TrueCrypt**, **VeraCrypt**, **BitLocker**, and **FileVault 2** (as stated in their latest official documentation), and **LUKS** can at least read partitions encrypted with them.

### Creating a LUKS encrypted partition via the GUI

As mentioned earlier, to make a **LUKS** encrypted partition via a **GUI**, I recommend using **KDE Partition Manager**. **GParted** doesn't officially support the creation of **LUKS** encrypted partitions, so if you are using **Xfce** or **GNOME**, you must still install **KDE Partition Manager**.

Apart from the settings mentioned earlier, there is nothing more to do. Consider only that a good password will be *at least 12 characters long*, including uppercase and lowercase letters, numbers, and a special symbol.

Once created, you will only need the passphrase to mount an encrypted partition. That's the only difference compared to non-encrypted partitions. If you add it to the `fstab` file, you must enter its passphrase (decryption key) every time your system boots. Regarding permissions – what we mentioned in the Mounting section is equally valid.

### Creating a partition via the terminal and optionally encrypting it with LUKS

Although easy, it requires a few steps compared to the **GUI** approach. They are defined separately in this section, numbered 1 to 4.

### Step 1 – Identifying an existing partition to modify

For blank new drives, read the following two sections only for information and continue from step *3.1*.

First, we need to know on which hard disk we will create the new partition. For this, we execute $ `lsblk -f`.

For this example, let's imagine that we currently have one old unnecessary partition of 50 GB that is *mounted*, and we will *remove* and *delete* it to create it again from scratch via the terminal.

For me, the previous command result is shown in *Figure 10.4*:

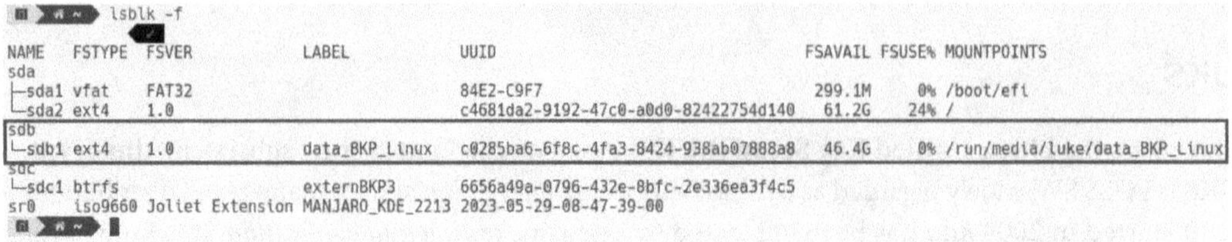

```
           lsblk -f

NAME   FSTYPE  FSVER           LABEL            UUID                                       FSAVAIL FSUSE% MOUNTPOINTS
sda
 ─sda1 vfat    FAT32                            84E2-C9F7                                   299.1M     0% /boot/efi
 ─sda2 ext4    1.0                              c4681da2-9192-47c0-a0d0-82422754d140         61.2G    24% /
sdb
 ─sdb1 ext4    1.0             data_BKP_Linux   c0285ba6-6f8c-4fa3-8424-938ab07888a8         46.4G     0% /run/media/luke/data_BKP_Linux
sdc
 ─sdc1 btrfs                   externBKP3       6656a49a-0796-432e-8bfc-2e336ea3f4c5
sr0    iso9660 Joliet Extension MANJARO_KDE_2213 2023-05-29-08-47-39-00
```

Figure 10.4 – lsblk result on the terminal

We must identify which partition to modify, or we *might format the wrong one and destroy its information*. This is one of the reasons why *currently mounted partitions cannot be modified* (reformatted, deleted, etc.), so the tools will always refuse to format partitions in use. For me, the partition is /dev/sdb1, mounted automatically at /run/media/luke/data_BKP_Linux.

## Step 2 – Deleting an existing partition

To **unmount** the partition (to be able to delete it), we need the mount path, listed at the end of the row in the previous listing, and the command is as follows:

```
$ umount /run/media/luke/data_BKP_Linux
```

We call $ lsblk -f again to check that the mount point has *disappeared* in the last column. We see now at the beginning of the same row that the target partition is sdb1. So, we call fdisk with it to modify a partition from the sdb drive without the additional numeric identifier (in this case, 1) as follows:

```
$ sudo fdsik /dev/sdb
```

Inside, we will be prompted for options. Typing m and pressing *Enter* will print the help. First, we have to use p to list all partitions. Then, we use d to delete a partition, and if we have more than one partition, we will be asked to *provide a number* (as when we have more than one partition, they are named after the main one, so sdb1, sdb2, sdb3, sdb4, etc.). If there are more partitions, refer to the number – for example, 3 – and press *Enter*. If we have only one, *it will be selected automatically*.

Finally, we are prompted for a command again; type w to *save the changes and exit*, then press *Enter*. We verify the deletion by executing $ `lsblk -f`. Know that `lsblk` *doesn't report unallocated space* but *only existing partitions*, so we know the deletion is successful if it is not listed anymore.

If you get an error message that the partition is *still in use* (although we have just successfully *unmounted* it) and the changes will be *applied upon system restart*, **restart** your system from the terminal by calling $ `reboot`. After restarting, the partition will have been deleted and not reported by $ `lsblk -f`.

### Step 3.1 (optional) – Creating a new partition table on a blank new drive

If this is for a completely *blank* or a new drive, you have to execute the interactive `fdisk` terminal utility and first create a new GPT table with the name of the drive, as in the previous step:

```
$ sudo fdsik /dev/sdb
```

If you haven't tried it, type m and press *Enter* to see all possible options. Then, type p if you want to *list any potential partitions on this drive* so that you are sure what you will wipe out, as partition creation makes any older content *unrecoverable*. To *create a new GPT table* type g, press *Enter*, and then type w to *save* the changes on the drive and *exit*.

### Step 3.2 – Creating a new partition on a drive

Call `fdisk` as before to double-check your partitions. Typing F and pressing *Enter* will provide us with a *summary of the unallocated space*. Now, create the *new partition* by typing n and pressing *Enter*. You will be asked for a partition number, which you can leave empty. Press *Enter*. Then, confirm the default for the first and last sectors by typing nothing and pressing *Enter* each time. If you are asked for the removal of a previous signature (in my case, `ext4`), type Y for *Yes* and press *Enter*. Finally, type w to *create the partition and exit* and press *Enter*. The result for me was as shown in *Figure 10.5*:

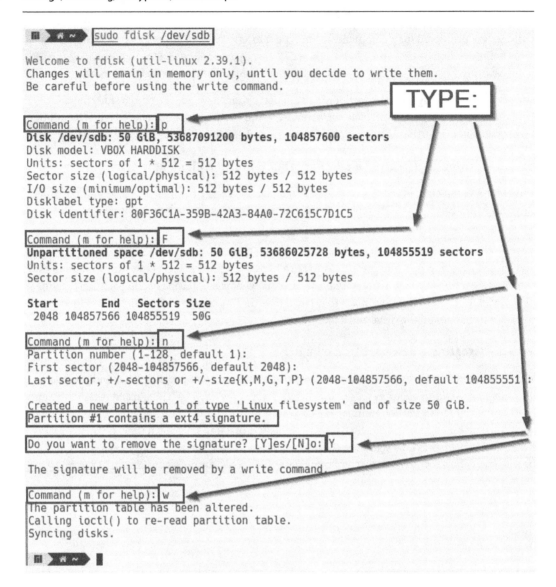

Figure 10.5 – Creating a new partition with fdisk

For the encryption case, a FS is added once an encrypted partition is opened with `cryptsetup` in *Step 4*. In this case, read the rest of this section for your information, and don't do it. In other words, continue with the rest of this section if you want to create a regular unencrypted partition.

Ensure you have the correct device node provided by `lbslk  -f`. For me, this is `sdb1`. Then, to create the FS, execute the following:

```
$ sudo mkfs.ext4 /dev/sdb1
```

Now, this partition *can be used*, but of course, you need to mount it to a location in your FS and change the *ownership* to your user via $ `sudo chown -v $USER:$USER /mnt/MountPoint`.

## Step 4 – Encrypting the new partition

First, check your partition file node with `lsblk -f`. In my case, it is `sdb1`. Now, set up **LUKS** encryption by executing the following:

```
$ sudo cryptsetup luksFormat /dev/sdb1
```

This command will ask you to explicitly confirm your intentions by typing YES in capital letters on the terminal and pressing *Enter*. Then, you have to *enter the passphrase twice* for the partition. The eternal reminder is here again – use *uppercase and lowercase letters*, numbers, and at least one special symbol. The whole *passphrase/key* should be *at least 12 characters long*. A great example is `Winnie1Pooh#2` – use similar words that are memorable to you, with symbols and numbers of your choice. The command will initialize the **LUKS** *header* on the partition and perform the **encryption**.

Next, *open the partition to be encrypted* with the following command:

```
$ sudo cryptsetup luksOpen /dev/sdb1 MyEncrPart
```

This will map the encrypted partition to a device called `MyEncrPart`.

Create a *new FS on the encrypted partition* by running:

```
$ sudo mkfs.ext4 /dev/mapper/MyEncrPart
```

Create *a mount point for the encrypted partition* by running:

```
$ sudo mkdir /home/luke/MyEncrPart
```

*Mount the encrypted partition*:

```
$ sudo mount -v -w /dev/mapper/MyEncrPart ~/MyEncrPart
```

Now, *change the ownership* of the created partition:

```
$ sudo chown -v -R $USER:$USER /mnt/MyEncrPart
```

Now, you can work with the partition.

Finally, when you are done working with the encrypted partition, *unmount it* with the following command:

```
$ sudo umount /mnt/MyEncrPart
```

Now *close the LUKS partition* with the following command:

```
$ sudo cryptsetup luksClose MyEncrPart
```

*The encrypted partition will be inaccessible* until you map it again with the `luksOpen` command.

If you want to mount it permanently via `/etc/fstab`, follow the instructions in the *Shrinking/resizing root or home partitions to free space* subsection. Each time the system boots, you will be asked for the passphrase.

# Backups, tools, rsync, and recovery

In this section, we will review what a **backup** is and what to consider when doing one. Having a basic understanding of the terms when choosing a backup tool and strategy is essential.

## What is a backup?

A **backup** is *a copy of specific data* in a separate location – on another hard disk (**HDD**), a separate **partition**, **cloud** storage, or a remote **server**. A **backup** ensures that any files will have a copy if your system suffers a **FS** or SW failure. These can be any *long-term projects*, documents, SW applications, art, and so on. While recovering the OS is easy via reinstallation, lost family photos may be unrecoverable.

We can even *back up the whole OS* with its configuration, the installed SW, and our personal files. However, backing up an installed OS or making a snapshot of a whole partition makes sense for servers. Considering that in case of a failure, Manjaro can be reinstalled in 15 minutes, we need to back up only our *personal data* and potentially *some custom configurations*.

Making a partition snapshot or a full OS backup requires significant space and is often slow. In addition, if we don't do periodic refreshes of such backups, their state will get older with time, and recovering them will put us in some old state of the system. This doesn't make sense for a Rolling Release distribution.

The types of backup actions available for us, regular users, are as follows:

- **Full backup** – this is to make a *full copy* of a given directory, which is *a slower task*. It requires as much space as the amount of data to be backed up.

- **A backup update** – once we have made one backup, we *only add* to it with *changes*. The required space corresponds to the changes in size, being a matter of a second for a few tens of files. After all, we rarely change thousands of files in a single day.

## Deciding what to back up

Most tools *don't support* **compression** as they preserve the full directory and file attributes, some including the extended ones, as explained in *Chapter 9*. Thus, to back up *30 GB* of `/home` data, you will need *30 GB* of space.

If you have large files you know don't need to be backed up, you should explicitly exclude them to save space and time. A good example is two virtual machines I use for experiments – each takes up over *25 GB* of space, and I don't need to back them up. **Steam games** can also be quite big, and considering that you can download them again (and that they need periodic updates managed by the Steam application itself), backing up *40 GB* just for one game can be pointless and slow. The **GUI** application **Filelight** is excellent for examining your data sizes, and the recommended tools *support manually selecting the directories to back up.*

I will just add that I store old data, big photos, and archive documents on a Synology **NAS** with **RAID**, as they are over 2 TB. I work with a 20 GB directory backed up daily.

## Backing up SW, rsync, and recovery

There are many backup tools; we will review the best available and list potential alternatives. Most are terminal-based, but several also provide a **GUI**, so making a backup is a matter of a **single** terminal command or a few clicks in a **GUI**. The only thing you need to *perform a backup* is your additional storage.

The reason to list the tools given here is that these are *considered mature and good by many users* – for each of them, I have found tens of fresh forum posts of people using them for a long time and across multiple distributions. I will not list the 70+ links here.

### The best tools

**rsync** is one of the best terminal-based *fast*, *reliable*, and *versatile remote* and *local* file-copying tools. Written in C, it is an effective single-threaded SW. The project is mature (developed since *1996*) and so good that many GUI and other tools are *based on it*. One of the strengths of **rsync** is its special optimized checks, so updating existing backups with changes is fast and effective. Transferring **cloud/ network** backups is also fast, as the data is compressed for transfer. After making a backup with it, *you can browse the files in the backup location.*

It is *preinstalled* on **Manjaro** and many other distributions. Once you define the correct command-line interface (CLI) arguments (reviewed later in the chapter), you can put the command in a script and trigger it manually or automatically. Scriptwriting basics and automatic triggering are reviewed in *Chapter 15*. From all my tests, this is the only tool that never failed for a whole system backup.

**Back In Time** is a **GUI** and **CLI rsync**-based tool. Its GUI is one of the richest. You must explicitly start the **root GUI** version from the main menu to *perform a full system backup*. It has never failed me when backing up personal files. Unfortunately, it is *not preinstalled on* **Manjaro**. Installing SW on a **Manjaro live USB** is very hard, so you can't use it for system recovery from there. On the other hand, it is excellent for use on a freshly installed Manjaro to recover personal directories from a backup. *It is my only and strongest recommendation for a GUI tool.*

### Other GUI tools

**Timeshift** is a **GUI-** and terminal-based tool preinstalled on **Manjaro**. Apart from **rsync** backups, it also supports **btrfs** snapshots for **btrfs**-formatted partitions. It has a simple **GUI**, can do periodic automatic jobs, and is famous and widely used. However, its simplistic GUI starts with a setup wizard, which lacks custom directories selection. Even after I found and configured the custom directories settings, **Timeshift** tried to make a full system backup. This tool failed me more than five times for a whole system backup.

**deja-dup** is a simple GNOME-based **GUI** for **duplicity** and **rsync**. It *failed for me once, and I'm not too fond of its lack of options* compared to **Back In Time**. Despite this, some people use and trust it.

**grsync** is another **GUI** for **rsync**, but it is also limited compared to the others.

### Other CLI tools

**rclone** is one of the most famous CLI tools *optimized especially for synchronization with* **cloud storage**. Its supported features depend widely on the *features supported by the remote location*. Inspired by **rsync**, it supports compression and is widely used and recommended.

The next five tools are good and worth checking out *for advanced users only*.

**duplicity** is a CLI tool based on **rsync**. It is widely used and supports multiple *cloud storage servers*.

**FSArchiver** supports *compression* and recovery on a different partition and **FS**.

**restic** supports a lot of **cloud** services and explicitly involves *differential backups*.

**borg**, with the full name **BorgBackup**, is a *great terminal tool* that optionally supports *compression* and *authenticated encryption*. It can backup securely over the internet via *SSH*, supports *resuming backups* and *deduplication*, and has many other features.

**Kopia** is another excellent tool offering *compression, incremental backups*, and *deduplication* like **borg**, and it can even serve as a frontend for some of the *cloud* storage options supported by **rclone**. For this, **rclone** also needs to be installed.

### Test results

I tested all the GUI tools and **rsync**. I also tested a whole system backup, as this is the *ultimate test* for a backup tool. Of all the tools, **rsync** was the only one that never failed for me. Considering that it is installed on Manjaro *by default* and is also *the tool* **Calamares** *uses for Manjaro installation*, I guess you see why it is the best for our purpose.

From the GUI tools, the only good one was **Back In Time**. Its only disadvantage is that it is not preinstalled, so you cannot use it from a Manjaro live boot. The preinstalled **Timeshift** failed many times.

## Common points

Remember that no matter which tool you choose, you **must test it**. This means *making a backup* and *restoring it on your machine*, ensuring everything *works correctly*. If we don't know it works correctly, how can we even consider making backups with it? One of the essential characteristics of **rsync** is that you can *browse the files after the backup*. While I can guarantee the quality of **rsync** based on over 100 tests, one incorrect argument may corrupt a backup.

System backups are complicated as they involve backing up the root directories and special files. As they are complicated to handle (and I have tested this over 50 times for this chapter), doing a system backup for a regular user system is pointless. They are also slow due to the number of files. As a result, if an *unrecoverable system failure* occurs, it is *much easier to* **reinstall** *the OS and* *recover only your personal files*. In addition, except if you are experimenting heavily with Manjaro, its regular usage has *never led to OS corruption* for me. Finally, for common failures, there are thousands of posts in the forum, and they are typically easy to solve, so don't rush into reinstallation if you have issues with Manjaro.

How can we test a simple **rsync** directory backup? Easy – open the Filelight GUI application and inspect the backed-up directory and its backup copy. If they are exactly the same size, you copied all the contents. Then, open the backup copy with your file manager and inspect the files, particularly the most recently changed ones. If the backup is successful, the latest modified dates will make sense. Open some of them, and that's all. Once you know your backup command is working, keep it in a script or a simple text file so you can execute it directly next time.

## rsync

Using **rsync** via the terminal is simple. Before starting the following commands, ensure *you own* the already mounted `BKP` directory. Otherwise, your backup will not work, and you will *get no warning*. Check this with `$ ls -l` or `$ exa -lah`.

Reading the full manual is overwhelming but can help in special cases. However, I recommend the many nice short guides on the web, just like mine here. The general form of using **rsync** is as follows:

```
$ rsync [-SingleFlags] [--SpecialFlags] [BackupFromDir] [--Options]
[DestinationDir]
```

For a regular backup, use the following command but replace `/home/YourUser/SomeDirectory/*` and the other paths with *yours* as necessary, and also *read the explanations that follow*. Keep in mind that the whole command goes on one line, but as it is long, it is spread over several here:

```
$ rsync -aAXHv --progress --delete /home/YourUser --exclude=
{"lost+found","/home/YourUser/SomeDirectory/*"} /home/YourUser/BKP/
BKP_home > BKP_18_Oct_2023.log
```

Let's split and explain the command.

-aAXHv – are the single flag options – a for all files and a recursive copy of subdirectories, preserving symlinks, permissions, the modification time, group ownership, the owner, and device files; A for preserving **Access Control Lists (ACLs)**; X for preserving extended attributes; H for preserving hard links; and v for verbose, to display the *progress* details. v is bold, as it is not required; once you have proven your command works correctly, you might not need the extra verbose report.

--progress displays per file the operation *time left* in detail and, if the process breaks, where this happened. It is also bold, as once you have proven your command works correctly, you might not need the extra report and can skip it.

--delete – tells **rsync** to explicitly delete from the archive location files removed from the backed-up local directory. In other words, your backup will be cleaned from old deleted files.

--exclude and the big string in curly braces { } list *directories to exclude*. This can be skipped if not necessary.

/home/YourUser/BKP/BKP_home is the destination, which I have mounted in /home/luke/BKP. As I want each backup to be in a separate directory, I've made a subdirectory BKP_home for this one.

> BKP_18_Oct_2023.log is the terminal redirection operator > (explained in *Chapter 7*) followed by a file name to put the whole log in a text file and not print on the terminal. This is also optional.

v and --progress make the execution slower, as they print an enormous amount of information in the terminal. If the terminal does not have set up an unlimited or at least large scrollback size (100K lines), you will lose it. Due to this, I strongly recommend redirecting to a log file if you use any of the two report options. The log is also good if you need to check results later. We don't need a log for regular personal data most of the time, as checking the overall result is done quickly with Filelight.

For me, backing up *4.2 GB* (over 50,000 files) without the v and --progress flags on one of my machines took *less than a minute the first time*. Each next update took less than a second for small changes. Of course, this depends on the amount and type of files, storage drive characteristics, currently running processes, and your machine's characteristics. Thus, please don't take it for reference.

To back up your whole /home directory, you can use this version (replace the bolded paths and potentially remove the unnecessary parts):

```
$ rsync -aAXHv --progress --delete /home/ --exclude={"/
lost+found","lost+found",".cache",".VirtualBoxVMs","/home/luke/BKP","/
home/luke/BKP5"} /home/BKP/BKP_OF_HOME > BKP.log
```

**For advanced usage**: If you want to back up a whole system, you need `sudo`, and it becomes a bit more complicated:

```
$ sudo rsync -aAXHv --progress --delete / --exclude={"/dev/*","/
proc/*","/sys/*","/tmp/*","/run/*","/mnt/*","/media/*","/var/
tmp/*","/var/cache/*","/lost+found","swapfile","lost+found",".
cache","Downloads",".VirtualBoxVMs",".ecryptfs","/home/YourUser/
BKP","/home/YourUser/BKP5/*"} /home/YourUser/BKP > BKP.log
```

For the last command, having your `/home` directory on *a separate partition* is not a problem. As long as it is *mounted*, it will be copied as well. What is essential is to remember to *exclude large folders of your choice* and the listed system directories (`/dev`, `/proc`, etc.).

Once your version of the arguments is ready, save it in a file, and it is strongly advised to also copy them in a backup location (e.g., an **email** or some form of **cloud** storage).

### Recovering with rsync

To recover with rsync, simply revert the `source` and `destination` with only one exclusion:

```
$ sudo rsync -aAXHv --progress –delete /home/YourUser/BKP/BKP1
--exclude="lost+found" /home/YourUser > BKP.log
```

Remember – **rsync** is a *highly optimized* **copying** *tool*; it copies files from one directory to another. You can *selectively copy/back up* anything you want. This program is also used to synchronize *data* or *selectively copy files* from hosts to **servers** and vice versa (including for **Unix**, **BSD**, and other **Unix**-like OSs).

# Summary

In this chapter, we delved into storage, covering partition management, mounting, extending, shrinking, and encryption techniques through both GUI and terminal. We also explored backups, covering **rsync** and its easy usage.

In the next chapter, we'll dive into network fundamentals, covering IPv4, IPv6, network adapters, and address scanning. We'll delve into network sharing with NFS and Samba, concluding with SSH and SSHFS as the best tools for secure remote Linux machine access.

# 11

# Network Fundamentals, File Sharing, and SSH

This chapter will continue our journey by investigating our local network connection. We will first check the basics of addresses, pinging, and network scans. Next, we will see how to configure the Linux **Network Filesystem** (**NFS**) for data sharing. We will explore its configuration on the client side for Linux, Windows, and macOS. Further, we will briefly check Samba as a potential alternative. In the last part, we will explore how to connect to our system via **SSH** and how to share files with it.

The topics in this chapter are as follows:

- Network basics

- Short network sharing introduction

- Sharing via NFS

- Sharing via Samba server

- **Secure Shell** (**SSH**) and working remotely

- Sharing via SSHFS

## Network basics

This section will examine the network basics by explaining how they work in *local networks*. An explanation of **protocols**, how the **internet** works, common network attacks, and how firewalls work is provided in *Chapter 12*. It also reveals how to connect securely and anonymously to the internet via a **Virtual Private Network** (**VPN**). What is important to know here is that a protocol is a set of rules for the network messages' contents and sequences. A message is a small data packet sent over the network. We exchange thousands of packets when we access servers, other computers, and websites. Without a protocol, we can't communicate over a network.

IP stands for **Internet Protocol**, and an **IP address** is the unique identifier for a computer in a network. A network, wireless or cable-based, connects many computers and devices. The internet is the ultimate network of networks, connecting small local networks with public networks and servers.

We need a **cable** or **wireless card** and a local **router** to connect to a local network. To connect to the **internet**, we need an Internet Service Provider (**ISP**), which provides internet connectivity to our router via cable or wireless connection. No matter whether we have 5, 10, or 500 devices connected in our local network, when they connect to the internet, they all connect via the **router** and the **ISP**.

Simple local networks offer no special services except connecting to the central router for internet access. In comparison, the internet offers connections to web pages, servers, web applications, and so on. Sometimes, in local networks, information sharing is done by sharing a directory with locally connected users. Other internet users not in our network have no access to these resources.

When local special services such as servers, local wiki pages, and others are provided to local network users, it is often called **intranet**.

In this chapter, we will learn two ways to share information with our local network, but first, let's see the basic terms. This chapter will also refer to starting some services, explained in detail in *Chapter 13*.

## IPv4 address

Until more than a decade ago, **IP addresses** were only **IPv4** (IP version 4), written as four numbers between 0 and 255 separated by dots, such as 10.58.236.17. This scheme provides approximately *4.2 billion addresses*, distributed between all countries worldwide and separated into five continental regions. There is a small percentage of reserved addresses in this whole range. Between *2011* and *2019*, all regions *gradually ran out of addresses*. This is called *IPv4 address exhaustion*, predicted long ago and analyzed officially at the beginning of the *1990s*. Nowadays, IPv4 addresses are used mainly for **Local Area Network** (**LAN**) cable connections inside a single building. The old international **IPv4** addresses still belong to those who bought and reserved them for their use. For new servers, companies now pay for **IPv6 addresses**.

## IPv6 address

In *1995*, the first steps were taken for the development of **IPv6**. In *1998*, a second improvement started its development. In *2003*, major universities worldwide gradually started using it experimentally, and **IPv6** also found its way into the **4G** mobile devices standard. The **Domain Name System** (**DNS**, reviewed later in this section) translates *human language* web addresses to numerical **IP** addresses, and has **IPv6** support *since 2008*. By *2011*, practically all major computer OSs and servers had *production-quality* **IPv6** capabilities.

**IPv6** provides *340 trillion trillion trillion* addresses written in the form of 8 groups of 4 hexadecimal numbers, each separated by a colon. Hexadecimal is a numerical system with digits from **0** to **9** and then letters from **A** to **F**. An example address is as follows:

```
2001:0db8:0000:0000:34f4:0000:0000:003d/64.
```

There are several ways to reduce this string length, the most important being *one double colon instead of the longest sequence of groups with all zeros*:

```
2001:0db8::34f4:0000:0000:003d/64
```

If there are more occasions of more than one group with zeroes, the leftmost is replaced by the double colon. As this is still long, *the second instance of all zeros in a section* can be *replaced with a single* 0, resulting in:

```
2001:0db8::34f4:0:0:003d/64
```

Finally, leading zeros from a group can be removed, and the initial address can be written as:

```
2001:db8::34f4:0:0:3d/64
```

This string is much shorter than the original; as it has six groups, we know the double colons represent two groups with zeros.

## Hostname

Your computer's **hostname** is *a name/label* that can (but not necessarily will) be used to *represent your machine on the local network*. You got it *automatically* or set it manually during Manjaro's *installation*. You can always change it if you wish. To see your current name, execute $ hostname in the terminal. To change it, execute:

```
$ sudo hostnamectl set-hostname NewName
```

Discovering a machine by its **hostname** is not covered in this book.

## DNS and WWW

The **internet** is the *network of networks*. **WWW** stands for **World Wide Web**, or the web pages hosted on different servers, which we access through the **internet**. Each web page has a **Domain Name** (such as www.google.com) and a corresponding **IP** address at which the page contents are located. The **DNS** translates the human language addresses (like google.com) to **IP** addresses. The three *w* letters are often omitted in a **web address**.

With the ping command, you may check whether you can reach a given address, such as google.com or packt.com. If the page server has not disabled ping requests, this will work. Try $ ping google.com, $ ping packt.com, or $ ping lwn.net. To test your **IPv6** connectivity to a server, add the -6 option: $ ping -6 google.com. This is the most frequently used tool to check our internet connection.

Note that **DNS** usually is not installed and configured in simple local networks. It is used globally for the internet and in big companies' **intranet** networks.

## Local network configuration basics

The next point is *how to get the address of our machine locally* for the **network** on which we will share. Again, whenever you are part of some local network (such as in your home or small office), sharing information is *limited inside this network*. You will need a so-called public IP to share it with everyone worldwide, so don't worry.

The local network connection is done via a **LAN** cable-connected **Ethernet card** or a **Wireless card**. Such cards are also called **network adapters**. To list all active *network adapters* on your PC, use the `$ ip -br -c a` command. Here, `br` stands for brief format, and `c` is for coloring. For me, the `ip` command reported the following:

```
$ ip -br - c a

lo    UNKNOWN 127.0.0.1/8      ::1/128
eno1  UP      10.51.8.166/22   fe80::6140:852b:fdea:3660/64
wlp3s0 UP     192.168.87.152/22 fe80::8ed8:84e3:6d3:d2b8/64
```

We see three rows, as I have three network adapters. The first one, `lo`, stands for the **loopback interface**. Via `lo`, we can *load a website hosted on our PC* to check it out (via the web page address `http://localhost`). **Localhost** is a **hostname** that calls the **IP loopback interface**. It is a network interface for us to our own PC and usually has the **IPv4** address `127.0.0.1`. The other address, `::1`, is my **IPv6 localhost** address. Both these addresses come from the default settings of the Manjaro network driver, following the network standards. This means they are the same (and by default present) for everyone.

Any line starting with `en`, letters, and digits after this stands for **Ethernet** – a local **LAN** cable *network card/adapter*.

The last line, with `wl`, letters, and digits after this, stands for **wireless**, as my mini-PC has both **wireless** and **Ethernet** network adapters. Manjaro KDE can connect simultaneously to both, improving the throughput and network performance. Also, whenever one of the interfaces has a partial connection loss, the network manager will switch to the other network adapter, and I will not even notice.

In the preceding `ip` report, `lo`, `eno1`, and `wlp3s0` are interface **IDs** for the adapters, which we can use in the terminal as arguments for different commands.

You can see that each of the interfaces has an **IPv4** and an **IPv6** address. The number after the slash / is the so-called **subnet mask**, which we will look into at the end of this section. To get more detailed information in the `ip` report, omit `-br`.

To get the list of adapters, you can also call the *obsolete* `ifconfig` command, which will show the same identifiers: `lo`, `eno1`, and `wlp3s0`. As it is *obsolete*, it is not installed by default on **Manjaro**. I mention this as *many online guides still don't mention that* `ifconfig` is obsolete. To get it, you must install the `net-tools` package.

## Pinging a computer in our local network

To **ping** a computer means to check on the network *whether a machine can answer on a given* **IP** *address*. This is unrelated to *whether the given machine shares something* by whatever technology. In other words, **ping** only *validates* the *existence of a machine* with a given address and a working *network connection* to it. It is a continuous command and needs to be stopped with *Ctrl+C*.

You can ping a machine using $ ping 192.168.87.152, or, if you know your network is connected via **IPv6,** use the following:

```
$ ping -6 fe80::5078:d1ff:fe30:9b9f%wlp3s0
```

The last part after the % sign %xxxxxx is *the name of the interface* (i.e., the *network card*) that we will use. In this example, this is my wireless network card ID. The -6 parameter explicitly requires the **IPv6** protocol to be used. Be extra careful of *single or double colons*! Writing the *wrong address will give you no answer.*

There are three things you need to be aware of. *First*, computers with special network settings or special set-up firewalls may not answer to ping, leaving the impression they don't exist. By default, however, the ping answer is *enabled* on any home PC.

*Second*, many web guides contain references to **ping6**, which is as *obsolete* as ifconfig. A few years ago, **ping6** *was merged into* ping.

*And third, local* **LAN cable** *networks might often use only* **IPv4**, like in my office and home. This is because it is a small local network, and many users are still unaware of **IPv6**. Nobody implements it for local and small networks because **IPv4** allows enough devices to be connected: *4.2 billion*. The *local network* then has its own **router**, and all devices connected to it access the external world through it. A **router** typically has one *worldwide viewable (discoverable)* **public IP**.

How do we determine whether our router connects via **IPv6** to the external world? This is easy – try the website https://test-ipv6.com/.

Since many modern mobile devices use **IPv6** for their **Wi-Fi** connection, even if the cable (**LAN**) network is only **IPv4**, **IPv6** is often enabled by default for *devices connected via* **Wi-Fi**. This is because the **4G** mobile device standard includes **IPv6** in its specification. However, not all **IPv6** devices can be pinged. To run an **IPv6 ping**, we will need some scanning covered later in this section.

## Router

A **router** is a network device that connects multiple devices requiring network connectivity and is usually connected to the **internet**. It is practically a small **LAN** microcomputer that manages our local network. Either our **ISP** (the company we pay for internet access) *provides us with one*, or we have *our own*. It is the entry point for the **internet**. The **router** uses many protocols to connect on one side to the internet and on the other to *all possible local devices* that use the network: PCs, printers, mobiles, smart TVs, laptops, tablets, IP cameras, and so on.

## Static and dynamic IPs

A **dynamic IP** is a network **router** service based on the **Dynamic Host Configuration Protocol** (**DHCP**) that *assigns* an **IP address** to any *new device* connected to its network. When we connect to the router via the network name and password, we get an **IP**. Say we disconnect from it and connect again after a week – we may often get a different (currently free) local **IP**.

A **static IP** is used when a **router** knows a given **IP** is *reserved* especially for one device. Based on the machine's unique *network adapter* ID, it keeps the **IP** *for this machine* between reconnections. This unique ID, named **Media Access Control (MAC) address**, typically doesn't change. In comparison, the **IP** is assigned by the router or network administrator. Many devices support editing their MAC address, but this is done only on special occasions.

The other way to have a static **IP** is to set up your router or server to have a reserved range of **static IPs** allowed on your network. In this case, you set the **static IP** in the given machine. Thus, the static IP is configured either in the router or in the device.

It is critically important that two machines in the same network *cannot have the same* **IP**. Once a machine is registered with an **IP**, connecting another machine with the same **IP** leads to connection errors.

We use **static IPs** to provide a long-term connection for other users to devices, which rarely change. Such are network printers, servers, or a home **NAS**. It is convenient; otherwise, each time the device address changes and we want to connect to it, we will have to reconfigure – for example, our printer setup.

When a device such as your laptop connects frequently to a network, it may get the same **IP** address each time. This doesn't mean it has a **static IP**. Some routers work this way, especially if only a limited set of devices connect to the network.

## Local and public IP

When a device connects to a local network, the local router assigns to it a local IP address, no matter whether static or dynamic. To illustrate this, I have created the schematic in *Figure 11.1*:

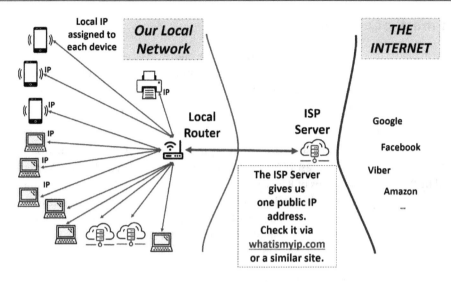

Figure 11.1 – Local network architecture example

To use a metaphor, say we have a big business building with 500 offices on 47 Sunshine Blvd., in Chicago. Its address is the external public address known to anyone worldwide, equivalent to the public IP by our ISP. The internal addresses consist of floors, companies, and people's names. They are known only by the building administration and are equivalent to the local IPs.

## Scanning the network

Scanning is a standard functionality of **IP** communication (and not hacking). For **IPv4,** you can download the **arp-scan** tool via `pamac` and run the following:

```
$ sudo arp-scan --localnet
```

**ARP** is the **IPv4 Address Resolution Protocol**. This command lists all **IPv4**-discoverable devices.

Scanning on an **IPv6** network is done with `ping` with the `-1` (lowercase L) option, an *explicit network adapter ID after it*, and a special broadcast address. In the following example, I use my **Wi-Fi** card. This command scans for *all* **IPv6** *devices* via the **IPv6 multicast address** `ff02::1`:

```
$ ping -1 wlp3s0 ff02::1
```

This call will list some **IPv6** addresses (at least the one of your router). Then, you can use one of them and test the **IPv6 ping** to it, adding the obligatory adapter ID at the end:

```
$ ping -6 fe80::5078:d1ff:fe30:9b9f%wlp3s0
```

Try the presented commands. The results are always interesting.

## Subnet mask of a network

Our last stop in this section is a **subnet** of a **network**, which defines *a range of IP addresses*. Understanding this is easier when you know binary numbers. If you don't understand them, check this link related to **IPs**: `https://www.networkacademy.io/ccna/ip-subnetting/converting-ip-addresses-into-binary`.

Now, let's consider a simple example with **IPv4**. The addresses in it have the format `0-255.0-255.0-255.0-255`. This means we have *4 groups of digits between 0 and 255*. The total possible number of addresses is `256*256*256*256` = *4.29 billion addresses*. Each of these 256 numbers is represented by 8 bits; thus, 4 groups of 8 bits give us 32 bits. The **mask** tells how many of the 32 bits are fixed and how many are free to change. If I want to tell a **user**, **server**, **firewall**, or software to work with all possible **addresses** starting with `156.144.xxx.xxx`, I will use a **mask** of *16 bits* written in the form `156.144.0.0/16`. This example is a subnet of all addresses formed by the first two numbers, `156.144.`, and completed by two other numbers between `0` and `255`.

As the first digits of the **address** provide its geolocation, `156.144.xxx.xxx` is located in the area of Chicago, Illinois, USA. Suppose we want a particular **server** to *be active only for this region*. We then set a *rule* in its **firewall** stating that requests *from any other address not within this* **mask** shall be *discarded*, providing an *allowed range of requestors* via the **mask** `156.144.0.0/16`. The *zeros* indicate that you don't care which number is in their position (but only when used with a **subnet mask**). For this example, *requests* to the **server** from an address such as `157.144.xxx.xxx` will be *discarded*.

Here is another example. Say we have a big company in a skyscraper 50 stories high with *25,000 devices* connected to its network. Instead of *exposing them all directly* to the internet, we want them all in a **local network**, which will be behind a **firewall**. Then, we will give the **DHCP** server a range of *65,535 addresses* by defining a rule to work with IP **mask** `178.62.0.0/16`. The network router will provide an IP within this range to each newly connected device. We will *limit all devices' local configurations to work only with addresses within this range by configuring their firewalls* to ignore other addresses. This would allow them to communicate only with local devices. Then, one of the *valid local* **addresses** within this **mask** will be *the address of our internet gateway*. In the local settings of each device *requiring internet access*, we set that internet requests go to the local network's gateway **IP** address. The gateway will have only a few *public/international* **IPs** visible from the internet, and all the traffic will go through them. Thus, instead of using *25,000 international* addresses and leaving our *local network devices unprotected*, we have a local **sub-network** based on a **mask**. Typically, the gateway also has a firewall, protecting our internal network. Though this example is realistic, real settings may differ a lot, depending on multiple factors.

Any home **router** *does the same* – it has no problem connecting at least 127 **devices** by default (and more if necessary) and provides a local **IP** in your network for each. You practically have a **subnet** hidden from the outside world behind a dedicated public **IP address** provided by your **ISP**.

An excellent explanation of **subnetting** is provided here: `https://www.freecodecamp.org/news/subnet-cheat-sheet-24-subnet-mask-30-26-27-29-and-other-ip-address-cidr-network-references/`.

Now, how will this work for an **IPv6** address? For **IPv6** addresses, the **subnet mask** is called a **routing prefix**, a **global prefix**, or sometimes only a **prefix**.

You have already seen an example **IPv6** address earlier in the chapter:

```
eno1    UP       10.51.8.166/22  fe80::6140:852b:fdea:3660/64
```

**IPv6** provides *128-bit addressing* and *trillions of trillions of possible* **addresses**. The **routing prefix** for IPv6 is not up to `32 bits` but up to `128 bits`, as we have *8 groups of 16 bits*, each written with *4 hexadecimal numbers*. In the decimal system, a 16-bit binary number is the range between *0 and 65,535*.

So, the `/64` **prefix** in the preceding snippet says that only *the first 4 groups* (16*4=64) are limited, and *the other groups can have any number*.

If you remember, the double colon means at least one group of zeros is hidden. As we *have to have eight groups*, the full address is as follows:

```
fe80:0000:0000:0000:6140:852b:fdea:3660/64
```

Here is an article on IPv6, including information on shortening and prefixes: `https://www.networkacademy.io/ccna/ipv6/ipv6-address-representation`.

## Short network sharing introduction

Sharing files over a **network** is essential in a home or company network. Before we dive in, it is vital to point out that *security-sensitive information* shall *never be shared over a network unintentionally* and without the *required protection*. Even with the improved security of modern-day systems, sharing data allows attackers to *exploit* potential **vulnerabilities** and steal information or modify your files. Of course, sharing your media or e-books on a home network, or sharing a development project in a *protected company internal network* is not risky, and is a typical scenario for such usage. In addition, in *Chapter 12*, we will review the most common types of attacks and explain how to set up a firewall, effectively mitigating most of them. To add to the topic, nowadays, **vulnerabilities** are *handled with updates ASAP*, and the public is also usually informed ASAP. In the case of Manjaro – as it is a **rolling release** distribution – you will get the updates and patches ASAP.

Over the years, many technologies have been created for this task. Here, we will review the **NFS** from **Sun/IETF** in detail. We will review two more at the end of the chapter: **Server Message Block** (**SMB** or **Samba**), initially from **IBM** and later developed significantly by **Microsoft**, and **Secure Shell Filesystem** (**SSHFS**). Before going on, we will check a few other technologies to have a basic idea.

**CIFS** is an obsolete technology from **Microsoft** based on **SMB** *1.0*. It was abandoned after most users didn't adopt it due to flaws, and CIFS *never reached standardization*. **Microsoft** then switched to **SMB** *2.0* directly and has supported it actively since then.

**AFP**, or **Apple File Protocol**, is used mainly with macOS and Apple devices.

**FTP**, or **File Transfer Protocol**, is an unencrypted protocol with lower security. It can be used in some cases but is not recommended.

**FTPS**, or **File Transfer Protocol Secure**, has potential vulnerabilities. Due to its incompatibility with firewalls, it is not a good alternative in the common case.

Many more file-sharing technologies exist, but **NFS**, **SMB**, and **SSHFS** are the most widely used. Most of the other technologies are dedicated to specific use cases.

We will discuss network protocols in detail in *Chapter 12*. Here, we will mention that a **protocol** defines *how data packets are organized and transferred* between computers on a network. Without a **protocol**, we cannot send data. Different protocols have different purposes, usage, and features. The most famous are **TCP/IP** and **HTTP(S)** – the basic **internet protocols**. For **NFS**, the important one is the **Open Network Computing Remote Procedure Call** (**ONC RPC**).

Both **NFS** and **SMB** have **security** features. However, it is essential that none of them is highly secure. Thus, if not on a trusted home or internal office network, they should not be used, especially if sensitive content is to be shared through them. They can be used on global networks only in a **VPN** environment or if the **Kerberos** network authentication protocol is set up.

In contrast, **SSHFS** is encrypted by default and provides higher security. However, multiple users report it is slow for high volumes of information with Windows clients. In addition, its default settings provide access to the *complete server filesystem*. There are ways to share only dedicated filesystem parts, but this requires special settings and is not recommended for the typical case. I will share links to guides on setting it up, should you need them.

Even though **SMB** is centered around Microsoft, as it is a standard network-sharing technology, it is not limited to MS Windows. Instead, there are Samba server implementations for practically any OS, and the same applies to **NFS** and **SSHFS**. In addition, **NFS** supports Microsoft's **Active Directory** access control technology and keeps up with syncing data between machines.

The mentioned protocols are **client/server**-based. A computer that *shares directories is always a* **server**, and *a computer browsing the directories is always a* **client**.

## Sharing via NFS

In this section, we will review the complete NFS setup and its access from client machines with any OS.

# A bit of NFS history and characteristics

**NFS** is a distributed filesystem protocol initially developed by **Sun Microsystems** in *1984*. **NFSv2** was released in *1989*, **NFSv3** in *1995*, and **NFSv4** in *December 2000*.

**NFSv4** has two more minor versions – *4.1* (released in *2010*) and *4.2* (released in *2016*). Version *4.1* added usage via global **wide area networks** (**WANs**), standard parallel **NFS** for improved network bandwidth usage, **UTF-8** filename and identifier support, an improved session model, and directory delegation. Version *4.2* added scale-out storage support (such as **NAS**), support for server-side copying (enabling cloning and file snapshots), space pre-reservation, support for sparse files with large blocks of zeros, and integration with **SELinux**.

**NFSv4.2** has been integrated into **nfs-utils** since *2016* and supported by the **mainline kernel** since *2017*. Even in *2023*, *all versions since 4.0 continue to get updates*. For version compatibility, the **NFS server** and the **client** attempt to connect as follows: try connecting on *4.2*, then on *4.1*, then *4.0*, and finally, *3.0*. Older versions are not supported and are **obsolete**.

A final point – **NFS** *filters users* only by **IP**. Hence, if you want a special authentication mechanism, use it inside a **VPN** *infrastructure* or combine it with the **Kerberos** protocol for network authentication.

# Enabling time synchronization for NFS

When sharing files via **NFS**, we will also share the *files'* **attributes**, which include the *last modified time* and any other **timestamps** *related to the server*. Consequently, to share with NFS, we *must enable* the same **time synchronization** *on the server and client sides*. Otherwise, software that relies on timestamps may not work correctly.

As Manjaro is **systemd**-based, it uses the `timedatectl` program to report and control the statuses of the clocks, and by default, no synchronization is enabled. Most Linux distributions use the **Network Time Protocol** (**NTP**) to synchronize their clocks over the **internet**.

First, we check the current settings with `timedatectl`:

```
$ timedatectl
               Local time: Tue 2023-08-10 11:26:24 EET
           Universal time: Tue 2023-08-10 09:26:24 UTC
                 RTC time: Tue 2023-08-10 09:26:26
                Time zone: Europe/Sofia (EET, +0200)
System clock synchronized: no
              NTP service: inactive
          RTC in local TZ: no
```

Then, we set the **NTP** *time sync* to `active` via:

```
$ sudo timedatectl set-ntp true
```

Further, we set the **hardware clock** to the *current system time*:

```
$ sudo hwclock --systohc
```

We finally run $ `timedatectl` again to check the `NTP service` status:

```
System clock synchronized: yes
              NTP service: active
           RTC in local TZ: no
```

## Sharing on the server side

Remember that the **NFS** server shares directories recursively. It is strongly advised to share a dedicated directory created for this purpose and to set its **permissions** (and those of its contents) accordingly. When the shared files are writable by all users, *backing up the stable states of the data periodically* to another partition, a NAS server, or cloud storage is always the correct approach.

No matter the distribution, clients need **nfs-utils** to access the shared data and mount it in their directory structure. Manjaro has this package installed by default, but clients may need to install it for other distributions. For **macOS** and **Windows**, we will check for potential solutions later.

We first prepare the directory we will share – in this case, `sharedDocs`. As it is in the **root**-owned `/srv/` directory, we use `sudo`:

```
$ sudo mkdir /srv/nfs
$ sudo mkdir /srv/nfs/sharedDocs
```

Now, we change its ownership with the following:

```
$ sudo chown -v -R $USER:$USER /srv/nfs/sharedDocs
```

Then, we put into it what we want to be shared. To create a test file, run $ `touch /srv/nfs/sharedDocs/testFile.txt`. Edit it with **nano** or **Kate**, write an example text, such as `Server wrote on Friday`, and then save the file.

Next, we must set the proper **permissions** for others, and apart from `read`, they will also need `execute` ones, or they will not even be allowed to run `ls` on the shared directory. We do this with the following command:

```
$ chmod -R o+rx /srv/nfs/sharedDocs
```

Now, edit `/etc/exports` with **nano** via $ `sudo nano /etc/exports`. To share with *everyone*, add the following line:

```
/srv/nfs *(rw,sync)
```

To share *only with specific clients*, instead of the *, add them by their **IP address** and use the following format:

```
/srv/nfs 192.168.10.65(rw,sync,subtree_check)
```

If you change the **IPv4** address to the **mask** 192.168.10.0/24, **NFS** will share with all users whose addresses start with the first three groups of numbers. The zero at the end of the address designates any number between 1 and 255.

Before *saving the file and exiting*, read the *comment section* in it. As with many other system configuration files, the lines starting with the # sign are **comments** and ignored by the software. The last comment line currently states Use `exportfs -arv` to reload. Once you save it, just run $ sudo exportfs -arv, to apply the changes. Complete information on all possible parameters in parentheses is provided on the **exports** man page here: https://linux.die.net/man/5/exports (or by calling $ man exports on the terminal). However, here are the basics of the most important parameters:

- ro – Specifies that the directory *may only be mounted* as **read-only**.
- rw – *Allows clients* to **read** and **write** to the shared directory.
- sync – Forces **NFS** to write data to the shared directory *immediately* instead of caching it in memory. As a result, changes are distributed over the network synchronously but a bit slower than without sync. This delay is negligible for regular usage.
- async – Allows **NFS** to *cache data in memory* before writing it to the shared directory. In other words, it ignores synchronization checks in favor of increased speed. Usually, you don't need this for daily work and *it is not recommended to be used*.
- subtree_check – Enables subtree checking, an **NFS** *security feature* that prevents clients from accessing files outside the shared directory, and the host does any required FS checks.
- no_subtree_check – Disables subtree checking, which increases the performance a bit.
- no_root_squash – Allows the root user on the client to have full access to the shared directory on the server, which is *dangerous* as it allows remote **root** users *the same privileges* as the **root** user of the **host machine**.

The following list presents some advanced options; we will cover **user IDs (UIDs)** and **group IDs (GIDs)** in *Chapter 13*:

- all_squash – Maps *all client* **UIDs** and **GIDs** to a single server **UID** and **GID**
- anonuid and anongid – Maps all the clients' **UIDs** and **GIDs** to an *anonymous* group (unidentified)
- insecure – Allows clients that don't support *privileged ports* to connect to the NFS server
- secure – *Requires* clients to connect to the NFS server using *privileged ports*

- `fsid=value` – Assigns a FS ID to the exported FS so that clients can refer to the FS by this ID

- `crossmnt` – Allows the client to mount FSs that are exported from the same server but are not exported as part of the file hierarchy rooted in the current FS

The last step is to enable the `nfs-server` **service** with the following command:

```
$ sudo systemctl enable --now nfs-server.service
```

The `enable` option marks the service to be started at boot, while the `--now` option is to force the service's immediate activation. We will review `systemctl` in *Chapter 13*.

To check running services related to **NFS** in general, execute the following command:

```
$ sudo systemctl | grep -i nfs
```

The result on my machine is as follows:

```
proc-fs-nfsd.mount                loaded active mounted   NFSD
configuration filesystem
var-lib-nfs-rpc_pipefs.mount      loaded active mounted   RPC Pipe
File System
nfs-idmapd.service                loaded active running   NFSv4 ID-
name mapping service
nfs-mountd.service                loaded active running   NFS Mount
Daemon
nfs-server.service                loaded active exited     NFS
server and services
nfsdcld.service                   loaded active running   NFSv4
Client Tracking Daemon
rpc-statd-notify.service          loaded active exited     Notify
NFS peers of a restart
rpc-statd.service                 loaded active running   NFS
status monitor for NFSv2/3
```

If `/etc/exports` is modified while **NFS** *is active*, we shall either restart `nfs-server.service` via `$ sudo systemctl restart nfs-server.service`, or call `$ sudo exportfs -r`, to re-export all shared directories.

If you have an active **firewall**, be sure to have allowed the **NFS** communication and not have *blocked* communication with the given client machines based on **IP** and/or protocols. This topic is reviewed in detail in *Chapter 12*.

## Essential commands for NFS status

`$ nfsstat -v` provides a detailed listing of **server** and **client** *calls*. Running it on the **server** side (while sharing a directory with NFS) gives the number of *accesses*, *lookups*, and other FS operations in a `Server nfs v4 operations` listing. It also provides statistics for transferred **TCP** and

**UDP** packets. On the **client** side, this command summarizes the requests to servers and client file operations, listed under `Client nfs v4`.

`$ systemctl status nfs-server` – checks the NFS server activity status.

`$ rpcinfo -p` provides information on **ports** *used* by **RPC**-related **protocols**. We can filter with `| grep -i nfs`. Currently, **port** `2049` is used for **NFSv4** and **NFSv3**. This information might be necessary for clients connecting via **NFSv3**, as it requires the `rpcbind` service.

Any firewall communication blockages on either the client or server side will *not be reported* in any of the previous status listings. The reason is that a firewall may not be detectable – it will just look as if your packets are lost, although you have ping between the machines. Thus, check firewall settings on the server and clients if you have such issues. We will review firewalls in *Chapter 12*.

## Getting access to a shared directory on the client side

If you have not done it yet, please enable the **timedatectl ntp** and **sync** options, as shown earlier. If this is not a Manjaro distribution and the package is not installed, install the `nfs-utils` **NFS** package.

Next, we have to enable the service targets:

```
$ sudo systemctl enable --now remote-fs.target
$ sudo systemctl enable --now nfs-client.target
```

We will look in detail at **systemd**, services, targets, and `systemctl` in *Chapter 13*.

For **Manjaro** and most **Linux** distributions, run the `$ showmount -e 10.51.9.30` command, replacing the given **IPv4** address with the one of your **server**. You will get the list of exports from the server like this:

```
Export list for 10.51.9.30:
/srv/nfs/sharedDocs *
```

Make a directory in which to mount the given shared directory:

```
$ sudo mkdir /mnt/sharedDocsFromServer
```

If you want to use a directory only temporarily, you can mount with the following:

```
$ sudo mount -t nfs 10.51.9.30:/srv/nfs/sharedDocs /mnt/
sharedDocsFromServer
```

The `-t nfs` option limits this mount to type `nfs`. Then, we have the server's **IPv4** address and the path from its `Export list`. We end the command with *the local path to mount on*. Check with `$ mount | grep -i nfs` for the `mount` result.

## Synchronization on the client side

Usually, after we have set the same type of time synchronization via NTP and due to **NFS**'s design, the directories shall be *synchronized almost instantaneously* on both the **server** and the **client**. However, the first time I created a file on either side, it didn't sync immediately. I had to *exit the directory*, go one level up with $ `cd ..`, and *enter it again* with $ `cd sharedDocsFromServer`.

The other way is executing $ `cd .` with *a single dot* in the current directory – this will *reload its contents*.

After my first glitch and *two hours of testing with two* **Manjaro** *machines*, everything worked like a charm for me. To test the synchronization, you can use `touch` to create multiple files simultaneously with $ `touch file{1..10}.txt` on one side, then list them on the other. To unmount the shared directory, execute:

```
$ sudo umount/mnt/sharedDocsFromServer.
```

## Mounting permanently on the client side based on a static IP

To do this, edit `fstab` on the **client** side, but remember that this is acceptable only for a machine with a **static IP**. The format to be used is as follows:

```
#From_where      To_where FS_Type Opt1,opt2,opt3    0    0
```

The line I added for my machine is as follows:

```
192.168.21.23:/srv/nfs/sharedDocs  /mnt/sharedDocsFromServer nfs
rw,sync,nofail,_netdev  0    0
```

In this way, each time you start the client machine, the shared directory will be mounted to the given directory (in my case, `/home/precisionTower`).

The `_netdev` option tells the `mount` command this is a network FS, and the system shall *not attempt mounting it* until the *network service is up and running*. The `nofail` option tells the system *not to interrupt the OS boot* if the FS is missing.

### If you change directory permissions on the server side

Suppose you update the shared directory's **permissions** (or any file inside it). After changing them, you may have to run $ `exportfs -r` or $ `sudo systemctl restart nfs-server.service` and potentially **remount** the shared directory on the client side. This is because the directory *contents are synchronized*, but the directory **permissions** are related to NFS internals, and changed **permissions** *are not synchronized*. To `remount`, execute on the **client** side:

```
$ sudo mount -o remount 10.51.9.30:/srv/nfs/sharedDocs /mnt/
sharedDocsFromServer
```

If you do this before the client machine has connected to the server, you will have no such problem. In addition, if you remove certain permissions, you are safe, as once you do it, even if the NFS clients have not updated them, they will not be able to perform the given operations. Still, they may have to remount to see the updated permissions.

## Automating all the commands

Once you have set up your commands correctly, putting them in a script makes sense, especially if multiple clients use *the same shared content* from one server. We will cover this topic in *Chapter 15, Shell Scripts and Automation*.

## Accessing NFS shared data by server name

You would need a **DNS** server for this, and the **NFS** server needs to be configured to work with it. This is not simple. So, for **NFS**, use the **IP** address to prevent complications.

## Accessing data via an encrypted connection

This is not simple, either. You need the **krb5** package or **lib-krb32** for 32-bit machines to work with **Kerberos**, available on all official Manjaro flavors by default. You can follow the **Arch Linux** guide at `https://wiki.archlinux.org/title/Kerberos`. I will not put information here because this would require a whole chapter. Sharing without Kerberos is *sufficient* for a regular **home** or **small office** network because you can limit access based on IPs, groups, and rights. If you work with sensitive data, an encrypted connection is obligatory. For personal administrator usage, a good alternative is **SSHFS**.

## A few more references

I have tried to cover all the basics, but if you need more information, here are a few good links for NFS:

- A good video: `https://www.youtube.com/watch?v=ZparikqAo3E`

- Generic **NFS** information: `https://wiki.archlinux.org/title/NFS`

- A few nice guides:

  - `https://techviewleo.com/configure-nfs-server-on-arch-manjaro-garuda/`

  - `https://linuxhint.com/install_configure_nfs/`

  - `https://www.ordinatechnic.com/web-cloud-server/1/Network%20Services/2/Network%20File%20System%20(NFS)/introduction-to-network-file-system-nfs`

## Windows clients

To enable the **NFS client** on a **Windows** *7, 10*, and *11* system, open *Control Panel*, go to **Programs |
Programs and Features**, then **Turn Windows Features on or off**. Select **Services for NFS**, check the
**Client for NFS** checkbox, and click **OK**.

Now, press *Win+R*, type `cmd.exe` in the pop-up box, and press *Enter* to open *Command Prompt*.
Run the following (and replace the **IP** with your **NFS server IP**):

```
> showmount -e 10.51.6.188
> mount \\10.51.6.188\srv\nfs Z:
```

That's all. You will have **mapped** the NFS shared directory to the `Z:` drive, and it will be *visible in*
**Windows Explorer**. You can also use other available letters for the mounted **network share**. From
the **Windows** perspective, this is a **network drive**.

Some guides explicitly propose using the additional `-o  anon` Windows `mount` option to match
your user to the **Anonymous GID** (for more information, see *Chapter 13*). But with or without, *if
the directory is shared with write permissions*, you will have no problem *creating and editing files and
directories*. You will have a **synchronous** mount by using the `-o  mtype=hard` Windows `mount`
option. However, if the server connection blocks, instead of dropping out, this will *block* **Windows
Explorer**; hence `hard` is not recommended for regular mounts.

To **unmount** the drive, execute in `cmd.exe`:

```
> umount z:
```

To automate the steps, you can *add them to a simple* **.bat** *file and execute it with administrative rights*.

To see all the available options for mounting, execute `> mount --help`.

After **restart**, you will *not see the drive*. To make the mounting *automatic*, you can add a **batch script** to
the **Startup Programs** to be executed *at startup*, as mentioned earlier – meaningful only for static IPs.

Windows has a nice additional option. When the **Windows NFS** client is *enabled* and if the **Windows
Network Discovery** feature is *turned on*, you will be *able to see the* **server** and *browse* it directly when
browsing the **network**. In this case, you don't need the mount commands, and the directory will always
be present in your network listing in the **Windows Explorer** file manager.

## macOS clients

Since I currently don't have a macOS machine, I cannot test it and describe it to you. As macOS is
Unix-like, most commands are similar to the Linux ones. Here are two good guides:

- `https://www.cyberciti.biz/faq/apple-mac-osx-nfs-mount-command-
  tutorial/`

- `https://knowledge.autodesk.com/support/flame-products/`
  `troubleshooting/caas/simplecontent/content/how-to-access-to-`
  `linux-nfs-share-mac-osx.html`

# Sharing via Samba Server

**NFS** is not encrypted. Its security is primarily based on *managing* **permissions** and shared directories access based on **IP**. As an alternative, **Samba** has theoretically a bit higher security, but the following link lists *all its current* **vulnerabilities** (tens more than in **NFS**): `https://www.samba.org/samba/` `history/security.html`. Of course, each found *vulnerability* is fixed ASAP, so considering it is an old and mature technology, it is OK to be used.

**Samba** is the default way of sharing on **Windows** machines. It is essential to note the following *security advice* when using **Samba**:

- Never share on a *public network* except if explicitly required. **Samba** opens specific ports on your *network connection*, which is considered a *potential severe vulnerability*. Of course, it is acceptable in your home, office, or VPN network.

- Share *only what you need to share*. If you share extra unnecessary files, apart from slowing down the indexing and access, you may provide a potential additional *vulnerability point*.

- Consider the **permissions** – provide **write permissions** only when *required* and in a *private network*.

- If it's not required explicitly, *don't allow anonymous guests*.

- When sharing with only *a few dedicated users*, explicitly set share access control via **user** and **password**.

I tested Samba in *December 2022* and *August 2023* (when its behavior and issues had already changed). In both cases, I had multiple issues, no matter the flavor, including when sharing files from Windows 11. That's why I decided it was pointless to provide detailed guides for a technology I simply could not make work.

Still, I have to mention that many forums currently report issues with the **AppArmor package** and **Samba**. AppArmor is a **Mandatory Access Control (MAC)** system, which is a kernel **Linux Security Module** enhancement *to confine programs to a limited set of resources*. While it runs, the Samba server is not provided network access. To turn it off, I had to execute $ `sudo aa-teardown`, then $ `systemctl disable apparmor.service`, then **reboot**. *Even after this, I couldn't access the shared directory* (including with a password). I could only see it as a listing. I performed all these experiments on updated systems.

Despite its issues, I mention **Samba** as, apart from NFS, it has been the other most widely used and actively developed technology for decades. Considering the number of guides and recommendations for it online, and the issues reports (including official ones), I assume its problems will eventually be solved.

The Manjaro wiki has a basic installation guide depending on your flavor. Once installed, you have menus in your File Manager. Here is the link: `https://wiki.manjaro.org/index.php/Using_Samba_in_your_File_Manager`.

You can also check this article on how to do it from a terminal: `https://techviewleo.com/configure-samba-share-on-arch-manjaro-garuda/?expand_article=1`.

### Sharing directories by ourselves via Terminal and all configurations

This is possible but requires some effort. The mentioned **Samba** issues are equally valid. If you need to do it, follow this guide: `https://forum.manjaro.org/t/root-tip-how-to-basic-samba-setup-and-troubleshooting/100420/1`. It also has short instructions on a **Python**-assisted solution and links to many related forum topics. You can also check `https://techviewleo.com/configure-samba-share-on-arch-manjaro-garuda/`.

Finally, we have the official **Arch Samba** configuration `https://wiki.archlinux.org/title/samba#File_manager_configuration`, and the **Arch** information on **AppArmor**: `https://wiki.archlinux.org/title/AppArmor`.

## Secure Shell (SSH) and working remotely

**SSH** is a *higher-level network protocol* on top of TCP/IP, which provides a *secure encrypted connection* between computers over a network. It replaced the old insecure **Telnet** protocol, which transmitted *plain text* data, making it *vulnerable* to eavesdropping and hacking. As mentioned earlier, network protocol basics are explained in *Chapter 12*.

**SSH** uses *public-key cryptography* to authenticate a remote computer and encrypt all data transmitted between two computers. This provides a secure connection, allowing logging in to a computer, transferring files, and executing commands remotely.

Many applications use **SSH** to communicate. In this case, however, we will talk explicitly about connecting from one machine with *any OS* to another *network-accessible Linux machine* and logging into it as *one of its registered users*. The first is called the **client**, and the second is the **server**. The server needs an **SSH** daemon, available in the preinstalled Manjaro **OpenSSH** package (`https://www.openssh.com/`). While there are some alternatives, **OpenSSH** is one of the most widely used and famous. It started as a project over 23 years ago and has high *security* and *functionality*. This package also contains the `ssh` client, `scp` for secure copying, `sftp-server`, and other tools. **OpenSSH** has a port for macOS clients. For **Windows**, we will use one of the most famous clients, called **PuTTY** (`https://www.putty.org/`).

Since we are discussing *logging in remotely via the terminal*, you can do *anything* you usually do from the server's local Terminal via a remote one. This includes *administrative tasks* performed with `sudo`, starting and stopping services, editing configuration, and so on.

One significant advantage of using **SSH** versus any GUI remote desktop application is that the second one transfers *image-related data* and thus needs more *processing power* and wider *network bandwidth*. As a result, it also has higher latency. SSH only transfers small amounts of commands and text information, so it is *much faster*. In addition, **SSH** will also work for servers without a graphical environment.

Executing the w command in a terminal prints the currently logged-in users. Without going into details, know that for **KDE Plasma** it prints two additional sessions for the graphical environment. Whenever a user is logged in remotely, their session will be added and usually designated as bash. We can have multiple SSH connections to a server from different clients, each reported separately by w.

## Enabling SSH on the server side

First, edit the sshd *configuration file* via:

```
$ sudo nano /etc/ssh/sshd_config
```

With the arrows, navigate to the line with the string #Port  22, delete the *hash sign*, and change 22 to a random port number of your choice *between* 49152 and 65535 (in my case, 53786). Press *Ctrl+s* to save, then *Ctrl+x* to exit nano.

Enable the SSH daemon:

```
$ sudo systemctl enable --now sshd
```

If you already enabled it some time ago, *changing the* **port** *will require explicitly restarting the service*, so in this case, execute:

```
$ sudo systemctl restart sshd.service
```

You can always check the service status with:

```
$ sudo systemctl status sshd
```

*Firewall configuration* and **ports** are reviewed in detail in *Chapter 12*. If you have enabled a firewall, you have to explicitly enable the corresponding **port** – in my case, in the **ufw** firewall, via:

```
$ sudo ufw insert 1 limit log 53786/tcp comment 'SSH port'
```

## Connecting from another Linux machine locally

First, execute the following in the Terminal:

```
$ ssh -p 53786 luke@10.51.8.142
```

In this command, -p is for the port specification; the next parameter before the @ symbol is the user, and after @, we have the target machine IP. Pressing *Enter* will ask you to confirm the *host's authenticity* the first time. You have to type yes and press *Enter*; then, you will be prompted to enter the user's password. Once you do so and press *Enter* again, a session will be launched as if you are in a terminal *on the machine you have just connected to*.

## Connecting remotely from the internet

What if I want to connect to my machine from Brazil, the US, or Europe? This is typically a complicated task, requiring editing your **router** configuration and sometimes even assistance from your **ISP**. I will describe shortly the theoretical way of doing it to explain the idea.

First, you need to know your router's public **IP** *address*. If it is not static (fixed), you need to be informed each time it changes. Your ISP shall not filter the SSH data messages (though rare, this is possible). As all local networks are behind *at least one home router*, it has to have *the given port configured as forwarded* so that, when accessed from the **internet**, it will *forward the packets to your machine*. As this depends on your **router's** brand, I cannot provide a guide on how to do it here – hundreds of brands and models exist. If you have purchased the router yourself, you will know its model and be able to find guides on how to configure it. If you have a dedicated **ISP router**, you may need to speak with the support if you don't have administrator access to it. Once you have the instructions, configure the correct port to be directly forwarded.

Let's imagine you successfully forward port 53786. Then, get your public IP (let's imagine whatismyip.com reports 10.212.4.15) and your machine's local IP reported by $ ip -br a (e.g., 192.168.3.17). In the router's configuration, you will forward 10.212.4.15:53786 to 192.168.3.17:53786. Now, you should be able to connect from all over the world to your machine via the following:

```
$ ssh -p 53786 user@10.212.4.15
```

There are more modern ways to do this, depending on your **router's** brand and model. Check your router's manual. It may do so via **NAT**-related functions such as **PPTP**, **L2TP**, and **IPSec** or, if more modern, directly via **OpenVPN**, **WireGuard**, or **IKEv2**.

## Connecting remotely from the internet through a VPN

**VPN**s are reviewed in detail in *Chapter 12*, so this paragraph requires reading the corresponding section to understand it. Still, I leave the explanation here to have some basic idea.

If the machine you want to connect to uses a VPN to connect to the internet, this is typically not allowed. The reason is that having *port forwarding* enabled on a VPN connection is considered a **security breach**. So far, I haven't found a way to connect to my machine from the internet through my **Surfshark** VPN access. On the other hand, with the **Surfshark VPN** enabled, I can directly access my machine via **SSH** on port 53786 from the **local network**. However, when I activated the **Kill**

**Switch** option on **Surfshark**, my local **SSH** connection got blocked entirely. In other words, if you experience issues with connecting, you have to check the **SSH** configuration on the server, then its **firewall**, and, potentially, the **VPN**.

## Connecting from a Windows machine

First, open `https://www.putty.org/` and find the **Download PuTTY** section. You will most likely need the **64-bit x86** version. If it doesn't work as the given PC has a different architecture, check whether it needs the 32-bit version or the 64-bit ARM one.

Install and then start the program. You will see the following default screen, and you have to fill in the **IP** and **port** details accordingly, as shown in *Figure 11.2*:

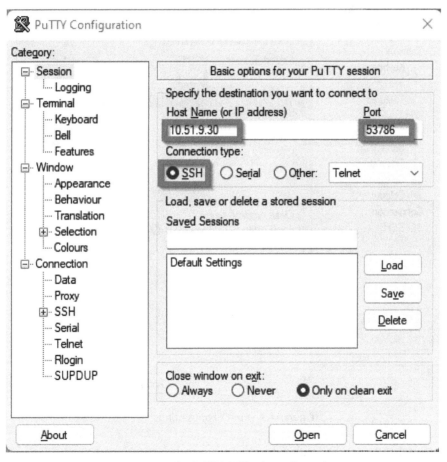

Figure 11.2 – The PuTTY GUI basic configuration

The first time you connect to your machine, you will be warned that *the host key is not cached for this server*. This just means **PuTTY** connects to it *for the first time*. Accept it, and then you will be asked first for the user and then for the password. Once you enter them, you will have a successful session.

Using the default font and view settings was uncomfortable for me. So, what you need to do for the current session is right-click on the top bar of the **PuTTY** window, select **Change Settings**, go to **Window | Appearance**, and change the **Font settings**. I've explicitly enabled the **Font quality** option **Antialiased** and then clicked **Apply**. Then, go back to **Session**, click on **Default Settings** in the list of saved sessions, and click **Save**, as shown in *Figure 11.3*:

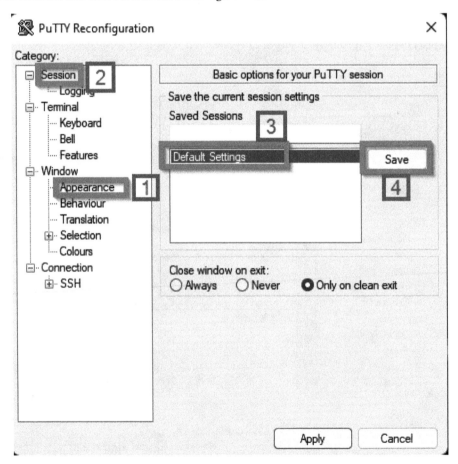

Figure 11.3 – PuTTY Font settings

If you also want the **IP** and **port** details to be saved, the next time you connect, you have to save them to the default settings in the same way just before starting the session.

# Sharing via SSHFS

**SFTP** over **SSH** (**SSHFS**) is a technology for remote access via the encrypted **SSH** protocol. It is currently implemented via **Filesystem_In_Userspace** (**FUSE**) and is comparable to NFS from a performance point of view. The performance is lower only for Windows clients for large amounts of data.

Compared to **NFS** and **Samba**, it mounts the server **FS** directly in a local **FS** location. Thus, in its standard usage, it is a way to *access the complete filesystem of a server* from a remote location. The presented simple example is applicable only for a non-root user. Mounting root-owned directories is not possible with the presented approach, thus you are safe. To mount root-owned directories, you need to perform additional configurations, which I will not provide, as it is more complicated and a severe security issue. In addition, you already know how to log in as a given user, and should it has sudo permissions, you are able to perform administrative tasks.

On the server side, we need to enable `sshd.service` via:

```
$ sudo systemctl enable --now sshd.service
```

On the client side, install the **sshfs** package. Then, create a directory to mount to. I have created the `fromSshfs` directory in /mnt/. Now, use the following format:

```
$ sudo sshfs yourUser@10.51.8.78:/home /mnt/fromSshfs
```

The command for me is as follows:

```
$ sudo sshfs luke@10.51.8.78:/home /mnt/fromSshfs
```

The first time you connect via **SSH**, you will be warned that *the authenticity of the host cannot be established*. You will have to type `yes` and press *Enter*. Then, you will have full access to the given server directory, but each time you perform some action, it will need `sudo` in front, like this:

```
$ sudo ls -l /mnt/fromSshfs/luke
```

Now, you can copy from or to the mounted directory. Practically, you have *full user access*. I guess you see why SSHFS is recommended only for selected users and inappropriate for common user share. There is one way to implement a restriction, but as it is pretty advanced, I will not write about it here. If you are interested, you can read this guide: `https://fedingo.com/how-to-restrict-ssh-users-to-specific-folder/`.

Apart from this, there are two more important notes.

First, the **SSHFS** has been an orphaned project since *June 2022*, subject only to critical bug fixes. However, as it is old and mature (in development since *2004*), using it is *safe*. In addition, its last maintainer has officially announced the search for a successor, and we hope to have one soon.

The second point concerns the official Arch article at `https://wiki.archlinux.org/title/SSHFS`. Reading it might answer additional questions. It also contains references to helper packages.

# Summary

In this chapter, we have covered the local network basics by explaining **IP** addressing, static and dynamic **IPs**, and ways to inspect our network configuration. Next, we reviewed the basics of network sharing, explaining in detail how to do it with **NFS**. We further looked into **SSH**, which is frequently used to access your machine remotely from a local network. We finally saw how to access a server's FS remotely via **SSHFS**.

In the upcoming chapter, we will investigate how the internet works and explain the structure of its protocols and ports. We will review the foundations of network security and explore strategies to counter common threats. A highlight will be a detailed guide on setting up the **Uncomplicated Firewall** as a great user-friendly security tool. The chapter concludes with a thorough review of VPNs with examples.

# 12

# Internet, Network Security, Firewalls, and VPNs

In the previous chapter, we explored the network basics, file sharing, and **SSH**. It is time for us to switch to one of the most significant features of **Linux** – **security**. After all, **Linux** is famous for being a notoriously hard system to hack, but hard doesn't mean *impenetrable*. Even the most secure fortress can be overtaken if certain weak spots are left unsecured.

We will continue looking at the **internet** and how it works by exploring networking models, packets, protocols, and ports. We will then dive into **network security** by exploring possible **attacks** and several *recommendations* that will keep you safe. Further, we will explore **firewall** types and then review one of the most famous of them in detail. The end of the chapter is dedicated to **Virtual Private Networks** (**VPNs**), the best way to stay *anonymous* on the internet, which raises your security to a cosmic level.

The sections in this chapter are as follows:

- How the internet works, network protocols, and ports
- Attacks, security advice, and firewalls
- Setting up your firewall from A to Z
- VPNs

## How the internet works, network protocols, and ports

In this section, we will start with the network models that are fundamental to the whole **internet** infrastructure. We will then dive deeper by explaining the protocol basics, as no data transfer can be realized without them.

## TCP/IP and OSI network models

After reviewing the networking basics in the previous chapter (**IP addresses, ping, DNS**, listing a few **protocols**, etc.), let's look at the overall networking model.

**TCP/IP** is an abbreviation of **Transfer Control Protocol/Internet Protocol**. It is the most widely applied network model and the current internet infrastructure base. You may have also heard of the **Open Systems Interconnection** (**OSI**) model. As **OSI** was introduced later, it was not directly and extensively applied, despite separating and presenting network communication better.

**OSI** has seven layers, each dedicated to a different *functionality*. Its upper layers are **software** (**SW**)-*oriented*, and the lower ones are **hardware** (**HW**)-*oriented*. They are presented in *Table 12.1*. The **bold** entries (in the *Layer* column) coincide, to some extent, with the ones in the **TCP/IP** model:

| Layer | Applied in | Function |
|---|---|---|
| **Application** <br> **(SW)** | SW | This is the level at which our mail client, browser, or Spotify applications work. They create the server data/service requests. When response data arrives, the applications use it to display web pages, play media, and so on. |
| Presentation/ <br> translation | | The translation layer packs the SW pure data (the useful data from the whole packet), takes care of compression, and additional encryption/decryption if necessary. |
| Session | | This logical level is responsible for session establishment and synchronization with the server from which the data is requested. It also provides security services if the connection is via an encrypted protocol such as HTTPS. |
| **Transport** | SW/HW | The transport layer is called the heart of the OSI model, as it provides services to the application layer and uses the services provided by the network layer. It is the middle point between the outside network and the local application. This layer takes care of the packets, traveling the whole route from your PC via tens of routers to a server, and then getting the delivered result packets. It ensures the proper message sequence. It also knows the local machine-opened ports. |

| Layer | Applied in | Function |
|-------|-----------|----------|
| **Network** | HW | Routers manage this level and assign IP addresses to data packets. Here, routing and packet transfer between network nodes are handled. Network nodes are multiple routers and servers, transferring the packets between us and their final destination. |
| Data link | | This layer is responsible for splitting datasets into small chunks. It then handles the low-level communication between a LAN cable or WiFi endpoints, such as routers and our PC. Each network node (a PC, network device, router LAN port, etc.) has a local data link connection based on its MAC address. |
| Physical | | This layer looks at the binary message form, sending and receiving bits. It is based on electrical signaling over wires for LANs or radio waves for WiFi. |

Table 12.1 – The OSI networking model and the functions of its layers

The **TCP/IP** model has only four layers, listed in *Table 12.2*. Those in **bold** are, again, the ones that coincide with the OSI model to some extent:

| Layer | Applied in | Function |
|-------|-----------|----------|
| **Application** | SW | This is the level at which our mail client, browser, or Spotify applications work, just like in the previous case. However, many high-level protocols such as SMTP, FTP, SSH, and HTTP are also constructed and processed here. |
| **Transport** | SW | The transport layer is responsible for low-level protocols, such as UDP and TCP, and packet delivery between two points on the network. |
| Internet | SW/HW | The internet layer handles the IP addressing and protocol and defines a packet's source and target addresses (e.g., a packet sent from 154.32.132.17 to 174.156.16.24). |
| **Network/ Link layer** | HW | This level is responsible for low-level data transmission, exchanged as electrical and WiFi signals between two points on a network. These two points can be at the two ends of a cable or two wirelessly communicating devices, such as your PC and router. |

Table 12.2 – The TCP/IP networking model and the functions of its layers

When a **data packet** has to be sent over the internet, *each layer adds additional data, often named* **header** data. We will start with a SW application that needs to send data (e.g., **SSH**, Spotify, or a browser), which prepares *some messages*. The **application layer** then adds the given *high-level* **protocol**. Further, the **transport layer** adds the assigned **port** number and the **TCP** or **UDP** parts. Finally, the **internet layer** adds the **IP protocol** information for the *sender* and *receiver*. The now-prepared packet is sent from your **network adapter** to the **local router**. The whole message structure (called a **datagram**) is shown in *Figure 12.1*:

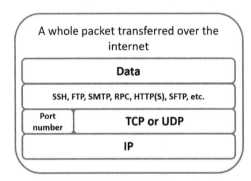

Figure 12.1 – The structure of a data packet sent over the internet

If we take a browser application as an example, the communication between it and a server occurs as shown in *Figure 12.2*:

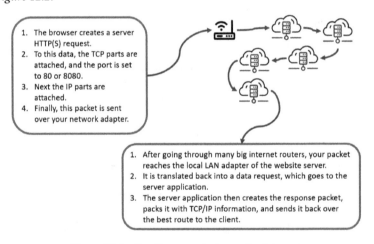

Figure 12.2 – Sending a packet over the internet

While all this happens *within milliseconds*, your **PC** waits for a packet to be received on a particular **port** (we look into ports later in the section). This means the **port** is **open**. Whether the data coming back is from the web server or a potential attacker, your computer will consume it. It will usually be *discarded* if it is *corrupted* or *unrelated* to any work in process. If it is correct, it will be accepted. All network applications then inspect the data and, again, discard it if it is malicious or corrupted.

To get a complete, beautiful web page with a loaded size of 50 kilobytes, a browser may exchange between 50 and several hundred packets with a server. Some may be corrupted or lost, and each such package will be *re-requested* each time this happens. Imagine your browser with 20 opened tabs, Spotify streaming music, your mail client, Zoom and Signal working, and so on. Even if you do nothing, at least 100 packets are exchanged *per second* between your **PC** and *multiple* **servers**. This number can reach up to millions of packets daily.

A **port** is a number defining the higher-level protocol data contained within a message. There are different **ports** for different applications and higher-level **protocols**. Most protocols use officially *dedicated ports*, while a few use unofficial ones. The **Internet Assigned Numbers Authority (IANA)** assigns the official ones for specific communication and usage. Let's take a look at *a few famous higher-level* **protocols** transmitted over the **TCP** or **UDP** transport protocols, as listed in *Table 12.3*:

| Port | Usage | Port | Usage |
|---|---|---|---|
| 20 and 21 | File Transfer Protocol (FTP) | 143 | Internet Mail Transfer Protocol (IMAP) |
| 22 | SSH communication, SCP, SFTP | 401 | Uninterruptible Power Supply (UPS) |
| 24 | Telnet (simple text messages) | 443 | HTTPS (secured HTTP) |
| 25 | Simple Mail Transfer (SMTP) | 530 | Remote Procedure Call (RPC) |
| 53 | DNS requests | 749 | Kerberos administration |
| 80 and 8080 | HTTP (for web sites) | 1194 | OpenVPN |
| 111 | ONC RPC (Open Network Computing Remote Procedure Call) | 2375 | Docker REST API |
| 115 | SFTP (secured FTP) | 4244 | Viber |

Table 12.3 – Some typical TCP/UDP ports used in network communications for dedicated purposes

**Port** numbers are from 1 to 65535, divided into **known ports** *from* 0 to 49151 and **dynamic/ private/temporary** ports from 49152 to 65535. Known ports are divided into **system ports** (or **well-known ports**) *from* 0 to 1023 and **registered ports** from 1024 to 49151. **IANA** manages the registered ports.

Nobody can and will stop you *from using whichever port you want* for a particular communication. Still, opening and using the **DNS** requests port to send and receive MP4 video clips would be *unwise* and *impractical*. Most **free and open source (FOSS)** projects/applications and commercial apps use the range from 1024 to 49152 for different long-term purposes. Refer to the following Wikipedia page for the hundreds of use cases (it includes historical usage from great names such as **Oracle, Unix, IBM,** and others). It explicitly states that the list is not always updated and is constantly changing: `https://en.wikipedia.org/wiki/List_of_TCP_and_UDP_port_numbers`.

When a **port** is **closed** on your PC, *even if packets for it are sent to your machine*, they are directly discarded. Thus, *a potential attacker can do nothing via the closed ones*. You cannot load a web page when ports 80 or 8080 are closed. As a result, at least these are open for web services to work.

## All types of network protocols and the ports for them

As you already know, the **internet protocol (IP)** adds **source** and **destination** *addresses* to your *data packet*. It has rare alternatives only on specific network types. This means that to send a packet from one PC to another, you will always use **IP** communication. The packets that will be transferred over it are not only **TCP** or **UDP** but also **SCTP, DCCP, IGMP, ICMP,** and others. In *March 2023*, the total number of transport-level protocols was *144*, as defined on this official **IANA** page: `https://www.iana.org/assignments/protocol-numbers/protocol-numbers.xhtml`. When a **packet** is received, the Linux Kernel network driver handles it and transfers it to the **TCP/IP** stack. Based on the **port** and **protocol**, the stack sends the data to the proper application. If you use the `$ cat /etc/protocols` command, you will see the same list as the one at `https://www.iana.org/`.

On top of **TCP** and **UDP** (and, again, very rarely others), you may have data based on *hundreds of higher-level* **protocols**, such as **SSH, FTP, SMTP, RPC, HTTP(S),** and **SFTP**. Each protocol adds additional information. If you don't have a driver and a working application to accept data over a given protocol, the data will be discarded even if packets are sent to you. The data is also discarded if you have a driver and the SW to accept them, but a **firewall** *blocks* the given port.

All **IANA**-assigned ports for the **TCP, UDP, SCTP,** and **DCCP** transport protocols are described on its official site: `https://www.iana.org/assignments/service-names-port-numbers/service-names-port-numbers.xhtml`. **TCP** and **UDP** *coincide completely* for these ports. There are only a few ports also assigned to SCTP or DCCP. For example, port 21 is designated for **FTP** connections over **TCP, UDP,** and **SCTP**. To connect remotely to a terminal on our PC, we use **SSH,** again transferred over **TCP, UDP,** or **SCTP** data packets, by default via port 22. Thus, if you *disable* port 22 for the lower-level TCP, UDP, and SCTP protocols, the higher-level one (**SSH**) will not work.

For some applications, such as **SSH**, you can *change the port*, which is frequently done for *security reasons*. Choosing *a random* port between 49152 and 65535 can be much safer, as regular attacks on port 22 for SSH will be ignored entirely.

## UDP versus TCP and others

**UDP** is designed to efficiently broadcast large amounts of data, no matter whether the **client** is listening, and will consume packets, irrespective of their *order* and *validity*.

**Conversely, TCP** is significantly slower for large amounts of data, as it has a larger **overhead**. Thanks to this **overhead**, it offers **packet** sequence management, error checks and corrections, reception acknowledgment, and retransmission requests. Additionally, it works only when a specific connection has been established between the sender and the receiver. As it ensures *data transmission and validity*, and many **routers** (particularly big international internet super-routers) have traffic management, *data congestion management* is applied for *billions of users*.

With all that said, **TCP** is *slower* and more *reliable*. **UDP** is *less reliable* due to only utilizing basic error checks, but it's *more efficient for large amounts of data*, including *streaming*. The most commonly used higher-level protocols are based on transport by **TCP/IP** or **UDP/IP**.

For reference, here are some other standard low-level protocols that are like **TCP** and **UDP** – **DCCP**, **SCTP**, **RDP**, **RRP**, **RUDP**, and **RSVP**. Some of them are used for streaming and games, and others to replace some web transmissions, but in general, they are limited to particular use cases.

## A few other essential protocols

In this section, we will mention a few other protocols related to the setup of the **Uncomplicated Firewall** (**ufw**). **ufw** is one of the common firewalls used on Linux distributions and is my recommendation to you, as it's a great, user-friendly SW. Take at least a quick look at this list:

- **IPSec**: A secure network protocol suite that *encrypts* the **packets** between communicating endpoints and is used in **VPNs**. It also ensures *authentication negotiation*, exchange of **cryptographic keys**, network-level peer authentication, data origin authentication, and other security techniques.

- **Authentication Header** (**AH**): This protocol is part of **IPSec** and provides a few special security additions.

- **Encapsulating Security Payload** (**ESP**): This protocol is again part of **IPSec**. In *tunnel communication mode*, it adds additional encapsulation (with authentication and encryption) for the payload data of an **IP** packet.

- **GRE**: This is a tunneling protocol developed by **Cisco**. It was used with the Point-to-Point Tunneling Protocol (**PPTP**) for **VPNs**, an older technology known for *certain security risks*. GRE can also be used together with **IPSec VPNs** to pass routing information between networks. Nowadays, it is still used with **Cisco** networking **HW**.

- **Internet Protocol version 6** (**IPv6**) provides the addressing for $3.4 \times 10^{38}$ (greater than 1 trillion trillion trillion) total addresses. It is not interoperable with **IPv4** (with addresses such as 192.168.12.17), but there are so-called *transition mechanisms* to ease this problem. It is required to communicate with any machine or server that connects to the internet via **IPv6**. Its addresses in short form look like this – 3012:dc8::8a27:370:7b34. For more information, check out *Chapter 11*.

- **IGMP** is used to establish *multicast group memberships*. A **multicast message** is sent from one server to *multiple clients*. Thus, streaming video data to 10 users on a local network is sent once to all clients instead of sending the same information 10 times. It is used only in **IPv4** networks, while for **IPv6**, the **ICMPv6** protocol is responsible for such tasks. **IGMP** can also be used for group gaming in local networks.

- **Address Resolution Protocol** (**ARP**) was developed in *1982*. It is a critical part of **IPv4**, as it discovers the so-called *link layer address* (the **MAC address**) of a network device. The **MAC address** is a physical address written in the form 0xFEAD6C34B254 (six groups of two hexadecimal digits). This is the physical address of your network card on the local network, which is necessary for communication with your local router. In **IPv6**, it is replaced by the **Neighbor Discovery Protocol** (**NDP**).

## The Linux basic network modules and a port-scan tool

This section will briefly discuss the basic Linux network modules so that you are aware of them. We will then see how we can check our system's open ports.

**Netfilter** is the framework that provides an interface for the **kernel network stack** to apply filtering and firewall policies. It has been part of the Linux kernel for over 20 years and has been in development since 1998. As a kernel module, it provides fast and highly effective network management. **Netfilter** practically is the **Linux firewall**. Higher-level SW firewall frontends, such as **firewalld** and **ufw**, just configure its options.

**iptables**, **ip6tables**, **arptables**, and **ebtables** are **Netfilter** *user-space modules* that configure filter rules for **IP** packets. They are all old and mature projects, the main one being initially released in 1998. All are actively developed but already have a successor combining them all, called **nftables**.

Together, all these tools allow not only filtering but also **Network Address Translation** (**NAT**) and **port translation**, which can serve as additional *security techniques*. They also provide logs for network traffic monitoring and analysis (and, consecutively, the identification and blockage of hacking sources).

You can check the complete list of Netfilter projects with news and releases at https://netfilter.org/projects/.

### Listing assigned ports and protocols on your machine

If you execute $ cat /etc/services, you will see the complete list of services with an **IANA**-assigned **port** and **protocol**. This listing simply shows the **IANA** assignments and not the **ports' open/closed** statuses. Unless explicitly used, all ports on Linux are, by default, *closed*. As the list is long, filtering with grep would be helpful. If we execute $ cat /etc/services | grep -i ssh, we will see *all services communicating over different protocols via* **SSH**. On one of my machines, I got the following result:

```
$ cat /etc/services | grep -i SSH
ssh              22/tcp
ssh              22/udp
ssh              22/sctp
sshell           614/tcp
sshell           614/udp
netconf-ssh      830/tcp
netconf-ssh      830/udp
sdo-ssh          3897/tcp
sdo-ssh          3897/udp
netconf-ch-ssh   4334/tcp
snmpssh          5161/tcp
snmpssh-trap     5162/tcp
tl1-ssh          6252/tcp
tl1-ssh          6252/udp
ssh-mgmt         17235/tcp
ssh-mgmt         17235/udp
```

### Scanning with Nmap for open ports on a machine

Many tools exist to do this, but we want the most widely used one from the Manjaro official repositories. This is **Nmap**. A graphical frontend for it, **Zenmap**, is available, but only as an AUR package.

Another frequently used tool is **ZMap**, which is *faster for some global network scans* but has had *only one beta release* since 2015 (in 2019). In comparison, **Nmap** is in active development and receives improvements and fixes.

To install **Nmap**, execute `$ pamac install nmap`. It works either with an **IP** or a *web server name* (and in the second case, it uses **DNS** to resolve the name to its **IP**). To run it for our PC, we need our own **IP** info. If you remember from the previous chapter, we get it by executing `$ ip -br -c a`. Then, we execute the following commands (and don't forget to replace the IP with the one of your machine):

- `$ nmap -p- 10.51.8.142`: The `-p-` parameter will trigger *a scan for all* **open TCP ports** on the given machine. In this case, I directly provided my IP. As the ports number thousands, the results will be only for **open ports**. You can also use the `127.0.0.1` address or the keyword `localhost` (it will also list TCP port `631` for the **Internet Printing Protocol (IPP)**. On one of my machines, the result was as follows:

```
$ nmap -p- 10.51.9.30
Starting Nmap 7.93 ( https://nmap.org ) at 2023-01-27 18:23 EET
Nmap scan report for 10.51.9.30
Host is up (0.000079s latency).
Not shown: 65528 closed tcp ports (conn-refused)
PORT       STATE SERVICE
111/tcp    open  rpcbind
2049/tcp   open  nfs
20048/tcp  open  mountd
33909/tcp  open  unknown
35735/tcp  open  unknown
49501/tcp  open  unknown
57621/tcp  open  unknown

Nmap done: 1 IP address (1 host up) scanned in 1.71 seconds
```

As I use the machine extensively (including for **NFS** shares), it has a few open ports.

- `$ nmap -p 20 10.51.8.142`: The `-p XXX` parameter will scan the **TCP** port XXX on the given machine and print *its status*. I have shown this with **port** 20.

- `$ nmap -p 20-40 10.51.8.142`: The `-p XXX-XXX` parameter will scan the port range XXX-XXX and print an *explicit status for each port in it* – in this case, from 20 to 40.

To scan **UDP** ports, you must add the `-sU` option and explicitly execute with `sudo`. The last command will then be as follows:

```
$ sudo nmap -sU -p 20-40 10.51.8.142
```

- `$ nmap -p- www.someWebSite.com`: This is used to scan a domain using its web address. It *can be a bit slow*, as it first needs to use the **DNS** service, and then make 65,535 requests to it (when scanning all ports, like this example). However, scanning commercial web domains in some states or countries is *illegal without permission, so please don't experiment on random websites*.

Use -h or man for the short and long help guides. Complete documentation is available at https://nmap.org/, but *for port scanning*, the link is https://nmap.org/book/port-scanning-tutorial.html.

This short *cheat sheet* is quite helpful: https://www.stationx.net/nmap-cheat-sheet/. There is also a nice short guide here: https://www.digitalocean.com/community/tutorials/nmap-switches-scan-types.

You can also try **Zenmap**, but the presented commands are easier and faster.

# Attacks, security advice, and firewalls

A **firewall** is a SW to filter or discard **network packets** and **port requests** from forbidden addresses or via closed ports. It can also discard valid packets with wrong credentials so they never reach your OS internals and applications.

Considering that for decades, many ports were **open** on **Windows** by default, without a **firewall**, this was like having an unguarded and half-opened vault – easy to hack.

Conversely, **Linux** distributions have *all* **ports** *closed by default* – this is why hacking them is extremely hard, and also why, despite **firewalls** being pre-installed on many distributions, they are often *disabled* by default.

In addition, the Linux kernel lives in a **separate namespace/memory**, accessible only by special means and *only from SW explicitly installed* **on the machine**. Its management also requires an administrator's password.

By default, any **Linux** distribution is installed with a user with *administrative privileges, who is never the* **root**. Despite being an administrator, this user must always use sudo to elevate their privileges to *change the system*. This all is by design, including in the filesystem, process authorization and management, and all possible system-critical settings.

*The final point against potential attackers* is that thousands of people constantly check for issues and vulnerabilities. As all Linux distributions are open source and share components, the community is immediately informed when any are found, and fixes are provided ASAP. Such fixes are very quickly distributed to all Linux users.

I guess you can see now why **Linux** is *rock-solid regarding security*.

## When is a Linux distribution in danger?

This is possible only if a *particular* **vulnerability** *related to network communication* is **exploited** in an attack precisely aimed at a machine known to have this **vulnerability**. For home usage, it is extremely rare and practically close to impossible.

People trying to hack into a **Linux** machine only target *organizations that store highly valuable information* on Linux machines or servers. Such attackers have immense technical knowledge. That's why such servers are behind private networks, with a HW firewall in front, an internal firewall on each big server inside the network, special security means, etc.

## When is a Windows machine in danger?

Even though **Windows** has immensely improved in terms of *stability*, *security*, and *efficiency*, there are still tens of ways to easily hack it, especially as it keeps backward compatibility with insecure technologies from the past. It has an integrated **firewall** and *security module* (which *will mitigate almost all common attacks*) but remains **vulnerable**. Direct proof is the hundreds of articles and YouTube videos on how to hack **Windows 11**; for **Windows 10**, there are already thousands of examples. Check out this fresh list from *July 2023 (the link is initially from 2018, but is updated during the years)*: https://www.techworm.net/2018/07/5-best-hacking-tools-windows-10.html.

## What about macOS?

It is considered equally hard to hack a **macOS** machine as it is to hack a **Linux** one. The reason is that **macOS** is **Unix**-like, just like **Linux**. In addition, **Apple** is famous for designing *all its products* and **SW** with the highest possible **security**.

## Common types of attacks

Here, we will review the main types of attacks. A firewall will protect us from those directly aimed at our machine. This section gives a basic idea; skip it if you already know them.

### Attacks to ports and open ports

A **firewall** will block any incoming traffic directed to **ports** not explicitly **open** to be worked with.

### Packets and communication requests with malicious content

Any **packets** considered *invalid* for certain types of communication will be **blocked**. In addition, any requests for services not explicitly allowed in the **firewall** *configuration* will be discarded.

### Invalid client and server requests

If a **firewall** has specifications for specific valid client or server requests, any such invalid requests will be discarded.

### DoS and DDoS attacks

**DoS** stands for **Denial of Service**, while **DDoS** stands for **Distributed Denial of Service**.

**DoS** happens when a given **network** is flooded with thousands or more valid and/or invalid packets. After a certain amount of such messages, a valid request between a server and a client cannot reach its destination. **DoS** has one source of invalid messages.

You can compare this to standing in the middle of a stadium surrounded by thousands of people at a concert and trying to shout to someone on the stage – they will never hear you.

**DDoS** happens when the invalid messages flood comes *from multiple points*.

**DDoS** or **DoS** *can't reach your OS internals* and are not dangerous to the *information* on your PC; *they just block your* **network**. However, this may lead to severe financial and market losses for commercial servers.

**DoS** can be *mitigated* when a network's entry point has a firewall and the source of incorrect messages is *marked as invalid* – this way, spam messages or packets never reach the internal network.

**DDoS** is *harder to mitigate*, as the sources are multiple. If they number tens and change their addresses, DDoS can become close to impossible to mitigate. However, its advanced form is also expensive and hard to execute, and has only been performed against big organizations for serious financial, political, or ideological motives.

### Ping floods

**Ping flood** and **Ping of death** are types of **DoS** attacks. A **firewall** will usually protect you from these. In a more sophisticated case, it will be harder to protect yourself, and you may suffer potential **DoS**. However, this is again never used against regular users but mostly against commercial networks.

### Brute-force attacks

This attack is carried out by a powerful computer, systematically trying all possible symbol combinations for a password. If the given system has no attempt counter, it theoretically allows such an attack to be carried out. To give a simple example, let's say you have a *4-digit* **PIN** for a device. This allows exactly `10*10*10*10=10,000` combinations to be tried out. For a combination of four English alphabet or digit symbols, the possible combinations are `1,679,616`. Adding even more symbols and a longer length increases the combination number exponentially. That's why a good password is *at least 12 symbols long*, contains *digits*, *small* and big *letters*, and *at least one special symbol*. The experts from Hive Systems researched the time needed for brute-force algorithms to crack a password here: `https://www.hivesystems.io/blog/are-your-passwords-in-the-green`.

When you make a few unsuccessful login attempts with wrong passwords, any Linux distribution and modern OSs usually block further attempts for a set time. In addition, the **Netfilter** packages on **Linux** have a **rate limiting** feature designed to counteract such attacks. **ufw** can set up **rate limiting**.

## Main advanced attacks that a firewall will not defend you from

For the curious among you, we will now review several advanced attacks. Remember that all these are rarely an issue *for any regular user*, thanks to **HTTPS**, **encryption**, and **Two-Factor Authentication** (**2FA**). In addition, if you combine these with the advice in the next section, you will never have such issues. If you are not interested in understanding them, skip the section.

### Man-in-the-Middle (MITM) attack

This attack occurs when a third party observes/listens to the *communication between two endpoints*. As they exchange user and password data, the third party copies it and can transmit messages to the server as if the client is doing it. Years ago, this was easy, as most websites used unencrypted HTTP instead of encrypted **HTTPS**. A **firewall** cannot protect you from this, as the sniffing happens *outside your machine*, somewhere along the packets' journey between you and the server. The easiest way to perform such an attack is to attach a *real network sniffing device* to your local network, which will listen to all transferred messages. In the case of **WiFi**, it is even easier, as you simply need *a listening device* that connects to an open **network** (one not secured with a password). In an ordinary case, there are two easy ways of mitigating this attack – using HTTPS and/or 2FA and, for some critical applications, the **TLS** protocol. *Advanced hacking techniques* can get around this mitigation but are *costly and never targeted at regular users*. When combining **HTTPS** with **TLS**, it is close to impossible to carry out a MITM attack. Most current websites use at least **HTTPS** with **2FA**, so with them, you are protected.

### Unauthorized access

This happens when somebody *has already stolen your credentials* (no matter how), and they access a given server with your credentials. **2FA** hinders such attacks entirely.

### Code/SQL injection

This only applies to *client-server connections*, which exchange interpreted code data such as **SQL queries**, custom **XML** data, **SMTP** (mail) **headers**, and other similar data packets. In this case, the malicious attacker must have already successfully performed some form of **MITM** attack. When attackers transfer data to/from the **client** and/or **server** as *a middle-man*, they inject additional data interpreted by them. Then, they exploit some SW *bug/flaw/vulnerability*, which harms the targeted machine or disrupts/falsifies the transferred data. This complex attack is usually mitigated using additional **encryption** techniques, **2FA**, specific custom-protocol queries, and so on. Again, this attack is only made against specific proprietary client-server systems, never against regular users, as they would not work with such data.

### Man-in-the-browser attacks

This was a well-known hacking technique against **Internet Explorer** and **Firefox** in the *first decade of the 21ˢᵗ century*. Since the turn of the *2010s*, these **Trojan horses** are rarely reported. They are called *Trojan horses* as most of them are preliminary installed malicious add-ons that can track and/or modify your data directly in the browser. In addition, these attacks were massively reported only for **Windows** and just once (many years ago) for **macOS**. Nowadays, theoretically, modern **smartphones** are *partially susceptible to them*, but as the security of both **iOS** and **Android** is already high, this is not a common concern.

## General good advice against all regular attacks

This advice guarantees your security up to 99%. It is based not only on my experience but also on standard **best practices** during the past 30 years of the digital age:

- *Never use low-quality email providers.* Several friends have suffered from stolen Facebook details and other credentials due to using very old email providers with low protection. Whichever your provider is, if it doesn't use **HTTPS**, additional security measures, or has a large enough user base to guarantee its security, it is a low-quality email provider. I will tell you of two widely used public providers with high security – **Gmail** and **Yahoo**. I also have an email server from my hosting company. They use Linux servers with additional firewalls, so I know nobody can hack them. Worldwide, there are *thousands of providers* in every country, both good and bad ones. When choosing one, search the internet for forums with regular user feedback; if there are reports of stolen information or bad services from a given provider, don't use it.

- *Never share passwords* over a *non-secured channel* or to people whose email provider is known for low quality.

- *Always use complex passwords.* They should include **lowercase** and **UPPERCASE letters**, **numbers**, and *at least one* **special symbol**. The length should be *at least 12 characters*. Never use common words or names related to your family. According to `https://www.hivesystems.io/password-table`, a 10-character strong password will take five months to crack. TSF says even 330 years here: `https://thesecurityfactory.be/password-cracking-speed/`. I would always consider the worst-case scenario, in which *5 months is good enough*. The problem is that shorter passwords that consist only of letters or numbers can be cracked almost instantly. That's why *12 characters* or more is best. Combining part of a saying, book, or a movie known to you like this is the best – `May1The%Force2`. It would be best to have such a password, especially for your email, as a lot of password-related information is usually sent to it. Of course, it would also be good for your PC, bank accounts, government platforms, and so on.

- *Never reuse the same complex password for critical services.* For example, using the same password for your email and bank account is dangerous. If your email gets hacked, your bank account is directly exposed.

- If you tend to visit random websites, *never provide your credentials to websites that look suspicious*.

- Consider all mail **phishing** attacks – this is the most common way for others to access your data, so be cautious with your emails. Any email offering *free money*, *vouchers*, or *prizes* from contests you never participated in is a perfect example of a **phishing** attack. And while it's pretty clear in those cases, problems arise with an email that looks completely normal. In *January 2023*, I got one that mimicked a **DHL** delivery parcel tracking message, as if I was expecting a package from an online store. As I use **DHL**, I was almost ready to click on the tracking link. However, hovering your mouse over a link (*hovering means going over it without clicking it!*) will show you the **URL** it leads to in the status bar of all modern **browsers** (if you haven't disabled the status bar). I saw a *strange link not from DHL behind the mimicked parcel tracking link*, so I checked my current packets, and I immediately knew this was a **scam/phishing** email. Any message that looks like it's from your bank but asks for your details is equally dangerous.

- *If working with sensitive information, transfer it only through an encrypted connection* – using **SSH**, a **VPN**, or **HTTPS**, but *never* **HTTP**. If a site that needs your sensitive information does not use **HTTPS** but simply **HTTP**, it is already highly suspicious, even if it is only because the web developer forgot to configure the proper settings. *Almost all widely used browsers* will depict a *locked padlock* (or similar symbol) just before the address of the web page you have loaded, informing you it uses **HTTPS**. Clicking on the icon will typically tell you whether the connection is secure.

### Securing your wireless router

Considering all the presented security advice, leaving your **router** with the standard/default **password**, or even worse, *without* a **password**, can be a severe problem. This means anonymous external users can directly log into your network and access all connected devices. In comparison, users must identify themselves when accessed externally (over the internet). This is only possible if **router** access from the **internet** is *enabled and configured*. Usually, *high-quality* **routers** provide at least basic **firewall** capabilities, which are *enabled by default*.

This leaves you with only a few things to consider nowadays. First, you have to *change the default* **administrator's username** and **password** to ones *known only to you* and, in addition, set up the access and security features of the router accordingly. Using well-known brands with *security features* is strongly recommended if you share sensitive information over your private network. This recent comparison (from *February 2023*) is an example of a few brands famous for higher security: `https://www.techradar.com/best/best-secure-router`. Follow the guides from each brand on how to configure them.

*The second thing you must do* is protect your **WiFi network** with a **strong password** (numbers, lowercase/uppercase letters, and one special symbol). This will guarantee nobody can easily hack it. The following link lists 20 wireless network hacking tools (updated as of *2021*, initially posted in 2016): `https://resources.infosecinstitute.com/topics/hacking/20-popular-wireless-hacking-tools-updated-for-2016/`. Here are 12 **WiFi** *hacking apps* for **Android** and **iPhone**: `https://techrrival.com/best-wifi-hacking-apps/`. If your **password** is *weak*, these tools will easily hack your network, except if your router has explicitly enabled additional **security features**. On most good routers, such are enabled by default. Stay calm – such tools are necessary to check network quality (for *penetration testing*), so it is great that we have them and they are publicly known.

In addition, all big brands constantly work on improving their routers. For personal needs, I had to buy a new router in *December 2022* and chose **Linksys**. After my explicit check in its admin panel, it immediately connected to **Linksys** servers and updated its firmware. Thus, *the third thing to do is to update your router SW once or twice per year*. If possible, *enable automatic updates*. Since these updates are *rare* and typically fast, they will not cause network interrupts. They are mainly for security and performance fixes, typically lasting less than five minutes.

## Firewall types

There are HW-, SW-, and cloud-based firewalls. In this section, we will cover **SW firewalls** configured *locally on our machine*. Again, this is only for curious people, so if you are not interested in the details, skip this section.

### Packet-filtering firewalls

**Packet filtering** means each packet's destination and origin IP address, type, port number, and network protocol will be checked *based on pre-configured* **rules**. This is the simplest type of **firewall** – *it will protect you from* the *most common attacks* and is easy to configure by default. The big downsides are that the **payload** (the useful data inside the packet) is not checked, the application layer data is not checked, and there are usually no logging or user authentication features. Finally, Access Control Lists are challenging to set up and manage.

### Circuit-level gateways

These **firewalls** work on the **OSI** session layer and check the **TCP** protocol *handshakes* between **clients** and **servers** in detail. This simple inspection is good to quickly deny time- or resource-consuming connections, but as it *lacks filtering*, it cannot be applied alone. It also often requires SW and network protocol tweaks.

### Stateful inspection firewalls

These are also called **dynamic packet-filtering firewalls**. They combine packet inspection with **TCP** handshake verification. They also have a table that monitors active and inactive connections, and any suspicious packets are filtered out directly upon arrival. Although they are much better than the previous two firewall types, they come with the cost of *more complex configuration* and *sometimes noticeably higher network latency*. This results in slightly lower internet access and speed performance from a user perspective. They are sometimes vulnerable to sophisticated TCP flooding attacks and cannot validate spoofed traffic sources.

### Proxy firewalls

These are one of the best types of **firewalls**, as they mask the client completely (i.e., your machine) and perform **deep packet inspection**, but they require a proxy server. Thus, they are frequently used as *a business solution*. They also lead to a latency increase and are pretty complicated to set up and manage.

### Next-Generation Firewalls (NGFWs)

The term comes simply from *combining all of the previous techniques with all the modern and latest options*. The underlying concept is good; however, it faces a few complications, starting with the price, slow deployment time, and a really complex configuration. These firewalls are necessary only for *expensive business* and state servers with sensitive information on them, which are targets for high-profile attackers.

## Setting up your firewall from A to Z

Now that we know all the basics, we will check out the two most common Linux distributions firewalls, including for Manjaro.

### ufw and firewall-cmd

**ufw** and **firewalld** are command-line interfaces for **iptables** configuration. As explained earlier, the firewall itself is the **Netfilter** package. Despite this, we call **ufw** and **firewalld** firewalls, as they are used for this purpose from a user's perspective.

Considering the great by-design Linux security, **ufw** is perfect as a **firewall** for a regular user. It includes a *hit-count feature* – that is, a **rate-limiting** configuration for brute-force attacks. **ufw** also offers a simple interface (compared to other tools), as its command syntax is short and clear by design. Its GUI frontend, **gufw**, offers *zone management* with different rules for each zone.

In comparison, **firewalld**, with its command-line frontend **firewall-cmd**, has multiple zones by default, which is great for frequent travelers' laptops. It doesn't provide a *hit-count-like feature*; you must manually configure it. Its command-line syntax is more *complicated* than **ufw**'s. The GUI addition to **firewalld** is called **firewall-config**.

Despite being different in arguments and handling, the main features of **ufw** and **firewalld** are similar from a practical point of view.

Although **iptables** can be used directly, this involves complicated manual configuration and is not expected from regular users. Manjaro **Xfce** and **GNOME** currently have **ufw** *preinstalled*, while **KDE Plasma** has *no firewall*.

An essential note – I have *never had a single case of viruses or any malware in over 10 years of Linux usage* (and on five different distributions). I also have never used a *firewall*, as if one is present, it is by default disabled. I'm one of millions of users with this experience. In contrast, as I am the person who supports my family's seven **Windows** machines, I have had to fix *malware* and *virus issues* hundreds of times in the last 20 years. In the last five years, this has been reduced to once or twice yearly (since my family has learned what not to do, and **Windows** security has improved significantly since **Windows 10**).

Despite this, for any **server**, local **network gateway**, or computer with *security-sensitive information*, it is *strongly advised* to have a **firewall**. As a simple example of the reason why, the moment I activated **ufw** on one of my machines and blocked the IP of another, **Nmap** and similar tools had no results when run from a second machine; even pinging was impossible. This guarantees security and privacy for a server, with the assurance that no random traffic will reduce the quality of its services.

In addition, using a **firewall** has one more benefit. With it, we can explicitly disable a particular SW's capability to send packets with data, no matter the reason.

My choice is **ufw**, as it is simpler; hence, I have prepared a complete guide for it. For **firewalld**, I have added a few links.

## Configuring ufw

**ufw** needs to be executed with `sudo`. The main options we will use are `enable`, `disable`, `allow`, `deny`, `status`, `list`, and `delete`. To start the firewall, execute the following:

```
$ sudo ufw enable
```

To active **ufw** on startup, execute the following:

```
$ sudo systemctl enable --now ufw.service
```

To disable it, use `disable` instead of `enable`.

Before going into the *configuration*, the simplest method to check whether a **firewall** *prevents a given application, service, or outbound connection from working* is to disable the **firewall**, test, enable, and test again.

For your convenience, an example configuration is available at the end of this section. As **ufw**'s commands need a specific order, the example provides it.

We will continue with the *commands* and their *syntax*. After each one, you will get either confirmation of whether the rule was successfully applied or error feedback if the syntax was wrong. Currently, **IPv6** is enabled by default, and for all **IPv6** applicable rules, you will get a second status after the **IPv4** one, like so:

```
Rules updated     # This is for all IPv4 rules
Rules updated (v6)    # This is for all IPv6 rules
```

You can use the `status` option to get a listing of your current configuration; add `verbose` for more details:

```
$ sudo ufw status verbose
```

Both man and related web articles use the terms **ingress filtering** and **egress filtering**. **Ingress filtering** refers to all rules for *incoming traffic*, while **egress filtering** refers to the ones for *outgoing traffic*. While it is clear why ingress filtering is used, *limiting outgoing traffic reduces the risk of potentially malicious SW communicating information from your machine to external third parties*. However, this needs to be done with care, as the complete denial of outgoing traffic will result in your PC and SW having no access to the internet.

Now, let's continue with a detailed explanation of the main `ufw` command options.

### Simple allow and deny

`allow`/`deny` will add a rule with *simplified* or *full* parameters. The *simplified* form only specifies *the port* and, optionally, the transport **protocol**. If *no transport protocol is specified*, **ufw** will apply rules for **TCP** and **UDP** but for no other. There will be brief information further in the subsections for the other supported and rarely needed protocols (**ah**, **esp**, **gre**, and **igmp**). Let's start with a few basic examples.

`$ sudo ufw allow 25/tcp`: adds a rule to explicitly allow connections via **port** 25. Not adding the protocol will allow **SMTP** (the standard protocol via port 25) over **TCP** and **UDP**, as listed for **port** 25 in `/etc/services`.

`$ sudo ufw allow https`: enables **HTTPS** on the standard **port** 443, as listed in `/etc/services` – via **TCP**, **UDP**, and **SCTP**. Adding a different port will cause an error.

`$ sudo ufw allow https/tcp`: enables **HTTPS** on port 443, as listed in `/etc/services`, but *only* via **TCP**.

`$ sudo ufw allow ssh`: allow **SSH** over **TCP/UDP**.

`$ sudo ufw allow ssh/tcp`: allows **SSH** *only over* **TCP**.

You can also use the short form for protocols and specify the direction:

```
$ sudo ufw allow out ftp
$ sudo ufw deny in http
```

With an **IP**, we work as follows:

$ `sudo ufw allow from 123.211.7.7`: explicitly allows all **TCP/UDP** traffic with this **IP**. Change `allow` to `deny` to block all traffic.

$ `sudo ufw allow from 123.211.7.7 to any port 25`: explicitly allows all **TCP/UDP** traffic with this **IP** on port 25. The `to any port` option tells ufw that *the user will provide* the **port** number to which the rule should be applied. Don't get confused – if you want to apply the rule to all ports, omit `to any port`.

$ `sudo ufw deny from 123.211.7.0/24`: explicitly denies all **TCP/UDP** *traffic* with **IPs** in this **subnet**. If you need more information on **subnet masks**, read the last part of the *Network basics* section in *Chapter 11*.

$ `sudo ufw deny from 123.211.7.0/24 to any port 8080`: explicitly denies all **TCP/UDP** traffic via port `8080` **IPs** for the given **subnet**.

## Longer forms of allow and deny

In a more detailed form, we can deny or allow specific **traffic** *to/from* a *dedicated network adapter*. One of the most frequent examples of this functionality is when we have one network adapter for *internet access* and *another for the local network*. Use $ `ip -br a` to get your network adapters' names (IDs). They will be provided as the first string on each line, like this:

```
$ ip -br a
lo      UNKNOWN 127.0.0.1/8 ::1/128
eno1    UP          10.51.8.166/22 fe80::6140:852b:fdea:3660/64
wlp3s0 UP        192.168.87.152/22 fe80::8ed8:84e3:6d3:d2b8/64
```

This example is the same as in the *Local network configuration basics* section of *Chapter 11*. We will use these identifiers in rules, as shown in the following examples:

- $ `sudo ufw allow in on wlp3s0 to any port 8080`: Here, `wlp3s0` is the name of my WiFi adapter; replace it with your identifier. This command will allow input network traffic via port `8080`, strictly for `wlp3s0`. Use `out` instead of `in` for output traffic.

- $ `sudo ufw deny in on wlp3s0 from 192.210.113.2`: Block all in network messages from this **IP address**, for `wlp3s0`. You can replace the IP with a **subnet mask** here.

The full syntax of `allow` can include the `proto` option and a transport protocol specification – `tcp`, `udp`, `ah`, `gre`, `esp`, `ipv6`, or `igmp`.

`tcp` and `udp` can optionally be paired with a **port** number, as we have already seen; if a port is not provided, it is checked in `/etc/services`.

`ah`, `esp`, and `gre` are valid *without a* **port** *number*. As explained in the previous section, `ah` and `esp` are part of **IPSec**, while `gre` was created by **Cisco** and is used with its **routers** and **network HW**. Usually, you will not work with them, but if you have to, use the following form:

```
$ sudo ufw allow proto esp
```

If an application has special network communication needs, its documentation shall specify the required **ports** and **protocols**.

As the man page explains, **ipv6** and **igmp** are valid for **IPv4** addresses and without port numbers. **IGMP** is used only over **IPv4** and is valid only when adding rules for an exact address, as follows:

```
$ sudo ufw deny proto igmp from 192.168.15.15
```

Here is another example:

```
$ sudo ufw deny in on eth0 to 192.168.15.15 proto igmp
```

This will deny all **IGMP** traffic to this address exactly and only via the `eth0` network adapter. We don't specify ports here; this is what the **man** page means by "*valid without port number*" for specific transport protocols different from **TCP** and **UDP**.

### Working explicitly with IPv6

To work with **IPv6**, it must be explicitly enabled in the ufw configuration (currently enabled *by default*). When **IPv6** is enabled, all applicable general rules will also be configured for it, and the *status* for *successfully accepted rules* will include an additional line for **IPv6**. If you need to check, execute `$ cat /etc/default/ufw`. At the beginning of the file, there will be a line `IPV6=yes`. Enabling and disabling **IPv6** is done via an editor, setting it to `no` or `yes`. Of course, to *disable* **IPv6**, you should have *a valid reason*. When disabled, the missing `(v6)` status will confirm that.

Blocking all **TCP** connections with an **IPv6** address can be done like this:

```
$ ufw deny proto tcp from 2001:db8:aa34:1823:12:bcd2:12:bcd2
```

### The limit rule and comments

**ufw** supports a *limiting rule* to limit the number of *login attempts* to 6 within 30 seconds. This blocks completely any standard **brute-force attack** methods. The typical usage is as follows:

```
$ sudo ufw limit ssh/tcp
```

As we saw in *Chapter 11*, using **SSH** on the standard port 22 *is not recommended*. If we have changed it (e.g., to port 53784), the `limit` option will be written like this:

```
$ sudo ufw limit 53784/tcp comment 'SSH port'
```

This example also presents how to set **comments**. The status output reports them.

### The default ufw policy and resetting

By default, if you enable **ufw**, it will be with policy as reported by status verbose:

```
$ sudo ufw status verbose
Status: active
Logging: on (low)
Default: deny (incoming), allow (outgoing), disabled (routed)
New profiles: skip
```

This means all *incoming connections* are not allowed. However, requests from your **browser** and *their* **responses** are *outgoing connections*. As your machine initiates them, such SW will work without problems.

routed refers to **routed packets**, used when your machine has two or more **network adapters**, and some network traffic shall be *directly re-routed from one to another*. This is usually done for PCs used as a local network **firewall** and **gateway**.

To change the *default* **ufw** *actions*, you must use the following format – $ sudo ufw default [policy] [direction]. For example, $ ufw default allow incoming is executed directly. The policy can be allow, deny, or reject. direction is as previously listed – incoming, outgoing, or routed.

Resetting is the easiest way to delete all newly added rules (switching back to default ones) and disabling the firewall in one step. To reset, execute $ sudo ufw reset.

### A routing rule and status example

For this, I will use a nice example from the man page. **Routing** means that incoming packets from one network card are directly transferred to another network card. In the man example, eth0 and eth1 are the network card IDs. The syntax is as follows:

```
$ sudo ufw route allow in on Card1 out on Card2 to AddressRangeCard2
from AddressRangeCard1
```

This translates to the following in practice:

```
$ sudo ufw route allow in on eth0 out on eth1 to 10.0.0.0/8 from
192.168.0.0/16
```

A ufw status report for this rule is based on the packets' direction, and the following statements are equal in status reports – Anywhere, any, 0.0.0.0/0 (**IPv4**), and ::/0 (**IPv6**).

The man page example shows how the following four commands will be reported from `status verbose` (considering it misses on purpose the `Rules updated` result messages). As you can see, the report is in a human-readable format, not in the commands' format:

```
$ sudo ufw allow in on eth0 from 192.168.0.0/16
$ sudo ufw allow out on eth1 to 10.0.0.0/8
$ sudo ufw route allow in on eth0 out on eth1 to 10.0.0.0/8 from
192.168.0.0/16
$ sudo ufw limit 2222/tcp comment 'SSH port'
$ sudo ufw status verbose
    To                     Action        From
    --                     ------        ----
    Anywhere on eth0       ALLOW         192.168.0.0/16
    10.0.0.0/8             ALLOW OUT     Anywhere on eth1
    10.0.0.0/8 on eth1     ALLOW FWD     192.168.0.0/16 on eth0
    Anywhere               LIMIT         Anywhere      # SSH port
```

## Order of rules and deletion

This section may look a bit complex, but it is *essential*. If you don't need to understand it in detail now, take a quick look and continue with the chapter. The penultimate subsection provides a default configuration sequence to make it easy for you.

When you add rules, they are inserted in the ufw configuration in the order of addition. To see this, execute `$ sudo ufw status numbered`. This will provide you with an exact identifier number for each rule. You can delete one by executing the following – `$ sudo ufw delete 7`, where 7 is the rule number. You will be asked for confirmation and provided a hint about the rule to be deleted. A proper error status will be printed if you write a non-existent number.

This *order is essential*. As the man page explains, "*Rule ordering is important and the first match wins. Therefore, when adding rules, add the more specific rules first with more general rules later.*"

The order is as follows:

1.    All rules *specific to* **IPs** shall be *first*.

2.    Rules for **subnets**.

3.    Rules for specific **high-level protocols**.

4.    Rules for certain **ports**.

5.    General rules for **TCP** or **UDP**.

6.    Rules valid *for both* **TCP** *and* **UDP**.

7.    For `igmp`, `gre`, `ah`, `esp`, or **IPv6**, add the rules at the end.

The last important rule is that all **IPv6** commands (if separately inserted by you) should be *at the end of the list*. Any generic port rule (such as disabling **port** 25 completely) will add the **IPv4** rule at the end of the **IPv4** rules; its automatic **IPv6** version will be added at the end of the second half of the list. This results in *two separate rules* – for example, for **port** 25. *Deleting one of the rules will not delete the other*, as it is a separate rule number. In this case, either manually delete the second rule by its number or use the direct rule deletion form:

```
$ sudo ufw delete deny 25
```

This form will *delete both rules*.

Let's look at an example related to **ordering**. If you *allow a whole* **subnet** (a more *general* rule) but then add a rule *to deny one IP address from this* **subnet** (a more *specific* rule), the *first one will be applied*, and then the *second* will be *discarded*, as the first range *already includes* the given **IP**.

There are *three* **workarounds** for this. *The first* is the insert 1 command, which will put a given rule on top of the list (*although it currently works only for* **IPv4** *rules!*):

```
$ sudo ufw insert 1 deny from 123.211.7.7
```

*The second option* is deleting more rules and *inserting them again in the correct order* (at the end or on top of the list).

*The third option* is having all your commands *in a shell script* to ease your work. Then, if something happens, you reset **ufw**, edit the script if necessary, and *rerun it with a single command*. We will look at shell scripts in *Chapter 15*.

### Logging

**Logging** is excellent for *monitoring connection attempts* and network traffic based on your rules. It is reviewed in detail in Chapter 13, while here we will briefly explain how to use journalctl particularly for ufw logs. It is usually applicable to **servers** and machines exposed to higher security risks. To get **logs**, you have first to enable them via:

```
$ sudo ufw logging on
```

The possible levels of **logging** are low, medium, high, and full, and the events they log are described in *Table 12.4*:

| Level | Logged events |
| --- | --- |
| low | Logs all packets blocked or allowed by defined policies |
| medium | This level adds the packets that don't match any policies |
| high | The medium log level without rate limiting, plus all packets with rate limiting |
| full | Complete logging without any rate limiting at all |

Table 12.4 – The ufw log levels

Any level above `medium` is not recommended by the **ufw man** page, as it may fill up your disks with too much information. The **man** page also states that even `medium` may generate too many logs on a *high workload* system (such as a **gateway** *server* or *main* private network **firewall**). This is why *the default level* is `low`. Changing the level is done via this command:

```
$ sudo ufw logging medium
```

In addition, you can activate per-rule logging, as by default, no logging is performed when a **packet** *matches* a **rule**. To log each packet matching a rule, add `log` after the `allow` or `deny` **policy** and the `in` or `out` **direction**, like this:

```
$ sudo ufw allow out log https
```

To enable full packet logging (i.e., *all* packet transfers based on a particular rule), use the `log-all` parameter:

```
$ sudo ufw allow out log-all https
```

In the case of using the limit rule, see the following:

```
$ sudo ufw limit log-all 53786/tcp comment 'SSH port'
```

For a dedicated network adapter, write it after specifying the interface:

```
$ sudo ufw deny in on enp0s3 log-all
```

In both cases, if you do this for *an existing rule*, it will *be updated*, relieving you from the burden of bothering about *the order*. In other words, *if you need to log explicitly for a particular rule* a few months after adding it, use the same command. Extract it from your text file or script with default configuration commands, and add `log` or `log-all` after the `allow`, `deny`, `in`, `out`, and optional `network adapter ID` keywords.

It is strongly advised to keep logging permanently enabled only at `low` level. `medium` is acceptable only if you have no severe network traffic. Depending on the network load, any **firewall/gateway** server must consider enough HDD space for bigger logs. Usually, the higher logging levels are enabled temporarily to check specific traffic.

Once you have enabled **logging**, there are two ways to read the logs – `journalctl` and `rsyslog`. The former is the **systemd** logging tool; Manjaro (like most major Linux distributions) is based on **systemd**, and *Chapter 13* discusses it in detail.

On Manjaro, currently, **rsyslog** *doesn't work for a combination of reasons* (related to **systemd**, **AppArmor**, **rsyslog** configuration, etc.). If you happen to come across articles related to rsyslog, I will post basic instructions at the end of this section, but for **Manjaro**, *please work with* `journalctl`. It is *unclear* whether the **rsyslog** issues will be solved or `journalctl` will remain the only log access tool.

## Log access via journalctl

As part of **systemd's logging service**, journalctl is the utility to *query* and *display* journald logs, both reviewed in detail in *Chapter 13*. **journald** stores log data in *binary format*, and journalctl *reads it*. Once you enable **ufw logging** (and if there are events to be or that were logged), you can check for logs by executing the following:

```
$ journalctl | grep --color -i ufw
```

As journalctl expects q to quit, when piped with grep, you must use *Ctrl+C* to exit the command. The lines that are of interest are long and look as shown next. For an *incoming* packet, I see on one line on my big monitor the following:

```
Feb 03 16:44:51 luke-HP440 kernel: [UFW BLOCK] IN=enp0s3 OUT= MAC=33:
33:00:00:00:01:88:d7:f6:8e:c1:9a:86:dd SRC=fe80:0000:0000:0000:5371:4
614:2f11:24e7 DST=ff02:0000:0000:0000:0000:0000:0000:0001 LEN=49 TC=0
HOPLIMIT=1 FLOWLBL=579081 PROTO=UDP SPT=62490 DPT=1434 LEN=9
```

Here's another one for an *outgoing* packet:

```
Feb 07 17:34:37 luke-HP440 kernel: [UFW AUDIT] IN= OUT=enp0s3
SRC=10.51.8.142 DST=8.8.8.8 LEN=62 TOS=0x00 PREC=0x00 TTL=64 ID=61200
DF PROTO=UDP SPT=43884 DPT=53 LEN=42
```

In these two reports, we can see *the date* and the machine on which the event was logged. Further, we have the **ufw** *status*, the transmission *direction* (IN or OUT), and the source (SRC) and destination (DST) **IPs** (**IPv4** or **IPv6**). From the other reported parameters, the protocol (PROTO), source port (SPT), and destination port (DPT) are always important. This is a good explanation of the **ufw** logs: https://askubuntu.com/questions/1116145/understanding-ufw-log.

The former simple journalctl extract produced a few hundred lines for me. Depending on the *number of rules* and *logged packets*, you might need a second piped grep at the end of this command.

The reason for the report length is that it goes back far. If you enabled **ufw** today and looked into the logs *after a month has passed*, it might be hard to find the proper information. For this purpose, journalctl has for date limitation the short -S and -U options (the long options are --since and --until). The following are a few examples:

```
$ journalctl -S "2012-10-30 18:17:16" | grep --color  -i ufw
```

Here, -S refers to since one particular date and time. If the time is not specified, 00:00 is assumed; if only the seconds are omitted, again, 00 is assumed:

- $ journalctl -S "Feb 07 18:23:01" | grep --color -i ufw – since Feb 7 this year
- $ journalctl -S "today" | grep --color -i ufw – yesterday and today are understood as well
- $ journalctl -S 08:00 | grep --color -i ufw – since 8 A.M. this morning

In the same way, you can add -U or --until, as shown here:

```
$ journalctl -S 08:00 -U 16:25 | grep --color -i ufw
$ journalctl -S "2023-02-01 08:00" -U "2023-02-05 16:25" | grep
--color -i ufw
```

-S and -U come from the time format specification of **systemd**, which you can check with $ man systemd.time. Refer to *Chapter 13* for more information.

### Log access via rsyslog

In the past, **ufw logging** needed **rsyslog** to work; for Manjaro, it is only available as an AUR package. Thus, you must have AUR enabled in Pamac's third-party packages, and then you can install it via $ pamac install rsyslog. After this, you can enable it via $ sudo systemctl enable --now rsyslog.service. You may also have to edit the **journald** configuration with $ sudo nano /etc/systemd/journald.conf – in it, find the #ForwardToSysLog=no line, remove the comment *hash sign*, and change the value to yes. Then, you should be able to find the logs in the /var/log/ufw path.

Unfortunately, this failed for me multiple times on KDE and Xfce, potentially because (at least since December 2022) there is an issue with a **rsyslog** *dependency*, and it *fails to activate* via **systemctl**. I did my best to find a solution, but as of March 2023, there was still none. The latest posts about the issue are from *January 2023*: https://groups.google.com/g/linux.debian.user/c/BNxAA2I1N_g?pli=1. As explained earlier, I have kept this rsyslog section, as even if these issues need time, they will either be resolved or announced as no longer supported. Strangely, after disabling **AppArmor** on **Xfce** the next day, the **rsyslog service** started successfully. It also worked on **KDE**, but in both cases, **ufw** logs were absent in the /var/log/ufw directory.

### The app list

**ufw** has an application profiles list, holding information for **ports** and **protocols** necessary for each application to operate. To list all currently available applications, execute $ sudo ufw app list.

**NFS** and **SSH** (which we reviewed in *Chapter 11*) are easy to recognize from the default ones. Executing `$ sudo ufw allow SSH` will allow the *corresponding standard port*. Unfortunately, this is again related to the explained *rules ordering*; if it is incorrect, you will get issues. Most existing predefined app profiles are helpful only for specific work, so I recommend working directly with corresponding ports instead.

### Editing configuration and advanced functionality

All **ufw** configurations are located in the `/etc/ufw/` directory. It's important to mention here that, as the official documentation explains, anything that **iptables** can do, **ufw** can be configured for. For more information, refer to the following links: `https://wiki.ubuntu.com/UncomplicatedFirewall`, `https://serverspace.io/support/help/osnovnye-komandy-ufw`, and, in particular for the config files, `https://manpages.ubuntu.com/manpages/bionic/man8/ufw-framework.8.html`. The second link even mentions **rsync**, which, to *work remotely*, needs **port** `873` open. Of course, you can search the internet for any more complicated cases, as **ufw** is widely used and, thus, covered in forums and web articles.

### A standard, good default configuration

After all these commands, you may ask, can you please give us an easy configuration for beginners? Yes, and here it is. It is tested (as is each command in this book). Consider the optional rules; if you don't need them, skip them while retaining the order:

- `$ sudo ufw default deny incoming`

- `$ sudo ufw allow from 91.198.174.33 to any port 873`: *Optional*, for remote **rsync** *connections*; add it only if needed, and change the **IP** to your **rsync** web server's **IP**

- `$ sudo ufw allow from 127.0.0.1 to 127.0.0.1 port 80 proto tcp`: *Optional*, explicitly allows `localhost` connections; it is frequently necessary to test web services on your PC.

- `$ sudo ufw allow in on eth0 from 192.168.0.0/16`: Optional, applicable if you run a network service to a specific subnet.

- `$ sudo ufw allow out http`: Enables **HTTP** in general.

- `$ sudo ufw allow out https`: Enables **HTTPS** in general.

- `$ sudo ufw allow from 91.198.174.33 proto tcp to any port 22`: Optional, used to allow SSH access on the standard **port** (`22`) from a specific **IP** address (not recommended; see the following command instead).

- `$ sudo ufw limit 53784/tcp comment 'SSH port'`: Enables **SSH** on a particular **port**; you can use any random one between `49152` and `65535`. If you have any other SW that needs input/output via *specific* **ports/protocols**, add them here.

- `$ sudo ufw deny in on eth0`: Now that we have the more specific rules applied, we want to *disable any other incoming connections*. Use `$ ip -br a` to get your network card identifier and replace `eth0`.

- `$ sudo systemctl enable --now ufw.service`: Enables the **ufw service**, including on **startup**.

- `$ sudo ufw enable`: Enables **ufw** in general.

- `$ sudo ufw status verbose`: Reads the status in general to see the default rules.

- `$ sudo ufw status numbered`: Prints *all commands* in a numbered list.

After applying your configuration, keep the commands *in a text file*. You will see in *Chapter 15* how to put them in a shell script so that you can execute them all at once.

Then, you must test any network/internet-related SW function (e.g., **Spotify**, **Zoom**, etc.). The internet is full of information on most **SW**. For **Spotify**, I read that by default, it works via port 4070; if it cannot use it, it switches to 443; if that fails as well, it will switch to port 80. As listing all possible internet-related SW ports here is impossible, every user should check for themselves. **Spotify** works fine with the former configuration.

The presented configuration *may result in some serious limitations*. For any peculiar issues, if you suspect **ufw**, first test disabling *it completely*, and if the given **SW** works, search the internet to see which **port** is required for it. If you cannot find information, read the **logs** and see which packets were blocked in the last few minutes. Enable exactly these ports with `insert 1`, and test again. Once you know the correct configuration, modify your commands following the ordering rules, `reset` **ufw**, and set the configuration again.

If, even then, the given **SW** doesn't work, delete the last two general `deny` rules and try again. However, in general, this is hardly possible to happen. If it still does, and you are sure you already know all the **ufw** basic configurations and have checked every other possible solution, ask for help in the Manjaro or other forums.

### The GUI frontend (gufw)

**gufw** is friendly and powerful for certain functions. However, compared to the simplicity of having a *single configuration script*, especially if you have some complex configuration to set, clicking all the options on a **GUI** is not ideal. Conversely, *it can be a great addition* to the basic **ufw** configuration. Let's see why:

- It supports **profiles**, has *three default ones* (home, office, and public), and you can add others. Switching them is easy and the changes are instantly applied.

- It supports adding *advanced rules per profile*, relieving you from bothering with complex command syntax.

- It *shows a current zone rules listing* in a comprehensive report.

- When adding a specific network adapter rule, you *get its network adapter identifier*.

- When adding a rule, you can add it *to whichever position in the current profile rules list*. However, be careful, as this may result in a different than expected order for specific reasons.

- You can control logging and quickly check the firewall with a single click.

- It supports *importing* and *exporting* **profiles** to transfer them between machines or installations.

- A report section lists the *applications* using the network, specifying the **protocols** and **ports**.

- It has *over 100 pre-configured profiles* for **applications** and **protocols**, available in a simplified rule-addition dialog. Use the *PgUp* and *PgDn* keyboard buttons to scroll its long list.

Conversely, there are a few drawbacks to consider:

- To work with it, you still need to know the networking basics from *Chapter 11* and the **ufw** basics from this section.

- If a rule is added by **ufw**, **gufw** can delete it but can't edit it.

- The **gufw** log is not related to the `journalctl` log. If you need to investigate logs, you still must use `journalctl`, as shown earlier.

If you use Manjaro on a **laptop** or a machine that needs *multiple profiles* (e.g., a **server**), **gufw** can come in handy. Refer to the following guide if you consider using **gufw**; it is one of the most recent and detailed ones: `https://itsfoss.com/set-up-firewall-gufw/`. Also, consider getting directly hands-on with it, as the **GUI** is straightforward.

## Configuring firewalld

You already know that the general functions of **firewall-cmd** and **firewalld** are the same as those of **ufw**. I will provide a few links for it here, as this is another frequently used firewall on many Linux distributions.

The official website for **firewalld** is `https://firewalld.org/`. It includes a nice *manual page with examples*. In the following link, we have a limited beginner's guide from **RedHat**: `https://www.redhat.com/sysadmin/beginners-guide-firewalld`. However, I found *the most detailed list with great examples*, divided by usage type, at `https://www.golinuxcloud.com/firewalld-cheat-sheet/`. Check these links out; they should be enough for any regular or specialized usage of **firewalld**.

# VPNs

In the context of all the information on hacking attacks and protocols, **VPN** is a technology to create a **secure, encrypted** connection *over the internet* between *a computer* or device and *a remote network*. This connection allows users to *access resources on the remote network* as they are directly connected to it, while also providing complete privacy and security. Any exchanged data *tracking, spoofing,* or *hacking* is completely mitigated thanks to the *encryption*.

What do we need for a **VPN**? First, at least one **VPN server** with both HW and SW firewalls. Second, *a SW client application*, which, when *loaded* and *connected* to the server, will pass *all internet and network requests and communication via the encrypted connection with the server.* An example is provided in *Figure 12.3.*

Suppose I reside in the UK but connect to a server in Germany. In that case, any local hacking activities will be next to impossible, as the potential attacker will only see the encrypted packets exchanged with the VPN server. Since it usually relays the traffic of thousands of customers, a potential attacker cannot easily track the packets of one particular user. Hacking the VPN server itself would not be possible thanks to the firewalls. When we add to this the fact that nowadays all internet services such as Gmail and YouTube connect via HTTPS, this traffic is impossible to hack or trace.

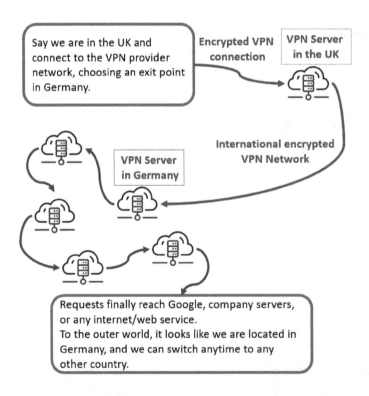

Figure 12.3 – Tunneling an internet connection via an encrypted VPN provider connection

As VPN networks are not easy to create, *several tens of providers* specialize in this. To mention a few, we have **Surfshark**, **ExpressVPN**, PureVPN, NordVPN, IPVanish, AtlasVPN, CyberGhost, and TunnelBear. All of them provide connection points in *tens of countries on all continents*. Only two VPN providers have clients for Manjaro – **Surfshark** and **ExpressVPN**. Fortunately, *both are considered the two best providers*. Any other provider will require downloading and configuring an AUR package (if there is such a package at all). The **Surfshark** client is available from the official Manjaro repositories, while for **ExpressVPN**, you have to download and install it from their website.

**VPN services** are available not only for PCs but also for *smartphones*, *tablets*, other *smart devices*, and practically *any OS*.

## The benefits for a single home user

Using a **VPN provider** has the following main benefits:

- We can bypass *internet censorship or geographic restrictions* to access content that may be **blocked** in certain *countries or regions*. This means that if certain content, movies, or streaming is available only for users from a particular country, you can connect via the VPN server of your provider within this country and access it.

- We can securely connect to public *WiFi networks* that may be vulnerable to **hacking** or **snooping**. This means that if I'm at the airport and connect via public WiFi, if I use a VPN, no attacker can trace my packets and internet communication.

- The last but equally important point is that *via VPNs, many political dissidents* have *secured untraceable access to internet services* to upload or download content without restrictions in their countries. As expected, this feature has a severe *downside* – many criminals use these connections to *perform illegal actions without being caught*. Many governmental and police network activities are also performed via VPNs.

## For a business use case

Let's imagine a big company in the center of London has 30 offices worldwide that must be connected virtually to the same network. We also want this connection to be private, not traceable from the public internet space. Then, we connect all devices in each office via a **VPN**. We connect the internet gateways of all offices via VPN servers, and all the communication goes through them. Every user has access to different resources inside the network, such as local or remote data servers, chat services, connection to client machines, and others. Usually, such companies have VPN servers on all continents with offices, for higher speed and network load distribution.

The solution from the previous example also allows us to apply **networking rules** for all users *in all locations* when they connect to the **VPN network**. This makes user management *centralized* and *easier*, compared to the alternative of setting up 30 different offices' internal networking rules.

For the last 14 years, I have used VPN networks with all of my customers when they share any internal company data with my teams, often including *confidential information* on prototypes and current developments. Several times, I even worked from airports, connecting securely and privately to the companies' networks.

## Countries in which using a VPN is illegal

Using personal VPN service providers such as Surfshark, ExpressVPN, TunnelBear, and others is forbidden or limited in a few countries. According to this article at `https://thebestvpn.com/are-vpns-legal-banned-countries/`, in Belarus, Iraq, North Korea, and Turkmenistan, these services are *banned*, while in China, Iran, Russia, Turkey, Oman, and the UAE, only government-controlled VPN services can be used. Some countries even ban proxy servers. There is some more recent information from **NordVPN** at `https://nordvpn.com/blog/are-vpns-legal/`, which adds Egypt and Uganda to the list. There are differences and periodic changes in regulations. If you reside in any of the mentioned countries, check the current situation; otherwise, you may face bans, fines, or even jail. In the rest of the world, using VPNs and proxy servers is, as expected, *legal*. The rules may differ for the business use case, so again, check the current regulations state if you have business with the mentioned countries.

## What are the potential drawbacks of using a VPN connection?

Due to *encryption*, *protocols* with additional *security overhead*, and traffic routing via dedicated servers, VPN connections usually are *at least a bit slower*. All regular user services, such as the ones provided by **Surfshark**, **ExpressVPN**, **TunnelBear**, and others, don't experience serious issues, as each provider has built a *worldwide server network* for the best distribution of traffic and high workloads. However, they cannot avoid the issues resulting from a user trying to access a server overseas or their email provider via a location 10,000 kilometers away. So, unless you need to avoid *country-specific content limitations*, choosing VPN servers *near your current location* is the best option. I currently reside in Bulgaria, and as we have direct internet optical highways to Germany, I have often used servers located there.

The company business example is a bit more complicated. This depends on the company *network infrastructure*, the allocated *network traffic capacity* provided by the company's ISPs, and the current *company servers and network load*. As a result, their speeds are often significantly lower than the local internet speed.

## Measuring VPN connection quality

Network connection quality (including VPN connections) includes several components, **speed** being *the most important*. We also have *latency, packet loss, bandwidth*, and others; however, *we can never reach high speeds if we have severe packet loss or bandwidth limitations*. To measure the speed of our internet connection, we can use tools such as `https://www.speedtest.net/`, which measures the speed to a distant network server. By default, the server is close to your location. You can also set it to test connections to *worldwide servers* and use it on many smart devices. It will measure both the download and upload speeds.

Usually, when you connect via a VPN provider, *any speed test results* will be *at least slightly lower*. You might experience issues with HD content streaming. This depends on multiple factors, and besides trying another VPN provider, there is rarely anything you can do. However, especially when accessing high-quality online video streams (such as an HD 1080p clip on YouTube), if you experience any delays or issues, dropping to a lower quality (such as a 720p resolution or less) might solve the problem.

In the business case, one of my last customers (a big international company with over 50 offices worldwide) had at least five VPN servers on each continent to connect to. Considering that they have over 25,000 employees, this kind of solution is expected.

## Is it possible to hack a VPN connection or a PC connected to it?

Although a VPN *immensely decreases the possibility of tracking or hacking your OS*, there are expert tools that, when used by *high-profile attackers*, make it *hypothetically* possible. However, if you *add a firewall* and a good browser (e.g., Brave, Opera, or Vivaldi), considering that we are using Manjaro, it is *practically impossible*. The most that a regular tracker can do is to detect that you are located somewhere that is not your actual location. After all, that is your purpose when using a VPN. Hacking a VPN connection requires expert tools and very sophisticated HW and SW, making the task both extremely difficult and expensive. Finally, specialists who can do this are scarce and highly paid. That's why VPN usage is so widely spread.

To make this hypothetical situation even harder, **Surfshark** offers a *MultiHop option*, which routes your packets via two VPN servers, and **ExpressVPN** also offers some traffic distribution. These techniques make any network tracking *impossible*.

There is only one exception to this. In the case of **Surfshark,** when it is activated with the default settings, local secure connections are not forbidden. Thus, connecting from another local PC via **SSH** is possible. Activating the **Kill Switch** option completely disables local traffic, including `ping` and **SSH**.

## Considerations when choosing your VPN provider

The following is a *key considerations list*, with a focus on **Surfshark** and **ExpressVPN** for some of them. Both are considered the best VPN providers and *cover all bullet points*:

- **Server locations**: A VPN provider with *servers in multiple locations* can help you bypass geo-restrictions and access content that may be blocked in your region. **Surfshark** offers 3,200 servers in 100 countries, and this number is growing. **ExpressVPN** has over 2,000 servers in 160 locations and 94 countries.

- **Logging policies**: Choosing a VPN provider with a *strict no-logs policy* is critical. This means they do not collect or store any data about your online activities.

- **Encryption standards**: Ensure the VPN provider uses *strong encryption standards*, such as **AES-256**, to protect your online activities from hackers and cybercriminals.

- **Connection speeds**: Choose a VPN provider that offers *fast connection speeds*, especially if you plan to *stream videos or play online games*. Both the mentioned providers are famous for excellent connection quality. However, if you intend to use the VPN for certain dedicated network-related streaming services, test it first. Some differences in the VPN providers' server locations and local **Internet Service Providers** (**ISPs**) may result in quality differences.

- **Privacy protections**: Check that the VPN provider has *strong privacy protections*, such as *a kill switch* that automatically disconnects you from the internet if the VPN connection is lost.

- **Cost**: Consider your budget and the needed features when selecting a provider. Both mentioned providers have the same monthly price (as of February 2023, 12.95 USD per month). In addition, both provide *extended plans with lower costs*. However, **Surfshark** offers a 60% cheaper plan for *a yearly subscription* (a total of 47.88 EUR for the first year) and even a longer plan of 59.76 EUR for 26 months. Both providers offer a 30-day money-back guarantee for their long-term plans.

- **Customer support**: Look for a VPN provider offering *reliable customer support*, such as live chat or email support, should you need help. Again, both Surfshark and ExpressVPN provide 24/7 support.

- **Simultaneous connections**: Check the ability to *connect multiple devices simultaneously* (your laptop, PC, mobile device, and some others). **ExpressVPN** supports a maximum of five *simultaneous VPN-connected devices*, while **Surfshark** offers unlimited *simultaneously connected devices*.

- **Ease of use**: Choose a VPN provider with an easy-to-use interface and client SW compatible with your devices and their operating systems.

**Surfshark** is based on the open source **WireGuard** protocol. **ExpressVPN** uses its **Lightway** protocol, which is also open source and was developed for the same reason as **WireGuard**. Interestingly, **ExpressVPN** supports the unique service of direct installation on some routers. One thing is evident from all possible reviews and comparisons – *both companies have been at the cutting edge of the best VPN services for years*. I have used both, and they are equally great. Still, more comparisons consider **Surfshark** to be *better and significantly cheaper in the long term*. Unless you need to install the VPN on a router (when you must use **ExpressVPN** and see the supported router brands), **Surfshark** is the better option. Despite this, I will also provide brief instructions for **ExpressVPN**.

*The official Manjaro team recommendation is* **Surfshark**, and it offers a **GUI**, whereas **ExpressVPN** has to be configured via the Terminal. Conversely, if you explicitly need Terminal interaction for **Surfshark**, you will have to install the `surfshark-vpn` AUR package and work with it. I haven't tested it, as the GUI application works perfectly, offers OS startup loading, and has an automatic connection option.

## Installing Surfshark and working with it

First, open `https://order.surfshark.com/`, choose a subscription, register, and pay. Then, open the **Pamac GUI**, search for `Surfshark`, and install the official package (a **GUI** module). Restarting your PC is recommended. Then, open your main menu, type `Surfshark`, and start it. You will have to enter your credentials (email and password), and then you will be presented with a list of servers on the left.

Load an *IP geolocation* site such as `https://www.whatismyip.com/` in your browser, and check your current IP. Don't close the page. Load `www.speedtest.net` in another tab, and check your current speed results on the default server.

Continue in the Surfshark GUI, where you have several options to connect. The first is the **Fastest location** option, which will guarantee the *highest possible speed*, if possible, via a server in your country. The fastest connection option is appropriate only if you want *simple security and privacy* but don't care about whether you are detected in the country you are currently in.

The **Nearest country** option will connect you to a different country closest to you from an internet connection perspective.

The other general option is *to choose a server location from the* **Locations** list and click on it. If you are already connected, *clicking directly on the other location* (or connection option) will automatically disconnect you from the current one and connect you to the new one. The **Surfshark** application is convenient and *flawless*, and I have tested it on all three official Manjaro flavors.

There are also the **Static IP** and **MultiHop** options. The first one allows you to connect to *the same* **IP** today, tomorrow, and whenever you connect. This can be useful if you always need your world-visible **IP** address to be the same.

The **MultiHop** option is a magnificent security addition, as it makes tracking your actions and packets even harder, since it uses two VPN servers for you to connect to. Sometimes, this potentially increases the speed as well.

The first time you connect, you can open `www.whatismyip.com` in a new tab to check for yourself that your **IP** is now different from the original one. Don't check this each time; **Surfshark** reports it anyway. A speed test check is worth it if you need a fast connection or more bandwidth (e.g., to stream HD clips).

Unless you need the highest possible security, using **MultiHop** is not necessary. **Static IP** is needed only for certain special occasions (and you are expected to know that you need it). Usually, you will need the fastest location, and *only if you need country-specific access* should you choose *a dedicated country*. The Surfshark GUI is shown in *Figure 12.4*:

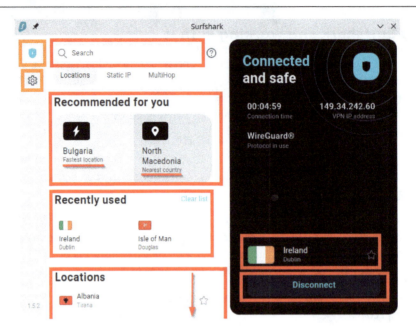

Figure 12.4 – The Surfshark GUI

On the left, there are two icons – the upper one, which looks like *a shield*, is for the shown VPN view. The status and the **Connect/Disconnect** button are marked on the right with red rectangles. The locations list is long; scroll down to see all the available countries.

When you click on the settings wheel (marked with an orange rectangle in *Figure 12.4*), you will see settings in three groups – **VPN**, **Application**, and **Account**. In the **VPN** group, there is a setting for the **Quick-connect** option, an **Auto-connect** option, the **CleanWeb** add and malware blocker, and **Kill Switch**. The Auto-connect option ensures that if you need a VPN constantly, it will always *automatically connect*. **Kill Switch** ensures that if the VPN connection drops, your regular internet access will be blocked until the VPN is restored. There is also a protocol choice option, but leaving it as **Automatic** is good enough, as the application will use **WireGuard** by default.

It is essential that when **Surfshark** is activated with the default settings, local secure connections are not forbidden. Thus, connecting from another local PC via **SSH** is possible. Activating the **Kill Switch** option completely blocks any local traffic, including `ping` and **SSH**.

In the application settings is the critical **Launch at startup** option; activate it if you need a permanent and constantly active VPN for your machine.

From my experience over the last 10 years, the **Surfshark GUI** is the best VPN I have used.

## Installing ExpressVPN and working with it

**ExpressVPN** typically doesn't offer a **GUI** for Arch-based distributions, although there are two unofficial ones in the **AUR**. I would *not use an unofficial one* for any *security-critical actions*. Here are the steps and commands to use **ExpressVPN**:

1.  Open `https://www.expressvpn.com/order`, choose a plan, activate it, and make the payment. The 30-day money-back guarantee is excellent if you want to test it. Once you pay, you will get your activation code.

2.  Go to the download page at `https://www.expressvpn.com/latest/1`. Choose **Arch 64-bit** from the drop-down list, as Manjaro is Arch-based. Download the package.

3.  Open the download folder, double-click, and install it. Reboot your machine – it is not required but still recommended.

4.  Open your browser, load an IP geolocation site such as `https://www.whatismyip.com/`, and check your current public IP. Don't close the page. Load `www.speedtest.net` in another tab and check your speed results; the test will take at least 30 seconds. Don't close this page either; keep the results in the browser.

5.  After you connect to the VPN service, load both pages in another tab again to check your changed IP and the speed result.

6.  Open the terminal and execute the following:

    ```
    $ sudo systemctl enable --now expressvpn
    ```

7.  Then, you have to enter your activation code, which is done via the following:

    ```
    $ expressvpn activate
    ```

8.  This command will prompt you to paste or write the activation code. Don't get confused that nothing is printed on the terminal when you do it; it is hidden for security reasons. So, paste it (*Ctrl+Shift+V* for KDE's Konsole and most Terminals), and press *Enter*.

9.  To list some possible connections, execute the following:

    ```
    $ expressvpn list
    ```

    Add `all` after the `list` parameter to see all possible connections. Choose one (in my case, I chose `uklo`, which is UK, London) and type this:

    ```
    $ expressvpn connect uklo
    ```

10. Now, recheck the `www.whatismyip.com` and `www.speedtest.net` results to see the difference.

11. To disconnect, type `$ expressvpn disconnect`.

12. To see all possible options, use `$ expressvpn -h`.

## What happens with my local IP address reports?

As I now have **Surfshark** enabled, if I go to www.whatismyipaddress.com, it will provide a random port in Germany (87.120.132.12), as I connected through a German server. In the meantime, $ ip -br a, reports that I still have my local IP address, 10.51.8.142. How is this possible?

This is so because any local traffic goes through my local network adapter with the same **IP** as before – having a VPN will not change your local internal network **IP**. However, the packets exchanged between *my PC* and *Surfshark's VPN servers* are *encrypted*, and no one can tell what information is transferred. They are further *tunneled via the VPN server*, the **IP** is *replaced*, and the request to servers, such as www.youtube.com, is sent from the *VPN server's address*, keeping my **IP** *hidden*.

$ ip -br a, also tells me that I have two more network adapters:

```
lo       UNKNOWN 127.0.0.1/8 ::1/128
enp0s3   UP  10.51.8.142/22 fe80::dcf3:e077:8206:24fc/64

surfshark_ipv6 UNKNOWN       fdbe:2bde:b490:9e47::/64
fe80::ce51:9d5:c000:ed23/64

surfshark_wg    UNKNOWN       10.14.0.2/16
```

None of them has anything to do with my German IP. This is so because the VPN service hides this information.

If you remember, your local intranet **IP** *is not your public* **IP**. You have an ISP that assigns to your router some *public IP*, and then $ ip -br a reports your *internal local network address*.

The two Surfshark IPs are necessary to identify your machine in Surfshark's internal VPN network. They are not exposed by any means to the internet. In addition, one is for IPv6 communication, while the other is additional for the **WireGuard** protocol.

## Setting up your own VPN connection

If you want to set up a *VPN connection* other users can join, there are several FOSS VPN server solutions that you can use. These solutions typically require technical knowledge and setup time, but they can provide you and your users with a secure private connection. In this case, **OpenVPN** and **WireGuard** are the newest and most recommended toolsets, available for direct download from the official Manjaro repositories. They offer both server and client applications. There are others, but I will not go deeper into the subject in this book. Here are two articles that are a good starting point: https://wiki.archlinux.org/title/OpenVPN and https://wiki.archlinux.org/title/WireGuard.

# Summary

In this chapter, we covered in detail the internet basics and network security by explaining OSI and TCP/IP models, network protocol usage, filtering, and attacks. We then reviewed the general advice to increase our security and the various firewall types. Then, we took a detailed look at ufw as a user-friendly firewall, widespread on many Linux distributions. Finally, we took a deep dive into VPNs and how to work with the Surfshark and ExpressVPN clients.

In the following chapter, we will continue with the main services manager on most Linux distributions nowadays – systemd. We will learn about its pros and cons, why it was chosen, and what we can do with it. We will also cover `journalctl` and daemons. We will continue with user management, as it is one of the greatest strengths of Linux by design. We will conclude the chapter by looking at groups, privileges, ownership, and superuser privileges, all inseparable elements of user management.

If you find the content so far valuable, I will be eternally grateful to you for a short book review on the platform you purchased it from. With this, you will help people interested in learning more about Manjaro and Linux.

# Part 4:
# Advanced Topics

The last part is dedicated to tasks that may be called advanced but can also be performed by intermediate users. A special case is troubleshooting, which is more of an intermediate task, but as it is logically related to system maintenance, I've put it in *Chapter 14*. So, don't consider the topics here hard. Instead, as they are again structured and with examples, keep on reading. It is exciting.

We start here with processes and daemons, two of the main building blocks of any computer system. We then cover services and systemd as, for Manjaro and most current desktop distributions, this is how we manage tens of functionalities. Our next stop will be system logs – the best place to learn what happened on our system. *Chapter 13* ends with an explanation of user and group management, ownership, and root privileges.

Manjaro is a high-performing and optimized distribution with excellent efficiency. Even still, maintenance is a topic to consider for specific cases, all reviewed in *Chapter 14*. We will cover filesystem cleanup and then continue with troubleshooting, both explained with examples and tools. It ends with reinstalling and keeping our personal data during the process.

*Chapter 15* presents shell scripts, particularly BASH, as it is the primary scripting tool used in the kernel and many distributions. It then unveils task automation with cron jobs and systemd timers, providing examples and detailed explanations.

To conclude the book, we will learn with simplistic metaphors what the Linux kernel is and how it operates. Without going into details, understanding process and kernel isolation will help you understand even more why Linux is the best operating system kernel ever. We will then briefly review kernel modules and continue with kernel versions. The next section presents Linux kernel release switching, easily performed on Manjaro thanks to its tools and release model. To conclude the book, we will briefly cover the RTLinux kernel.

This part has the following chapters:

- *Chapter 13, Service Management, System Logs, and User Management*
- *Chapter 14, System Cleanup, Troubleshooting, Defragmentation, and Reinstallation*
- *Chapter 15, Shell Scripts and Automation*
- *Chapter 16, Linux Kernel Basics and Switching*

# 13

# Service Management, System Logs, and User Management

In the previous chapter, we reviewed the **basics of the internet**, explaining network models, packet sending, and protocols used. Further, we delved into network attacks and how firewalls and 2FA provide security. We then thoroughly reviewed **ufw**, one of the best user-friendly firewalls. The last part of the chapter was dedicated to **VPN**s, explaining how they increase our anonymity and privacy to the highest and covering the usage of the best public **VPN** provider at the moment, Surfshark.

In this chapter, we continue with the **systemd** service manager, which manages the Manjaro *system startup and services*. We will also cover the topic of **daemons**, a widely used term in the Linux world. Our next stop will be the multiple active TTY devices and how to access and use them. We will further explore **journalctl**, as it is the primary way to track what happened on our system by exploring its log. The last major topic is **user management**, focusing on groups and ownership.

The sections in this chapter are:

- Processes, daemons, and systemd
- Linux virtual TTY consoles
- Journalctl and system logs
- User management, groups, ownership, and root privileges

## Processes, daemons, and systemd

In this section, we will review processes and several ways to explore the ones running on our system. We will then continue by covering daemons, which are special processes running permanently on our system. Finally, we will cover systemd, the service manager on Manjaro and most major Linux distributions. Many people view it only as a service manager. However, it serves as a *system and service manager*, and we will see why and how.

# Processes

A process is any task started automatically by the **operating system** (**OS**), like an SSH server, or manually by us, such as a web browser. Each newly opened terminal runs as a separate process. If you copy a file, it is performed in a separate process. There is a constant running process for the events from your mouse – movements and button clicks, and another process for your keyboard input. More complex programs start several processes. For example, internet browsers always have one primary process, multiple child processes (at least one for each opened tab), and potentially others for additional background tasks. Without multiple child processes, such complex programs would work very slowly.

As we have already seen in *Chapter 7*, each process has a *unique* **Process Identifier** (**PID**). When the system starts, the motherboard BIOS loads first, then there's the bootloader setup, and then the bootloader starts the Linux kernel. When the kernel has loaded completely, it starts *the* **systemd init** *process, the first process on Manjaro* or *any other systemd-based Linux distribution*. **init** stands for **initialization**, and I will use its short form when appropriate.

It is essential that Linux (like any desktop OS) is a **multitasking** OS, which means it manages multiple processes in parallel. Processes can have multiple child processes (also named **threads**). The Linux kernel manages all these processes, providing them with resources, such as sections in RAM and processor time slices. The time slice period lengths are dynamically set and ranges from several tens of microseconds up to 100 milliseconds. This way, hundreds of processes can run together on just a few CPU cores. We will review essential kernel basics in *Chapter 16*.

The **systemd init** has **PID** 1, performs basic initialization and setup, and then starts all other processes to load the fully fledged OS. Each next process has a **PID** *incremented by at least 1*. The first few hundred processes are executed from the **root**. Most of them are small, fast tasks, taking between several microseconds and several hundred milliseconds. Nowadays, we usually work on multi-core processor systems, and many processes are executed in parallel when possible. Others require the **initialization** of specific OS subsystems, so they are started in a specific sequence. When we log in, a second copy of systemd is started to manage all user-specific initialization and processes. This second copy is the first process your user starts, and it is performed with the `/usr/lib/systemd/system --user` command. This command can be seen in most process explorer tools, some of which reviewed here.

Any process will go to sleep when it has no work, and it will be awakened to work and take CPU time only when necessary. If the process is not a constantly running task, then once it finishes, it disappears from the list of current processes. To get a simple snapshot of all currently running processes, use the `ps` command like this:

```
$ ps -1A
```

The -A option specifies to list all processes. The -l option provides 14 columns with details, including the following:

- **UID**: The ID of the user who started the process (**root** has UID 0 and the first regular OS user has UID 1000)
- **PID**: The process ID
- **PPID**: The parent PID (the PID of the process that started this process)
- **PRI**: The process priority
- **SZ**: The process size in kilobytes
- **TTY**: The terminal name the process is running on (if any; otherwise, a question mark)
- **TIME**: The amount of CPU time the process has used
- **CMD**: The command that started the process

The output of ps can be modified immensely by using different options; check its man page.

To investigate the currently running tasks in real time, use the top or htop process managers. Write their name in the terminal and press *Enter*. The help for top is shown when you press *H*; for htop, press *F1*. To quit them, press *Q* or *Ctrl+C*. If you compare their views, you might think they are different, but this is because they use different default *sorting* and *presentation* settings. The keyboard arrows or *PgUp/PgDown* keys will scroll through the processes list for both.

In htop, however, we can use mouse buttons and scroll. It also adds coloring and a system resources summary at the screen top.

The htop main menu options are found at the bottom of the screen. Clicking on any of them will activate the options; pressing *Esc* will exit them. One more press of *Esc* will hide the menu; choose any *F1* to *F10* option and press *Esc* once to show it again. The htop default view is sorted by CPU%, and as it refreshes every 1.5 seconds, it changes constantly. Press *Shift+N* to sort by PID and wait for 1.5 seconds for the view to refresh. View changes are saved in your personal view configuration file, located at ~/.config/htop/htoprc. Deleting this file will reset the htop view to the defaults.

We will look at some keyboard combinations shortly, but the setup option *F2* deserves a closer look first. There are several categories in it:

- **Display options**
- **Header layout**, to modify the top section layout
- **Meters**, to select what is displayed in the top section
- **Screens**
- **Colors**

In **Display options**, it is helpful to set accordingly **Tree view is always sorted by PID**, the several Hide and Show options, **Enable the mouse**, **Update Interval** (change it via the + and - *keys*), and **Hide main function bar** (I prefer to set this to 0, to have it always enabled). Use the *spacebar*, *arrows*, and *Enter* keys or the *mouse* to select/deselect these options. Selecting **Screens | Main** is important to change the columns in the main processes listing. When you select with the arrows or mouse from the **Available Columns** list, press *Enter* to add it to **Active Columns** for a given screen. In **Active Columns**, use *Enter* (or the F2, F5, and F7-F10 keys as described in the bottom menu) to reorder the columns.

The following are the other essential options to check. For most of them, we must wait for one update interval to see the effect. Also, note that the tree view changes the listing significantly and is sometimes incompatible with some options:

- *F1*: Help. Read it and try the operations. Taking a screenshot helps.
- *F3*: Searches for a process name; press it again for the next result.
- *F4*: Filters the view by process name. Try it on a process that is mentioned several times.
- *F5*: Toggles the tree view, showing the parent-child process relationships.
- *F6*: Sorts by one of the columns. Select one and press *Enter*.
- *I*: Inverse sorting, by whatever criteria.
- *Shift+Z*: Pauses/resumes the update process.
- *Shift+H*: Hides/shows the user process threads (doesn't apply to all user process threads).
- *Shift+K*: Hides/shows the kernel threads (doesn't apply to all kernel process threads).
- *Shift+N, P, M, T*: Sorts by PID, CPU%, MEM%, or TIME, respectively.
- *#*: Hides/shows meters.

*Figure 13.1* shows how htop looks for me, with only some columns selected, kernel threads enabled, and sorted by PID:

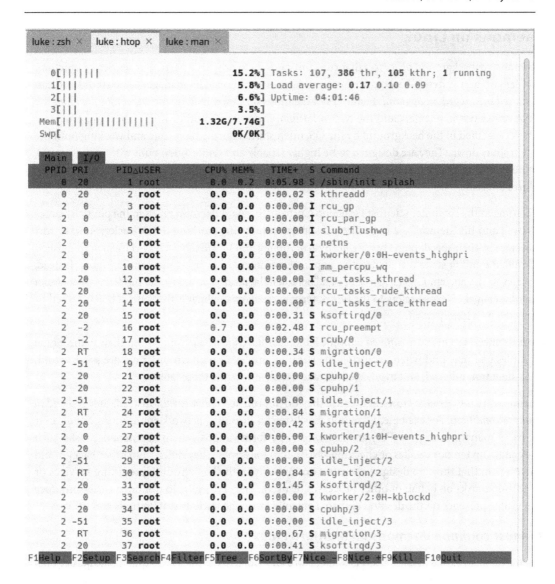

Figure 13.1 – Example of htop listing processes by PID, with kernel threads enabled

Play with the htop options. With htop, you can discover many child-parent process relationships, see when a process has blocked your CPU, and understand which processes are started by root and which by your user. htop (or top) is essential for inspecting any Linux distribution.

# Daemons on Linux

**Daemons** are background processes that run continuously to provide a particular service or function. On Manjaro and **systemd**-based distributions, daemons are normally managed by **systemd** and can be *started or stopped on demand*. Examples of daemons on Linux include the *Bluetooth daemon*, the ALSA *sound system daemon*, and the *NetworkManager daemon*. In other words, these are processes or tasks executed in the background by the OS, often started during boot time and working until the system shuts down. They are designed to be highly reliable and stable, often built with various error-handling and recovery mechanisms to ensure they continue running even in the event of system failures or errors.

According to the computer scientist Fernando Corbató, the word **daemon** refers to the physicist James Maxwell and his "demon" – a hypothetical supernatural being working in the background to sort molecules by their speed. In the book *Unix System Administration Handbook* by *Evi Nemeth* (published in *1989*), the author states that no negative implication is intended by using the term "daemon." Instead, it relates to the ancient Greeks' concept of a "personal daemon," similar to the modern concept of a "guardian angel." He also refers to the Greek term *eudaemonia*, which is the state of being helped or protected by a kindly spirit.

**Daemons** are *suffixed* with the letter d or have *the whole word* within their name. Some examples you may find when exploring the running processes with ps or htop are **sshd**, **crond**, **dbus-daemon**, **udisksd**, and **ntpd**. We will take a look at some of these later.

**Daemons** are *not periodic processes* because they do not run at *fixed intervals*. Instead, they *respond to events or conditions*. For example, a **daemon** may listen for incoming network connections, handle user logins, or manage print jobs. They are also used to perform various system maintenance tasks, such as cleaning up temporary files or checking system logs for errors. They differ from other background processes in that they are designed to run continuously in the background, providing services or performing tasks on behalf of other processes or users. They typically do not have a user interface or a visible presence on the desktop and are controlled through various system tools and utilities.

## *15 most common daemons on Manjaro Linux*

The following list is not extensive but will help you know some of the most common **daemons** working on Manjaro. Please note that not all of them will necessarily be present on your system. For example, on one of my workstations, a wireless network adapter is not present; hence, the **wpa_supplicant** daemon is not running, as during the initial installation, the **mhwd** module didn't detect such an adapter, and so didn't enable it:

- **systemd**: This is the parent process for all other processes on the system and is responsible for managing system services.

- **NetworkManager**: This daemon manages *network connections*, including *Ethernet*, *Wi-Fi*, and mobile broadband connections.

- **cupsd**: Manages *printing*, including printer discovery, driver installation, and *print queue management*.

- **sshd**: Provides secure remote access to the system over the **SSH protocol**.

- **crond**: Schedules and runs *periodic tasks and scripts*, such as backups or system maintenance tasks.

- **avahi-daemon**: Provides zero-configuration networking (**zeroconf**) services, including network discovery and service announcement. Zeroconf is a set of **protocols** that allow network devices to automatically discover each other and communicate without requiring manual configuration or a central server.

- **dbus-daemon**: This one provides inter-process communication services between different processes on the system.

- **udisksd**: Manages *storage devices* on the system, including *filesystems mounting and unmounting*.

- **ModemManager**: This manages mobile broadband connections on the system, including modem detection, configuration, and dialing.

- **wpa_supplicant**: This manages *Wi-Fi connections* on the system, including authentication, encryption, and network discovery.

- **ntpd**: *Synchronizes the system clock* via the network with an **NTP** time server.

- **systemd-journald**: Manages *system logging*, including log storage and retrieval. We will take a detailed look at it later in this chapter.

- **dhcpcd**: Manages **Dynamic Host Configuration Protocol (DHCP)** network configuration.

- **sddm**: This acronym stands for **Simple Desktop Display Manager**. It is a daemon that initiates the graphical login screen after the OS loads.

- **colord**: Manages color profiles and color calibration on the system, including monitor calibration and printer color management.

When we use $ `ps -1A` to list all running processes, the list will also contain all daemons. Search a given process name online to tell whether it is a daemon, as each daemon is documented.

As daemons are managed by **services**, controlling them on Manjaro is primarily done via **systemd**. Some of them offer additional configuration utilities, but again, you have to check their documentation to see whether a given daemon has such programs in its package.

## systemd introduction and a bit of history

A **service manager** is typically used to start, stop, or modify a service – e.g., Bluetooth, network, NFS, an HTTP server, or other **services. systemd** does this in a centralized and unified fashion. However, the startup process of a Linux-based distribution is quite complicated, and systemd was *created* primarily to improve Linux systems initialization and abandon the old **SysVinit**. It added parallel processes, reduced shell usage, and generally made the `init` process (i.e., *the first process* the Linux kernel starts) *organized* and *unified*.

**SysVinit** was the **Unix** `init` process during the *1980s*. It was introduced when multi-core complex modern computers were quite rare. *Complex* in this context means a system with multiple subsystems – multiple HDDs, USBs, Bluetooth, sound and graphics cards, support for network and wireless interfaces, PCI devices, and so on. In the case of laptops, it includes touchscreens, touchpads, and many more. This means that our PC needs to start hundreds of drivers and modules during `init` so that, when loaded, each device can be used immediately. **SysVinit** uses a sequential startup process, where each service must wait for the previous service to finish before it can start. This results in long startup times and sometimes system instability.

As a result, in *2009*, several **RedHat software** (**SW**) engineers and tens of other developers joined forces and started working on a new and better **init** and **service manager**, which they called **systemd**. Its first release was on *March 30th, 2010*. In comparison to **SysVinit**, **systemd** was a lot better in many ways, some of which are:

- systemd can *start services in parallel*, which speeds up the boot process

- systemd can monitor and *restart* services that *crash or terminate unexpectedly*

- In addition to *services*, systemd supports *socket activation*, user *session management*, and resource control

- systemd can also manage timers, sockets, mounts, devices, and more

It is crucial to point out that **systemd** *practically controls (or at least influences) all essential services*. It *didn't provide a good way of* **modularizing** management and separating it entirely from `init`. As **modularity** is a core value for the Linux community, this spawned a conflict.

Say you want to create an audio-related application that is to be started during system `init`. You must ensure it starts after the audio drivers and consider all **systemd**-service-related characteristics. So, to develop the application, you *must* have a basic understanding of **systemd**, or else you will make severe mistakes. This is why many developers voted against the widespread adoption of **systemd**.

For several years, there were debates about whether it should be widely adopted; the most notable debate occurred in the **Debian forum**. One of the leading **Debian** developers even *resigned* from the project after a few years of discussion on the topic. He stated that the discussions exhausted him completely.

However, **systemd** is constantly developed, and thanks to the mentioned improvements and the mitigation of some initial flaws, it was slowly accepted as the best alternative. The total number of **contributors** over the last *13* years has grown to *over 1,500*. Over the years, they have added multiple modules, and at the moment, **systemd** is the most widely used system manager in most Linux distributions. In *2019*, the **Debian** community voted for a second time to keep **systemd**, but it was officially concluded that they would look for alternatives.

**systemd** has been adopted as the default *system and service manager* by **Fedora** (in 2011), **openSUSE** (in 2012), **Arch** Linux (in 2012), **Manjaro** (in 2013), **CentOS** (in 2014), **RHEL** (in 2014), and **Ubuntu** and **Debian** (in 2015).

**Slackware** is the only major distribution that has not introduced **systemd**.

More than a few major distributions derivatives and custom distributions support other init systems. Some examples include **Devuan**, **MX Linux**, and **antiX** (all three are **Debian**-based), and the completely independent **Void Linux**, **GoboLinux**, and **Alpine Linux**. These have different init systems, such as **OpenRC**, **runit**, **Dinit**, and **s624**. There are even more variations. I will not go deeper into the subject as it is not our primary focus. Instead, let's get deeper into **systemd**.

## systemd configurations, units, and targets

**systemd** controls OS initialization, services, logging, and shutdown via configurations and units, all described in text files. The specifics of each of them are described in the following two sections.

### Configuration files

In the main directory, `/etc/systemd/`, we will find the `.conf` files used for the basic configuration of several functionalities. I will provide at least a short description for each of them. Most settings are commented out with a hash sign # at the beginning of the line, showing their *default* values. To change those defaults, you must *uncomment* the line by removing the # and changing the value. Remember that changing the settings in these files will often require at least a *service restart or even a reboot*. Also, remember that these files do not provide options for all possible settings related to the given functionality configuration. You often edit the given settings and then modify others that are unavailable in the `.conf` files via commands. systemd-related characteristics are modified via `systemctl` and other **systemd** commands.

The `.conf` files are helpful as they enable us to quickly see a list of settings, whether default or deliberately set, using the `cat` command. They contain many default *service-related settings* and define the overall *system behavior*.

In the same directory, `/etc/systemd/`, there are three subdirectories – `network`, `system`, and `user`. They contain `.service`, other unit files, and subdirectories with files that contain systemd configurations for specific services. For some of them, the files are located in other filesystem locations, and there are only symbolic links in these directories. We will look into some of them later.

As of March 2023, the `.conf` files are as follows:

- `coredump.conf`: This file configures the system's **core dump** settings, including the core dump files maximum size and where they are stored. A **core dump** file saves the memory image of a crashed process; it is created to help developers diagnose the cause of the crash. It contains information about the *process state* when the crash occurred, including its memory contents and *register values*. As these files may contain security-sensitive data, they will be accessible only to authorized users.

- `homed.conf`: This file configures the **systemd-homed** component, which provides *home directory management* and related features.

- `journald.conf`: This file configures the systemd **journal**, *a central logging service* that records system activity and events. The related `journal-upload.conf` and `journal-remote.conf` files contain the configuration for uploading **journal** data to a remote server. Locally, we use the `journalctl` command to access the `journald` logs.

- `logind.conf`: This file configures **systemd-logind**, which manages *user logins* and provides *session management services*. **systemd-logind** allows administrators to configure various settings related to *user sessions*, such as their *idle timeout*, *power management*, and *login restrictions*. For example, the *idle timeout* setting determines how long users can be inactive before their session is automatically logged out. The *power management* settings allow administrators to specify how the system should behave when users close their laptop lid or press the power button. The *login restrictions* settings also allow administrators to specify which users can log in to the system, and from which devices or networks. We usually use the `loginctl` command to modify these settings.

- `networkd.conf`: This file configures **systemd-networkd**, which provides *network configuration management*. It allows administrators to configure various settings related to **network interfaces**, such as *static IP address* or enabling **DHCP** for dynamic address assignment. Additionally, administrators can use this file to configure *routing tables* and set up **DNS servers**. We usually use the `networkctl` command on the terminal to manage some of these settings.

- `oomd.conf`: This file configures **systemd-oomd**, which monitors *system memory usage* and kills processes that consume too much memory.

- `pstore.conf`: This file configures the **pstore** persistent storage filesystem, which stores kernel errors when it "dies" (after reboot or power-off).

- `resolved.conf`: This file configures **systemd-resolved**, which provides *network name resolution services* to local applications.

- `sleep.conf`: This file configures the system's *sleep behavior*, including what happens when the system *wakes up from sleep*.

- `system.conf`: This file configures the global **systemd** *system settings*, which include timeouts for start/stop/abort, watchdog, logging settings, and others.

- `timesyncd.conf`: This file configures **systemd-timesyncd**, which provides time synchronization services over the **Network Time Protocol** (**NTP**). We usually control them via the `timedatectl` command.

- `user.conf`: This file configures the user-specific **systemd** settings, including the user's preferred locale and time zone.

## Unit files

The unit files describe systemd unit configurations and have different extensions for each unit type. Most are in `/usr/lib/systemd/system`; you can list this directory with `exa` or `ls`. To list *all unit files* with the state and preset, execute `$ systemctl list-unit-files`. To list only *loaded* **systemd** units with their *load* and *activation* status, execute `$ systemctl list-units`. You can **scroll** not only *up* and *down* but also *left* and *right (with keyboard left and right arrow keys)*, as the report is usually wider than the terminal on *small monitors*! Here are a few words on each of these types:

- **Service units** (`.service`): These define and manage system services. They typically contain the *service name*, description, and *dependencies*. If necessary, they can contain a command or script path that should be run to start or stop the service and any necessary arguments for it. systemd services include network daemons, web servers, databases, and others, which run continuously in the background. Some of their key features are *automatic restarts*, *dependency management*, and *logging*.

- **Socket units** (`.socket`): These units manage inter-process communication sockets that systemd can monitor and activate. For example, the `dbus.socket` file is the **D-Bus** message bus socket unit. **D-Bus** messages are used extensively by desktop environments like **GNOME** and **KDE** to provide desktop notifications, sound events, power, and network management. To list *all active systemd* **sockets**, you can use `$ systemctl list-sockets`.

- **Device units** (`.device`): These represent devices recognized by the kernel. For example, `dev-sda.device` is the **unit** for a *disk drive*. To list all active devices, use `$ systemctl list-units --type=device`. You will see **units** related at least to `sda`, as this usually is the first hard disk connected to your system. The different partitions will be listed as separate devices (e.g., `sda1`, `sda2`, `sda3`, etc.). You will also see your *network, Ethernet, soundcard*, and other *devices*. The last thing to note, which is always impressive for newbies, is the great amount of *TTY devices*. **TTY** is an abbreviation for *teletypewriter*, which basically means a terminal. Each **TTY** device is a *virtual terminal*, and by providing multiple active **TTY** devices, Linux easily supports *multiple desktops* and *terminal sessions* simultaneously. We will review them later in the chapter. Executing `$ exa -lah /dev` will list a file for each active device in this directory. Keep in mind that this listing will hold many more devices than those listed by the `systemctl` command. This is because everything in Linux is a file; thus, many other applications create device files in the `/dev` directory.

  We don't use the `.device` file to get more information about a device. Generally, `.device` files are not meant to be written or changed by users and administrators. We use the `$ sudo systemctl show sda.device` command *to list details about a device* managed by **systemd**. Replace `sda.device` with any name from the device unit list's (from the `UNIT` column) to see details about a specific device.

- **Mount units** (`.mount`): These files represent mount points in the filesystem hierarchy of local (such as `sda`), kernel, network, or virtual filesystems. They are located in `/usr/lib/systemd/system` on Manjaro. List them with `$ systemctl list-units --type=mount`.

- **Automount units** (`.automount`): These represent automount points that trigger on-demand mounting. List them with `$ systemctl list-units --type=automount`.

- **Swap units** (`.swap`): These represent swap devices or files that are activated or deactivated. A typical example is the swap space of your PC. It is an additional partition on your hard disk used when RAM is insufficient. An example could be a weak older PC with only 1 GB of RAM that has some additional space on the HDD allocated as a swap partition. Then, when you open a browser with 20 tabs and the PC needs more RAM, the swap partition is used, as 1 GB would not be enough for the PC and the browser's needs. Of course, this is slower compared to having more RAM. There are automatic generators for such files, and you will often not find them easily in the filesystem.

- **Path units** (`.path`): These represent paths in the filesystem hierarchy that **systemd** can monitor and (de)activate other units when certain events occur. For example, `gpm.path` is a way to monitor the status of the **General Purpose Mouse (GPM)** daemon and automatically start or stop it. This ensures that the mouse is always available and properly configured and that the GPM daemon is started and stopped in a controlled manner.

- **Timer units** (`.timer`): These represent timers activating other units based on calendar or monotonic (periodic) events. For example, `fstrim.timer` is the unit for running a periodic trim operation on SSD hard disks. `pamac-mirrorlist.timer` is a timer unit that periodically updates the repositories *mirror list* for the Pacman package manager on Arch-based Linux distributions, including Manjaro.

- **Slice units** (`.slice`): These represent groups of processes organized hierarchically for resource management purposes. For example, `user.slice` is the unit for grouping user processes under a common parent, and `machine.slice` manages the resources provided to virtual machines and containers at system level.

- **Target units** (`.target`): These represent groups of other units that can be used for synchronization or logical grouping. For example, `network.target` is the unit that represents the system's state when network services have been initialized and are ready to use. It is a collection of units and services required to provide network connectivity, which includes network interfaces, IP addresses, routing tables, and other network-related settings. This unit is typically used as a dependency by other systemd units and services (that require network connectivity to function correctly), such as a web server service.

Whether a given unit will be present and/or named in one way or another depends at least partially on your distribution, flavor, and the team's choice of basic tools. Each distribution team configures things accordingly. Usually, the team will not put *flavor-specific* or *graphical-environment-specific* settings in systemd's configuration. However, it is still possible for some systems to optimize and improve the boot of a given distribution.

Again, to see all currently present systemd units, call $ `systemctl list-unit-files`. Then, to get information for each of them, execute $ `systemctl show name.extension`, replacing the last part with a proper unit name from the previous listing. Here is an example: $ `systemctl show shadow.timer`.

As mentioned earlier, on all Manjaro flavors, all unit files are located in `/usr/lib/systemd/system`. Thus, in addition to $ `systemctl list-unit-files`, you can use an `exa` or `ls` directory listing, and then use the files there as an argument to $ `systemctl show`.

Furthermore, to get information for **systemd**, call $ `info systemd`. To get more information for a given unit, call $ `info systemd.timer` or any other extension. To get more information for the units in general, call $ `info systemd.unit`. You can replace `info` with man in the previous commands and potentially export to HTML.

Though I do my best to describe the components as well as I can, complex systems have different perspectives to be described. Thus, an excellent article to read is this one from Linode: `https://www.linode.com/docs/guides/what-is-systemd/`. When looking for information, it is always good to first look the `info`/`man pages`, and then for articles. When you look for articles, always check that they are up to date; otherwise, you might end up with outdated information. An example of the last case is this **DigitalOcean** article, which provides good but potentially obsolete information from 2015: `https://www.digitalocean.com/community/tutorials/understanding-systemd-units-and-unit-files`. As **systemd** has at least a few releases per year, reading material that old is definitely not the best option.

### Standard targets and more information

Knowing the standard targets is essential. Each standard target represents a specific system state in which a set of services are running. From a basic systemd design stand point, the most important ones are as follows:

- `rescue.target`: This is the lowest level when systemd starts, which includes mounted filesystems, basic services running, and a rescue shell on the main console.
- `multi-user.target`: When this target is loaded (which is obligatory after `rescue.target`), the system has multiple services loaded and running. It supports all registered users to be logged in via the terminal but not via a graphical environment.
- `graphical.target`: The graphical target adds all dependencies of the graphical environment. Thus, when loaded, the PC has a clickable GUI to log in.
- `default.target`: This target is often an alias for `graphical.target`.

There are also the `emergency.target`, `halt.target`, `reboot.target`, and `poweroff.target` standard targets.

To see the hierarchy of default.target, we can execute $ systemctl list-dependencies on the terminal. For Manjaro, this will show us all the units that need to be loaded to reach the default.target state and their dependencies. The following shows partial results from this command:

```
default.target
 ├─sddm.service
 ├─surfsharkd2.service
 └─multi-user.target
   ├─cronie.service
   ├─cups.path
   ├─cups.service
   ├─dbus.service
   ├─ModemManager.service
   ├─NetworkManager.service
   ├─pkgfile-update.timer
   ├─rsyslog.service
   ├─sshd.service
   ├─systemd-ask-password-wall.path
   ├─systemd-logind.service
   ├─systemd-user-sessions.service
   ├─ufw.service
   ├─vboxservice.service
   ├─basic.target
   │ ├─-.mount
   │ ├─tmp.mount
   │ ├─paths.target
   │ ├─slices.target
   │ │ ├─-.slice
   │ │ └─system.slice
   │ ├─sockets.target
   │ │ ├─cups.socket
   │ │ ├─dbus.socket
   │ │ …
   │ │ └─systemd-udevd-kernel.socket
   │ ├─sysinit.target
   │ │ ├─dev-hugepages.mount
   │ │ …
```

If we call the previous command with the additional argument graphical.target, we will see the same dependencies as those of default.target. You can use list-dependencies on *any unit*, which will show you its dependencies; for example, try $ systemctl list-dependencies NetworkManager.service.

## Analyzing systemd's init sequence

When we start our PC, the BIOS looks for the kernel bootloader located in /boot/efi and runs it. Once the bootloader finishes the basic hardware (HW) setup, it runs the Linux **kernel**, which, when ready, starts **systemd** in root context. After **systemd** initializes itself, it starts running the hundreds of OS root initialization processes until it reaches the default.target state. As all this happens in a few seconds, we may potentially see only partial logs. Thus, to analyze what happened during OS init, we must use the **systemd** tool systemd-analyze.

We have several systemd instances – one for the root and one for each logged-in user. The *user-specific* systemd instances are started at each user login to manage its *environment-specific* services. When inspecting with htop, combined with tree view (*F5*) and showing/hiding kernel/user processes with *Shift+H* and *Shift+K*, we can see which processes are started from the root and which from our or other users. To find the second systemd instance, filter with its name and look for the /usr/lib/systemd/systemd --user command. I will not go deeper into this as it becomes too complex. The important thing for us is that systemd-analyze has the additional argument --user, to access the following information *for our user*. It essentially tells systemd to extract information from the *user* systemd instance.

Here are several basic usages of it:

- $ systemd-analyze: Without options, systemd-analyze only provides a basic summary of the firmware boot, bootloader boot time, kernel init, userspace init, and graphical environment startup time (i.e., the time elapsed until the moment when you see your login screen). Remember that these times depend entirely on your configuration (CPU, RAM type and speed, storage type, motherboard, chipset, and systemd settings). Thus, the times you see in the following code snippets are not an example of good or bad performance. On a new HP laptop with a fresh Manjaro KDE installation, my total time was 20.891 seconds. In comparison, a virtual machine lacks the HW delays and reports a total load time of 12.016 seconds. On an older PC with Xfce, it reports the following:

```
Startup finished in 14.481s (firmware) + 14.237s (kernel) + 2
min 462ms (userspace) = 2 min 35.670s
graphical.target reached after 25.284s in userspace.
```

To access your *user* session startup time, call $ systemd-analyze --user. It reports the following on the older PC with Xfce:

```
Startup finished in 449ms (userspace)
default.target reached after 449ms in userspace.
```

- Though I have used `systemd-analyze` a lot and rarely had any issues, there is one issue encountered on multiple distributions for years. It is related to `plymouth.service`, which shows your machine/distribution splash screen logo during boot. It has a somewhat tricky unit configuration. When it is wrong, calling `systemd-analyze` results in a `Bootup is not yet finished` error with a short report after it. To fix it, the easiest way is to change the kernel boot configuration, which can be done in seconds.

  First, edit the grub file via `$ sudo nano /etc/default/grub`. In it, find among the first few lines the one containing the `GRUB_CMDLINE_LINUX_DEFAULT` configuration, which on one of my machines is currently set to:

  ```
  GRUB_CMDLINE_LINUX_DEFAULT="quiet splash udev.log_priority=3"
  ```

  If yours includes the words `quiet` and/or `splash`, delete them, then press *Ctrl+S* and *Ctrl+X* to save and exit. Now, execute `$ sudo update-grub` and then `reboot`. If your loading logs were missing and you had only a logo, you will now have a limited Manjaro kernel loading log, and the previously mentioned `systemd-analyze` error will not appear.

- `$ systemd-analyze blame`: The `blame` option asks the `analyze` tool to display the init time per unit; in this way we can see the separate start times of each service. As explicitly emphasized in the `systemd-analyze` man page, this listing provides times only for services with `Type` value other than `simple`. Those that are of `Type` `simple` are considered fast and therefore unimportant. It is essential to know that the longer startup times of some processes can be due to waiting for other slow-to-start processes. Again, as stated in the previous bullet point, all these results depend heavily on your HW, so don't take the following partial result example from one of my virtual machines for reference:

  ```
  2.968s man-db.service
  2.224s NetworkManager-wait-online.service
  1.604s pkgfile-update.service
  1.060s systemd-random-seed.service
   301ms dev-sda2.device
   270ms pamac-cleancache.service
   165ms ufw.service
   148ms udisks2.service
   130ms user@1000.service
   129ms ModemManager.service
   108ms ldconfig.service
    97ms polkit.service
    95ms systemd-journal-flush.service
    93ms systemd-udevd.service
    88ms logrotate.service
    79ms systemd-tmpfiles-clean.service
    77ms dbus.service
  ```

  To see this analysis for your *user* processes, add `--user` at the end.

- $ systemd-analyze critical-chain: The critical-chain option displays a tree of time-critical units for a specified unit or the default target (usually the graphical environment). The following is an example of its results:

  ```
  systemd-tmpfiles-setup.service @1.372s +32ms
  ```

  The time shown *after* the @ character is when it *started*, relative to the systemd start time. The time after the + character is optional; when present, it is the time *taken to start the unit*. However, note that the output might not always be accurate due to socket activation and parallel execution of units. Additionally, this option only shows the time spent in the activating state. It doesn't cover units that never went through this state, such as device units that transition directly from inactive to active. Finally, it doesn't provide information on jobs, especially those that timed out. Despite this, it is suitable for basic checks. You can use it directly, as shown above (try it out; it is worth it), or with some of the units listed with the previous commands, like this: $ systemd-analyze critical-chain systemd-journal-flush.service. Add --user after the critical-chain option to speak to the *user* systemd instance.

- $ systemd-analyze plot > systemd_boot_sequence.svg: This is a great option, providing us with an **SVG** graphic with a name of our choice. It can be opened in any web browser, and search will work on it due to the **SVG** format. This graphic shows all init *systemd units*, *highlighting* the ones with *longer start times* (and providing for them, in particular, their start time). A partial example is shown in *Figure 13.2*:

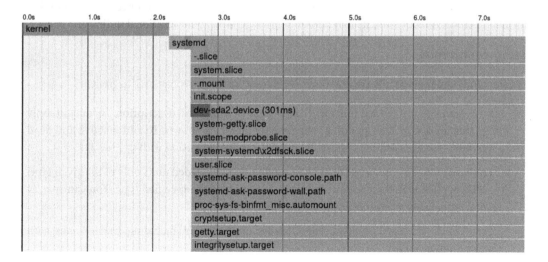

Figure 13.2 – Partial example of the systemd-analyze plotting function

Add `--user` after the `plot` option for the *user* systemd instance.

- `$ systemd-analyze dump > systemd_dump.log`: This option will provide a complete systemd current state log (the filename is our choice), with a full description of each unit and all its parameters. It contains information on CPU, memory, dependencies, and much more. As the log is exceptionally long (over 30,000 lines on my virtual machine), inspecting it with an advanced GUI text editor like **Kate** is the best option. Otherwise, it is hard to find the necessary information.

- `$ systemd-analyze unit-paths`: This option lists all paths from which unit files may be loaded. Add `--user` after this and the next option (about files) to get the *user* systemd instance results.

- `$ systemd-analyze unit-files`: This option lists all unit files and symlinks for them.

There are many other options – check out the `$ systemd-analyze --help` options. For complete information, check the `info` or `man` pages. The official Arch Linux page may also be helpful: `https://wiki.archlinux.org/title/Improving_performance/Boot_process`.

## The most important systemctl command options

Now that we know what services are started and when, let's see a summary of the most important `systemctl` options. Remember, you can add `--user` to each of these to manage **userspace** services. Without it, the commands default to the **root** systemd instance. Also, remember that most of this information comes directly from its magnificent `man` page, with some additional testing and checks.

### Listing commands

The commands and their details are as follows (I omitted writing `systemctl` in front):

- `list-units`: List loaded units with a load/active status and description. By default, only units that are active, have pending jobs, or have failed are shown. To add inactive and dead units, add `--all`. Without `--all`, I currently see 148 units, and with it, I see 324.

- `list-units -t device|mount|target|...`: List loaded units only of the given type. `--all` applies as in the previous example, but only for the specified type. You can specify several types using a comma-separated list.

- `list-unit-files`: List all unit files installed on the system, including disabled ones, with state and preset values.

- `list-sockets`: List socket units currently in memory, ordered by address.

- `list-timers`: List timer units currently in memory, ordered by next timer elapse.

- `list-automounts`: List automount units currently in memory, ordered by path.

## Status and list jobs

The commands and their details are as follows:

- `status [UNIT]`: Without additional options, this shows a hierarchical tree view of all started services, with the most recent journal log value. We get a detailed listing for each unit if we add `--all`. If we add a unit name, we get the detailed status of only that unit, with the 10 most recent journal log values. This function displays the runtime status but is not used to evaluate the loading status, as to extract the information, it will load the given module (if it is not loaded).

- `list-jobs`: List jobs in progress. After loading Manjaro and logging in, it will report no jobs are running (except if you manually started one), including with `--user`. This option helps find stalled systemd jobs after initialization.

- `systemctl --failed`: List failed units. If you have some, then take their name and get more information with `status [UNIT]`.

## Enable, disable, (re)start, stop, and more

The commands and their details are as follows (I omitted writing `systemctl` in front):

- `start <UNIT>` and `stop <UNIT>`: As expected, used with a unit name, but both commands also accept lists of unit names.

- `restart <UNIT>`: `stop` and `start` a given unit, even if it is in a stopped state. This command will not flush all temporary resources, such as the file descriptor storage facility. If you need to do that, use `start` and then `stop` separately.

- `enable <UNIT>` and `disable <UNIT>`: Here, `enable` creates a set of symlinks and then invokes a systemd configuration reload, to have the unit loaded and ready to start, but *without* calling `start`. To also immediately start after that, add `--now`. `disable` does the same but may remove additional symlinks. If `--system`, `--user`, `--runtime`, or `--global` is specified, that enables the unit for the system, the calling user, the system session, or all future logins of all users, respectively. In the last case, the *systemd daemon configuration* is not reloaded and must be done manually or via `reboot`.

  `disable` has one special action here: all units listed in the unit's `[Install]` section, in the setting `Also=`, will be disabled as well.

- `reload <UNIT>`: Reload *only* the *service-specific unit configuration*. In the case of Apache, this will reload Apache's `httpd.conf` in the web server, not the `apache.service` systemd unit file. If you want systemd to reload the configuration file of a unit, use `daemon-reload`.

- `daemon-reload`: Reload the systemd manager configuration. This will rerun all systemd unit generators (see $ `man systemd.generator`), reload all unit files, and recreate the entire dependency tree. While the daemon is being reloaded, all sockets systemd listens to on behalf of the user configuration will stay accessible.

- `log-level [LEVEL]`: Without an argument, this prints the current systemd log level. If an optional `LEVEL` argument is provided (as listed in the next section, in *Table 13.2*, specifically the `Code string` column), it sets the level.

### Options to get information about a service

The commands and their details are as follows:

- `show [UNIT]`: Without additional options, this provides the system's general properties. Adding a unit name or job ID gives their specific properties. By default, empty properties are suppressed; they are shown if we add `--all`. Add `status` to make the output human-readable. Test the combinations; it is worth it.

- `list-dependencies [UNIT]`: This shows (in a tree) the units required by a specified unit. It recursively lists units following the `Requires=`, `Requisite=`, `ConsistsOf=`, `Wants=`, and `BindsTo=` dependencies. If no units are specified, `default.target` is implied.

## Linux virtual TTY consoles

We explained the basics of the `/dev/` directory in *Chapter 9*, and earlier in this chapter, we saw a significant amount of **TTY devices** while listing `.device` units. We can see even more **TTY**-related devices when listing the `/dev/` directory itself (I will not go into details). These are virtual kernel consoles to which we can connect. Each is a text-based shell environment, supporting all the commands we use in our terminal emulators. On one of them, we always have the graphical environment we log in to, which later switches to our desktop. While most are accessible only through special tools, Manjaro and most Linux distributions usually configure most *F1–F12* keys for accessing them with keyboard combinations. These configurations differ a bit between different flavors and distributions. The main buttons are left *Ctrl+Alt*, and the additional is *FX keys*, where *X* designates the keys from *F1* to *F7*. On Manjaro Xfce, we have the main graphical session on `TTY7`, accessed with *Ctrl+Alt+F7*, while *Ctrl+Alt+F2* puts `TTY2` on the screen. *Table 13.1* presents the current (as of *September 2023*) configuration of each official flavor:

| Combination | Xfce | KDE Plasma | GNOME |
| --- | --- | --- | --- |
| Ctrl+Alt+F1 | Not configured | Kernel boot log | Login screen |
| Ctrl+Alt+F2 | TTY2 | Desktop session | Desktop session |
| Ctrl+Alt+F3 | TTY3 | TTY3 | TTY3 |

| Combination | Xfce | KDE Plasma | GNOME |
|---|---|---|---|
| Ctrl+Alt+F4 | TTY4 | TTY4 | TTY4 |
| Ctrl+Alt+F5 | TTY5 | TTY5 | TTY5 |
| Ctrl+Alt+F6 | TTY6 | TTY6 | TTY6 |
| Ctrl+Alt+F7 | Desktop session | Not configured | Not configured |

Table 13.1 – Linux virtual console configurations on official Manjaro flavors

You only have to enter a `username`, and then its `password`, and you will log in to the given TTY terminal as the given user. It is not an additional session, which means you share the same resources as your main session. You can also log in as root. Essentially, you have multiple doors to the same room.

Try the combinations. If you press the wrong one, nothing will happen. As no graphical/GUI environment is loaded on them, they are faster to work with than desktop sessions, especially on slower and older machines. Also, you can change the terminal session with a single key combination when working on several jobs.

## journalctl and system logs

In this section, we will review `journalctl` in detail and then briefly examine the **rsyslog** alternative and the **dmesg** separate kernel log tool. Finally, we will look at the **KSystemLog** GUI tool, as its filtering and sorting functions provide easier and more effective log exploration.

On Linux, hundreds of events are logged at each boot, during normal operation, and on power off. All system changes at runtime write log entries, the kernel also logs messages, and many daemons and processes write logs on different events. The resulting logs can be big, so its messages are provided only when requested. **journald**, the systemd tool suite daemon, unites and handles them so we can get all the information at once. It uses a binary format, so we cannot read its data directly; instead, we use the `journalctl` command (which we have already used for **ufw** in *Chapter 12*). **journald** supports data compression to optimize memory usage and upload-to-server to get logs remotely.

On Manjaro, and by a general rule of thumb on most Linux distributions, all log files are located in the `/var/log` directory and its subdirectories. The journald log is long – after a few months (and with enabled **ufw** logging), I already have over 300 MB on one of my machines. On Manjaro, it is stored in multiple files in the `/var/log/journal/` directory. To get a list of all log files, execute `$ journalctl --header | grep -i "File path"`. Depending on its settings, journald can keep log messages indefinitely. The `--help` and man pages are (as with all systemd tools) great.

## Inspecting logs with journalctl

Before going on, we must explain the essential journald log entry characteristics. It has the format `<Date,Hostname><Unit>[PID] [Message]`, with color coding to indicate priority.

The `Date,Hostname` bit is self-explanatory. `Unit` can be `systemd` or any of the **systemd units**, the `kernel`, or any process/daemon, such as `dbus-daemon`, `kwin_x11`, `rtkit-daemon`, `kernel`, `plasmashell`, `pulseaudio`, `user@1000.service`, `audit`, `at-spi-bus-launcher`, `surfsharkd.js`, `snapd`, `anacron`, `CROND`, and hundreds of others. `Message` can be any string; we will look at a few examples later.

We often see `systemd` as `<Unit>`, even though the given message is generated by some service or daemon under its management.

The log entries priority is crucial, and depending on your terminal coloring settings, you will see the most critical events in different shades of red. *Table 13.2* presents all the priorities, with `0` being the most critical:

| Number | Level/Priority | Code String | Meaning |
| --- | --- | --- | --- |
| 0 | Emergency | `emerg` | The system is unusable |
| 1 | Alert | `alert` | Action must be taken immediately |
| 2 | Critical | `crit` | Critical conditions |
| 3 | Error | `err` | Error message |
| 4 | Warning | `warning` | Warning message |
| 5 | Notice | `notice` | Important information |
| 6 | Info | `info` | Regular information messages (the most frequently used priority) |
| 7 | Debug | `debug` | Message for developers |

Table 13.2 – journalctl and dmesg log message levels

The `code string` is used as a parameter for `journalctl`, `dmesg`, and other related log viewers.

Here are a few examples:

```
Sep 02 11:01:11 luke-virtualbox kernel: BIOS-provided physical RAM
map:...
Sep 02 11:01:11 luke-virtualbox systemd[1]: Mounting Kernel Trace File
System...
Sep 02 11:01:14 luke-virtualbox dbus-daemon[463]: [system] Activating
via systemd: service name='org.freedesktop.home1'
Sep 02 11:01:14 luke-virtualbox sddm[580]: Setting default cursor
```

With thousands of such messages accumulated over months, it is easy to get lost, so let's see the options to make our lives easier. Remember to use the *Shift+PgUp* and *Shift+PgDown* buttons to navigate in the terminal.

Here are the most frequent uses:

- `$ journalctl -f` will continuously emit log messages *live* as they come in. This way, we can start a program to see whether it triggers some logs or indirectly triggers other process logs. Exit with *Ctrl+C*.

- We use the `-S` or `--since` and/or `-U` or `--until` options followed by a date/time to filter the output by a specific time range. Here are a few examples:

  - `$ journalctl -S yesterday -U today`: To view all logs from yesterday. We can also use the time specifier `tomorrow`. They all refer to time before and after 00:00. We may also use only `-S` or only `-U`.

  - `$ journalctl -S "2023-03-29 12:37:46" -U "2023-03-30"`: To view all logs between two dates. If the time part is omitted, it is set to `00:00:00`.

  - `$ journalctl -S "12:37" -U "13:46"`: To view all logs between two times on the current day.

- `$ journalctl -b`: To view all logs from a specific boot. A boot is one PC start; we have the next boot when we restart it. We can add a number, in particular, if we use negative numbers, we can access the previous boots. Executing `$ journalctl -b -0` will show log messages from the current boot, it is the same if we miss the `-0`. If we use `-1`, we will get the logs from the previous boot only. If we use `-2`, we will get the messages from the pre-previous boot, and so on.

- We can filter by log priority like this:

  - `$ journalctl -p warning` will provide us with all warning messages.

  - We can use priority `Code strings` or numbers, as listed in *Table 13.2*. Thus, `$ journalctl -p 3` shows all error messages.

  - `$ journalctl -p 2..4` works as a range, showing all critical, error, and warning messages.

  - We can combine priority with any date format we looked at earlier, like this: `$ journalctl -p err -S today`.

- $ `journalctl -u dbus.service`: Here, `-u` is for filtering by **systemd** unit. It will often work equally without the extension; in the `dbus` example, you can skip `.service`. Sometimes it will need an extension explicitly, such as for `boot-efi.mount` events. Use the $ `systemctl list-units` names, but try them with and without extensions. As for a few other examples, simply using `systemd` will not work; however, `systemd-journald`, `systemd-logind`, and `systemd-udevd` will work. We really need to know what we are looking for with this option.

- Adding `--system` will provide logs from system services and the kernel.

- $ `journalctl _PID=1` will give us logs explicitly from systemd (as we have specified **PID** 1). We can change the **PID** to any other one that works on our system and has log entries. The easiest is to try with some of the root processes with lower **PID** numbers. Remember that this does not always apply to user processes, as they vary in their PID.

- $ `journalctl _COMM=udisksd` will provide logs from a **process** with this name. We can use `systemd` as it definitely has logs. However, many processes will not write logs, in which case we will get `-- No entries --`.

- **Coloring**: By default, `journalctl` uses color coding to highlight different message priorities. Green is used for `debug` messages, blue for `info` messages, yellow for `warning` messages, and red for `err` messages, but these depend also on the terminal. I find the output most readable on **Konsole** with `zsh`, running on the KDE Plasma flavor.

- The following options can give us *insights about the available* journal log *fields*. First, use $ `journalctl -N` to print all *field* names used currently in all journal entries. Then, call $ `journalctl -F SYSLOG_IDENTIFIER` to see all existing entries for processes with logs. Now, we can use the `-t` option with any of them like this: $ `journalctl -t krunner -t konsole -t kaccess -t kernel -t NetworkManager`. The `-t` option is *special* – it can be used *multiple times in one command*, but when used so, it might have time-ordering issues when accessing many logs from longer periods.

- The final option we will look into is `-k`, or in its extended form `--dmesg`. This will show *only kernel messages* (by default, only from the current boot). It is based on the `_TRANSPORT` *field* with the *value* `kernel`.

To conclude, we can combine most options (as expected). Use `man` or `info` for more details.

## journald.conf

> **Important note**
> This subsection is for *advanced* users.

The journal configurations are in `/etc/systemd/journald.conf`. Currently, a recommendation is provided in this file to create a subdirectory, `/etc/systemd/journald.conf.d/`, for custom configurations and put an additional `.conf` file there. The `journald.conf` man page explains this in detail and provides a few other possible directories for custom configuration files. It is strongly advised to read it if you need custom configurations. If more than one custom `.conf` file resides in all listed directories, all `.conf` files will be sorted and processed lexicographically. When processed, they will override any changes in the default configuration. In addition, if a parameter is set to different values in those files, it will take the value in the last sorted custom file.

Let's first briefly look at all parameters and then explore an example of custom configuration:

- `Storage=`: Defines the storage mode used by journald. The options are **volatile**, **persistent**, **auto**, and **none**. The default value, **auto**, selects the most appropriate mode based on the system configuration, which, for a regular PC, is **persistent** in the `/var/log/journal` directory.

- `Compress=`: Enables or disables log compression by setting it to **yes** or **no**. By default, it is enabled to save disk space.

- `Seal=`: Enables or disables journal sealing with **yes** and **no**. By default, it is set to **yes**, and `journald` seals (encrypts) the journal files to prevent tampering or modification.

- `SplitMode=`: Specifies the log file split mode with the **uid** (the default) or **none** option. The **uid** option creates a separate log file for each user ID. Thus, messages specific to another user will not be listed when using `journalctl`. This option is related to `MaxLevelWall`, presented later in this list.

- `SyncIntervalSec=`: Sets the interval between journal file sync operations to disk and applies only for `error`, `warning`, `notice`, `info`, and `debug` messages. The default value is five minutes. Any `critical`, `alert`, or `emergency` message forces a flush to HDD (a write to the disk).

- `RateLimitIntervalSec=` and `RateLimitBurst=`: Configures the rate limiting applied to all messages generated on the system. During the time interval defined by `RateLimitIntervalSec=`, if a service logs more messages than `RateLimitBurst=`, all further messages are dropped until the next interval. A message about the number of dropped messages is generated. This rate limiting is applied per service, so two or more services that log do not interfere with each other's limits. The default is 10,000 messages in 30 seconds. The time specification for `RateLimitIntervalSec=` may be specified in the following units: `s`, `min`, `h`, `ms`, and `us`. To turn off rate limiting, set either value to `0`.

- `SystemMaxUse=`: Defines the maximum journal disk space for system logs.

- `SystemKeepFree=`: Specifies the minimum amount of free disk space that must be kept for system logs. You can investigate the contents of `/var/log/journal` to see the total number of files and space taken from journald logs since installing your Manjaro with **Filelight**. In the terminal, check with `$ journalctl --disk-usage` and `$ tree -L 3 /var/log/journal`.

- `SystemMaxFileSize=`: Defines the maximum single journal file size; by default, it is set to no value, which means no limit.

- `SystemMaxFiles=`: The maximum number of journal files stored for system logs, currently set to 100.

- `RuntimeMaxUse=`: Defines the maximum disk space for runtime logs. These are kept only in RAM and not saved as a file. This parameter, together with `RuntimeKeepFree=`, `RuntimeMaxFileSize=`, and `RuntimeMaxFiles=`, is applied only when `Storage=volatile`.

- `MaxRetentionSec=`: Specifies the maximum log files retention time – in other words, whether files older than a certain amount of time will be deleted.

- `MaxFileSec=`: Defines the maximum time to store entries in a single journal file before changing to a new one. For Manjaro, it is set to one month by default.

- `ForwardToSyslog=`: Enables or disables the forwarding of log messages to the system log via the traditional **syslog** daemon. By default, on Manjaro, it is set to **no**.

- `ForwardToKMsg=`: Enables or disables the forwarding of log messages to the kernel log buffer kmsg. By default, on Manjaro, it is set to **no**.

- `ForwardToConsole=`: Enables or disables the forwarding of log messages to the console. By default, on Manjaro, it is set to **no**.

- `ForwardToWall=`: Enables or disables the forwarding of log messages to all logged-in users. By default, on Manjaro, it is set to **yes**.

- `TTYPath=`: Defines the TTY device used for logging console messages when `ForwardToConsole=yes`.

- `MaxLevelStore=`: Sets the maximum log level for storing messages in the journal – currently set to **debug**, meaning all messages are stored.

- `MaxLevelSyslog=`: Sets the maximum log level for forwarding messages to the system log – currently set to **debug**, meaning all messages will be sent if `ForwardToSyslog=`**yes**.

- `MaxLevelKMsg=`: Sets the maximum log level for forwarding messages to the kernel log.

- `MaxLevelConsole=`: Sets the maximum log level for forwarding messages to the console.

- `MaxLevelWall=`: Sets the maximum log level for forwarding messages to all logged-in users. By default, on Manjaro, it is set to **emerg**; thus, any user will explicitly get emergency messages even from the kernel, whether they are a regular user, limited user, or superuser.

- `LineMax=`: Defines the maximum line length for log messages. Currently it is set to **48k**, which is absolutely enough. There are rarely messages with more than a few hundred symbols.

- `ReadKMsg=`: Enables or disables reading of the kernel log. On Manjaro, it is set to **yes**, so we can see kernel messages via `journalctl`.

- `Audit=`: Enables or disables auditing of system events. The auditing Linux kernel subsystem is responsible for logging security-related events and system activity, which can be used to detect and investigate security breaches, system misuse, or other unexpected events.

Now, with all these options, let's say that we want to change some of the settings. First, we create a subdirectory with $ `sudo mkdir /etc/systemd/journald.conf.d`. Then, with nano or Kate, we create a file in this directory named `custom_journald.conf`. In it, put the values you want to change like this:

```
[Journal]
Storage=persistent
Seal=no
SystemMaxFiles=50
```

Now save the file and then execute the command:

```
$ sudo systemctl restart systemd-journald.service
```

This is enough to apply the new configuration. Remember that the values presented here are only examples, not recommended ones. For regular usage, it is expected that you will rarely have to change the default configuration.

## Kernel ring buffer messages and dmesg

The **kernel ring buffer** contains logs generated by the kernel, including boot messages, driver information, and HW events. These messages, called **kernel messages** or **kernel logs**, can be read by various system utilities, most frequently the **dmesg** command. The word **ring** means it is a circular buffer that can store a fixed number of messages. When the buffer is full, each new message overwrites the oldest one. As this buffer is in RAM, **journald** periodically reads it and stores its messages on our HDD.

We will review the Linux kernel in *Chapter 16*. However, it is essential to note here that it is not limited to basic process scheduling. It also has integrated services based on subsystems for filesystems, device drivers, user management, virtualization, interprocess communication, timers, graphics, sound, cryptography, etc. The messages written to the kernel ring buffer include information about the kernel's interaction with all these subsystems, as well as error messages and system events warnings. Many subsystems may have limited functionality to write messages to the kernel log. Still, most of them write at least some data, so this log can be very helpful for troubleshooting system issues and diagnosing HW or SW problems. In addition, the kernel log is also helpful for monitoring system performance and resource usage, and for kernel modules and drivers debugging.

Though this log can be inspected both with `journalctl` and the GUI tool **KSystemLog** (presented in the next subsection), it is important to know what we can do with `dmesg`, the standard command-line utility for inspecting the kernel log. Here is a list of its most frequent uses with some options:

- `$ sudo dmesg`: Displays kernel messages since the last boot, sorted by time. The time format is the number of seconds since boot time, with an accuracy of microseconds. It looks like this:

  ```
  [    6.101671] input: PC Speaker as /devices/platform/pcspkr/
  input/input7
  ```

- `$ sudo dmesg -T`: Used to change the timestamp to human-readable format. However, we lose the microsecond-level accuracy:

  ```
  [Tue Apr  4 10:30:15 2023] input: Logitech Wireless Keyboard
  PID:4023 Keyboard as /devices/pci0000:00…
  ```

- `$ sudo dmesg -c`: Clears the kernel ring buffer. It is strongly advised not to do this!

- `$ sudo dmesg -l err,…`: Displays only kernel messages with the specified severity level(s) by using the codes from *Table 13.2*.

- `$ sudo dmesg | grep -i sda`: Displays only kernel messages containing the string `sda`, case insensitive. You can use any string to grep, remembering that the coloring will be cleared. You can also pipe with a pager like **less** for longer logs.

- `$ sudo dmesg -w`: This option will first list all messages since the last boot with the accurate timestamp. It then keeps the terminal blocked, showing you each new message arriving in the kernel ring buffer in real time. Press *Ctrl+C* to quit.

There are more options; check the `--help` and `man` pages. Most of the time, I prefer to work with **journalctl** or **KSystemLog**, as **dmesg** is limited, and we usually need more than the kernel messages.

## Other log systems and the KSystemLog GUI tool

There are also other log systems in the Linux world. For starters, we have the **syslog** tool, first introduced in the early *1980s* as part of BSD Unix. In *2004*, it got a successor named **rsyslog**, which added improved performance and features. **systemd-journald** can also get **rsyslog** messages, but on Manjaro, it is not configured to do so by default. In addition, as practically the whole system management is performed by systemd, **journald** contains all the information we might need. You can find more information on **rsyslog** at `https://wiki.archlinux.org/title/rsyslog`.

Without going into detail, here are a few other loggers and utilities: **syslogd** and **syslog-ng**, **logrotate** (installed by default on Manjaro), **Logwatch**, and **ulogd**. Typically, you will never need them on Manjaro.

There is one more critical system we shall mention – the **Xorg** log, which is the log of the graphical desktop environment. **Xorg** is an open source implementation of the **X Window System**. It provides a display server that implements the **X11** protocol and *allows GUI applications to communicate with the HW and graphics drivers* to display graphics on the screen. It is installed by default on all official Manjaro flavors. **Xorg** is the most widely used such implementation. The Xorg server messages are also handled by **systemd-journald**.

Finally, having a GUI tool with good filtering is often helpful. Manjaro offers **KSystemlog for this**, which is available in the official repositories. As additional tools don't have access to root-related resources, to access system logs, they need *root* privileges. On KDE and Xfce, it works flawlessly. On GNOME, it must be started from a root (su) terminal session via $ ksystemlog & to get the kernel logs. The su session is reviewed at the end of the chapter.

Now, let's say a few words about KSystemLog. By default, it will show you the **journald** log. At the top of the messages table, you have filtering capabilities by Date, Unit, Message, or the default All, which will search all fields. You can also select priorities. There are buttons for directly switching to the kernel or **X.org** logs. You can add tabs and even more logs. The Logs menu opens a rich selection of additional logs for multiple tools that are worth checking out. It is an excellent tool. *Figure 13.3* shows its default look:

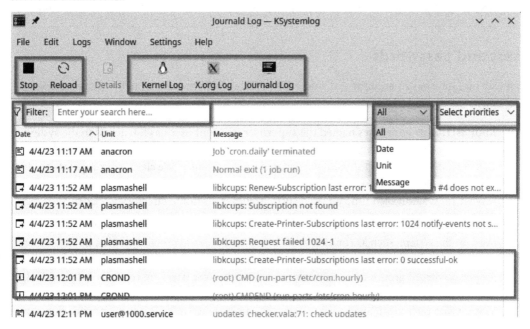

Figure 13.3 – KSystemLog screen view with options

# User management, groups, ownership, and root privileges

In any OS, user management is an essential part of system management. Many years ago, computers were too expensive, so teams within organizations and families had to share one, with different accounts for each member. Historically, less experienced members had fewer privileges to prevent them from installing inappropriate applications or harming the system. User management also encompasses parental controls, so parents can make sure their children don't get access to certain content and applications. In companies, user management ensures the company's confidential information security, protects against malicious SW installation, and provides employees with only the resources they need without full access to critical data. In addition, as remote Linux virtual environments are frequently used in companies nowadays, user management is essential for allocating security privileges to different users.

In this section, we will cover the different types of users and their privileges, how to manage them (and the groups they are in), and how to edit configuration files for users and superusers. It is easy once we know a few basic concepts, commands, and configuration files.

> **Important note**
>
> All presented commands can lead to unexpected effects and security issues when used without the necessary care. Be sure to double-check your intentions and also test the resulting behavior.

## Users and passwords

Each user on Linux has an **account** and a **user ID** (**UID**, a whole number from 0). There are several types of users/accounts in Linux:

- **Root user**: Also sometimes named the **superuser**, the **root** user has full access to the system, is unique, can perform any task, and has **UID** 0 and **GID** 0. This account is intended for *system administration* and should *only be used when necessary*. We elevate our regular user to its privileges with `sudo`, which applies only to the command provided directly after it. That's why we must add it each time we need such privileges. Its home directory is `/root`.

- **Regular users**: These are standard user accounts with *limited privileges*. They can perform basic tasks on the system, such as running applications and accessing their own files. By default, on Manjaro, they have no access to system files and cannot install applications; they are added to proper privileged groups and use `sudo` to do this. Regular users are typically assigned **UID** `1000` and bigger. Their home directories are located inside `/home`. When created, by default, a group with their username is created, and they are assigned to it. This is their primary group.

- **System accounts**: These are used for OS components that don't require root privileges and/or need to be limited to only access dedicated memory and other resources. They typically have UIDs between `1` and `99` and no home directory or shell.

- **Service accounts**: These are similar to system accounts but are dedicated to managing specific services' resources. Their UIDs are typically from `100` to `999`.

The **UID** is used by **root** to *identify users and determine their access level to services and resources.* The other important identification is the **group ID (GID)**, a numerical identifier for groups of user accounts. The root account always belongs to **GID** 0. There are multiple groups; we will look at some in the next section.

The passwords on Manjaro are encrypted with the **SHA-512** hashing algorithm. It has additional cryptographic **salting** applied, which makes passwords resistant to multiple attacks, including brute-force and dictionary-based attacks. **Salting** is the *addition to the password of random symbols specific to your PC and Manjaro installation* before the **SHA-512** hash is calculated. The resulting hash is saved in the /etc/shadow file, accessible only by **root**. Each time the root password is used, its hash is calculated and compared to the saved one. This solution was integrated into Arch Linux (then subsequently in Manjaro) more than 10 years ago, so it is rock solid and tested.

Now let's take a look at a few configuration files and manuals so that you know where to find the relevant information:

- $ cat /etc/passwd: This is the list of *all user accounts* with basic information. By default, our user can read it (so we don't need sudo) but cannot change it without root privileges. Lines within it are formatted as follows:

  ```
  username:password:UID:GID:comment:home_directory:login_shell
  ```

  The comment, home_directory, and login_shell parameters are *optional*. As the **passwords** are *saved in the shadow file*, we always have the letter x for them.

- $ cat /etc/login.defs: This holds the configuration control definitions for the shadow package. Like most Linux configuration files, lines starting with a hash sign # are commented. Here, we have only some of all possible configuration variables, and many are even commented out.

- $ man login.defs: This man page holds all possible parameters for /etc/login.defs, along with a detailed explanation of each of them.

- $ man "shadow(5)": This man page explains basic password parameters. The number in parentheses is necessary, as the default call $ man shadow will redirect you to the man page of a library for developers titled shadow(3). The *quote marks* are necessary as the string has *special symbols* (the parentheses, ()); with " ", we indicate that *this whole string* should be passed as an argument to man. Otherwise, the shell would try to interpret the data in parentheses as part of a command.

Many commands in Linux come with developer manuals. Thus, for some, we need an additional identifier, like "shadow(5)", to access the user's manual. Don't think that if we have one with 5, we might have all the natural numbers before this – we don't. If you're wondering how I found out about this manual, when calling the default man page without the additional parameter (5), there were *references for further reading at the end*. In addition, the same reference to shadow(5) is present at the end of the login.defs man page.

In addition to the `/etc/passwd` file parameters, passwords may have the following optional parameters::

- *Expiry date*: The date on which the given user account will be disabled, measured in days since `01.01.1970` – set by `useradd` or modified by `usermod`.

- *Inactive days limit*: The number of days a user password may be active, requiring the user to change it once reached – set by `useradd` or modified by `usermod`.

- `PASS_MAX_DAYS`, `PASS_MIN_DAYS`, and `PASS_WARN_AGE`: Three variables in `/etc/login.defs`. Here, `MAX` defines the maximum number of days after which password change will be forced, and `MIN` is the minimum number of days between password changes. The `WARN` variable controls how many days in advance the user will start getting warnings that their password will soon expire and needs to be updated. These are their default values.

In the following subsection, we will briefly review the `useradd`, `passwd`, `userdel`, and `usermod` tools, and a few others for *groups*. To repeat: for more information, use the man or info pages for each – they are detailed and usually explain everything you need to know. Almost all belong to the `shadow-utils` package, which has an up-to-date man pages summary at `https://www.mankier.com/package/shadow-utils`.

### Managing users

We start with the `useradd` command for adding users. Know that to activate an account after using it, you must use the `passwd` command presented after it. *A username can contain only upper and lowercase letters, digits, underscores, or dashes* and cannot begin with a dash, `-`. The following are the most common uses of `useradd`. Put the new username at the end when you combine options:

- `$ sudo useradd john`: Create a new user named `john` with the default options set by `login.defs`. Currently, on Manjaro Xfce, a new home directory will not be created, as the `CREATE_HOME` variable is commented in `login.defs`.

- `$ sudo useradd -m john`: Add a new user and *create* a home directory with its name in `/home/`.

- `$ sudo useradd -m -d "/home/NewUserSpecificDir" john`: Create a new user and *specify a custom home directory*.

- `$ sudo useradd -M john`: Create a new user and *don't create a home directory* for it, even if the *system-wide setting* from `login.defs` (`CREATE_HOME`) is set to **yes**.

- `$ sudo useradd -s <shell> john`: Create a new user and set the user's login shell path. Without this option, the system will use the `SHELL` variable specified in `/etc/default/useradd` (explore it with `sudo cat`). If that is also not set, the field for the login shell in `/etc/passwd` remains empty.

- $ `sudo useradd -U john`: Create a new user and a group with the same name, and add the user to this group, making the group its primary one (currently the default behavior on Manjaro). The default behavior (if the `-g` and `-N` options are not specified) is defined by the `USERGROUPS_ENAB` variable in `login.defs`.

- $ `sudo useradd -g <group> john`: Create a new user and add it to an *existing group* identified by **name** or **GID**.

- $ `sudo useradd -e <expiry_date> john`: Create a new user and set its expiry date, specified in the format `YYYY-MM-DD`.

- $ `sudo useradd -f <inactive_days> john`: Create a new user and define the number of days *after password expiration during which the user can update their password*. The value is stored in the `shadow` passwords file. An input of `0` will disable an expired password with no delay. An input of `-1` will blank the respective field in the `shadow` passwords file.

- $ `sudo useradd -r john`: Create a system account. System users are created with no aging information in `/etc/shadow`, and their numeric identifiers are chosen in the `SYS_UID_MIN-SYS_UID_MAX` range, as defined in `login.defs`. A home directory will not be created for this user, regardless of the defaults in `login.defs` for `CREATE_HOME`.

Next, we continue with the `passwd` command. Remember that *we must use it after adding the user with* `useradd`:

- $ `passwd`: When used without arguments it works for the currently logged-in user. We are first asked for the current password and then asked twice for the new one. To change the password of any other account, we must use `sudo` or a special account with administrative privileges.

- $ `sudo passwd john`: Sets a new password for the given account. You will be asked twice for it.

- $ `sudo passwd -S john`: Provides the password status for the given user.

- $ `sudo passwd -S -a`: Provides a full status for all users with the output format `[username] [status] [last_change_date] [minimum_age] [maximum_ age] [warning_period] [inactivity_period]`. The last four are provided in days and displayed with `-1` if disabled. `[status]` is interesting:

  - `L` means a *locked password*, so the user cannot log in until it is unlocked.

  - `NP` means *no password*. The given user cannot be logged in with a password as such doesn't exist.

  - `P` means a *usable password*.

- $ `sudo passwd -l john`: Locks the password for the account, but note that the account itself remains. It is active but can't be logged in to with a password. However, with an SSH key (if one is set), it can be logged in to.

- $ sudo passwd -u john: Unlock the password for the given account, reverting it to its value before using the -l option.

- $ sudo passwd -e john: Expire the user's password immediately, forcing them to set a new one at the next login.

- $ sudo passwd -d john: Delete the user's password; they will not be able to log in anymore.

To delete a user account, use the userdel command. Here are its most common uses:

- $ sudo userdel john: Delete the given account without removing its home directory and personal data. This works only if the user is not logged in.

- $ sudo userdel -r john: Delete the user and their home directory, files, and mail spool. This works only if the user is not logged in.

- $ sudo userdel -f john: According to the man page, with the force (-f) option, the user will be deleted even if logged in. However, this is not guaranteed and may lead to unexpected issues. *Do not use this option*.

For more options, refer to the man page and this article: https://pimylifeup.com/userdel-command/.

The last command we will review in this subsection is usermod. Some of its options partially resemble useradd, as this command changes *existing users'* options. However, they are different, so *don't count on this*! Here are a few common examples of its use:

- $ sudo usermod -d /home/user2 john: Changes the home directory associated with the user to /home/user2. If the –m option is given (before or after the whole string, -d /home/user2), the contents of the current home directory will be *moved* to the new home directory. When moving, the new home directory is created if it does not exist. If the current home directory *doesn't exist*, the new home directory *will not be created*.

- $ sudo usermod -e 2023-12-31 john: Sets the user expiration date.

- $ sudo usermod -f 30 john: Sets the inactive password lockout period for the user to 30 days.

- $ sudo usermod -l user2 john: Changes the username john to user2.

- $ sudo usermod -L john: Locks the user by putting an exclamation mark in front of the encrypted password.

- $ sudo usermod -U john: Unlocks the user by removing the exclamation mark from the encrypted password.

- $ sudo usermod -s /bin/zsh john: Changes the login shell associated with the user to the specified one.

- $ sudo usermod -u 1015 john: Changes the **UID** associated with the user to 1015 and accepts the new value only if it is not in use!

- $ sudo usermod -g someGroup john: Changes the primary group associated with the user to someGroup.

- $ sudo usermod -G group1,group2,group3 john: Sets the groups the given user belongs to. The groups *should all exist*, and when more than one is provided, they should be separated with commas and without spaces. If the user belongs to another group that is not on the list, they will be *removed from it*!

- $ sudo usermod -a -G group1,group2,group3 john: Adds the user to the listed groups without removing them from other groups. You can also use the combined options format -aG.

- $ sudo usermod -c "Some User" john: Changes the comment associated with the user to Some User.

## Groups

In Linux, groups are collections of users with common privileges. Each group is assigned a unique name and a **Group Identification (GID)** number. Groups can be used to manage files and directory access and to grant users elevated privileges.

When your user was created during the Manjaro installation, a group with your user's name *was created automatically*, which is its primary group. The regular users' **GIDs** start from 1001. The root user belongs to the root group with GID 0 and is only to this group. Regular users can be added to multiple groups. In addition, we have *system groups* created for special devices, daemons, and others. To list all groups, execute:

```
$ cat /etc/group
```

The format in this file is as follows: [name] : [password] : [GID] : [user1,user2,user3…].

[password] is a *legacy* field, and nowadays it's rarely used. When set, the group password is necessary to add users to the group. The last parameter in the format is a comma-separated list of users belonging to the group. In the same manner as for single users, the groups have a *shadow file*, /etc/gshadow. Its man page is called with $ man gshadow.

Of all the groups, the most famous is wheel. It dates back to Unix times when several users needed to share root privileges. As this was considered a security issue, trusted users with administrative privileges were added to the wheel group. Currently, on all Manjaro flavors, the user account created during installation will be added to it by default, which is *why you can use* sudo. We will see later how to remove the need to enter the **root** password after sudo, which is possible *but is not recommended to be permanently used!*

As we know, *everything on Linux is a file*, and every file on Linux is owned by a **user** and belongs to a group provided by `ls -l` or `exa -lg`. In addition, each file has permissions, as we have seen in *Chapters 7* and *10*. The permissions are always for *read, write, and execute*: first for the *owner*, then for its *main group*, and then for *others*. This way, system administrators can manage access to any file or directory. As *devices are also files*, such permissions apply to network devices, HDDs, keyboards, Bluetooth, and so on.

### Managing groups

Managing groups is easy with a few commands. Here they are:

- `$ groups`: Lists the groups your user is added to. This will skip all system groups. With `sudo` or from a root session, this command will print only the `root` group.
- `$ sudo groupadd developers`: Creates a new group called `developers`.
- `$ sudo groupadd -g 1566 developers`: The additional `-g` option sets a **GID** explicitly, which must be *unique and not in use*.
- `$ sudo groupadd -r newSystemGroup`: The `-r` option creates a new system group, normally *reserved for daemons and special purposes*. It should not be used for regular users.
- `$ sudo groupadd -U user1,user2,user3 newGroup`: The `-U` option adds users via a comma-separated list, as in this example.

To remove a group, use `groupdel` as follows:

- `$ sudo groupdel group1`: Delete the given group and all entries that refer to it. If this is the primary group of any existing user, you will be warned and the command will fail. You must remove the user or change its primary group to another existing one before you remove the group. In addition, you should manually check all filesystems to ensure that no files remain owned by this group.
- `$ sudo groupdel -f group1`: The force parameter will remove the group even if a user has it as their primary group. It is strongly advised *never to use it*!

To manage the group password, use `gpasswd` as follows, noting that only the `-A` and `-M` options can be combined:

- `$ sudo gpasswd group1`: Creates a password for `group1`
- `$ sudo gpasswd -a user1 group1`: Adds a user to `group1`
- `$ sudo gpasswd -d user1 group1`: Deletes a user from `group1`
- `$ sudo gpasswd -r group1`: Removes the `group1` password
- `$ sudo gpasswd -R group1`: Restricts the access to `group1`

- `$ sudo gpasswd -A user1,user2,user3 group1`: Sets the given users as group **administrators**

- `$ sudo gpasswd -M user1,user2,user3 group1`: Sets the given users as group members

To manage a group as an **administrator**, we use `groupmod` and `groupmems`:

- `$ sudo groupmod -U user1,user2,user3 -a group1`: *Adds* the listed users to `group1`

- `$ sudo groupmod -U user1,user2,user3 group1`: Sets the listed users to `group1` and doesn't keep the old users

- `$ sudo groupmod -n newGrName group1`: Renames `group1` to `newGrName`

- `$ sudo groupmems -a user2 -g group1`: Adds a user to `group1`

- `$ sudo groupmems -d user2 -g group1`: Deletes a user from `group1`

- `$ sudo groupmems -l -g group1`: Lists the group members

- `$ sudo groupmems -p -g group1`: Purges all group members

The `groupmems` manual states that we can use it to manage the group members of our main group without `sudo`. However, on Manjaro, this *doesn't work as expected* for my group `luke` with my user `luke`. As having a user and a group with the same name can lead to confusion on the command line, I recommend not using it for your own group and without `sudo` (even if this works in the future or on some other distribution).

If you need to work with multiple users and groups, you usually are the administrator and set the names accordingly for both groups and members. In this case, users will rarely need to use `groupmems` by themselves. Instead, it is better to use the `useradd` command options, `groupmod`, and `groupmems` with `sudo`.

## Changing file or directory mode, ownership, and group

We briefly reviewed the **GNU** `coreutils` package commands `chown` and `chmod` in *Chapter 7*. Here, we will review them in more detail, together with `chgrp`.

> **Important note**
>
> These commands are expected to be used only on your files! With `sudo`, you can change any file, and you should double-check your intentions and the resulting behavior to avoid unexpected effects on your system and SW!

## chown

chown changes a file or directory user and group **ownership**. It has the format:

```
chown [OPTION] ... [OWNER] [:[GROUP}} FILE_OR_DIRECTORY
```

Whenever you want to apply it to a directory and its contents, add -R as OPTION. Here are its most common uses:

- $ sudo chown john /home/SomeDirOrFile: Changes the *owner* of SomeDirOrFile to john. You can also use a **UID** value instead of the name john. The group will not be changed.

- $ sudo chown root:administrators /home/SomeDirOrFile: Changes the *owner* to root *and its group* to administrators.

- $ sudo chown john: /home/SomeDirOrFile: Changes the *owner* of SomeDirOrFile to john. When *a colon and no group is given*, the group is set to john's *primary group*.

- $ sudo chown -R user7:grRegular --from=john:administrators /home/ SomeDir: Changes the *owner* of SomeDir to user7 and the *group* to grRegular *recursively*. The optional --from=john:administrators is *a filter*, which means the change will be done *only to files and subdirectories owned* by john and the group administrators. The filter can also be written only with a user (john) or only with a group (:administrators).

- The filter option from the previous example can also be applied for *single files* when the -R option is not specified.

- For the -R option, there are additional modifiers for traversing symbolic links (shown in the following sub-list). By default, symbolic links are *not followed*. These three options *cannot be combined*, and if more than one is provided, only the last one is applied:

  - H: If a command-line argument is a symbolic link to a directory, traverse it

  - L: Traverse every encountered symbolic link – this will not change the link ownership itself

  - P: Do not traverse any symbolic links (the default)

An example would be $ sudo chown -RL :someGroup ../Directory5. In this case, we will recursively change the group (and not the user) of all files in this directory, *following all symbolic links*. The target directory is specified with a relative path to the current directory – in this case, it means "First go one level up, then execute the task for Directory5 in it."

By default, **symbolic links** ownership will not be updated; instead, only the pointed content will do – whether it is a directory or a file. If you want to update the links themselves, you have to add the -h parameter. For a file, an example would be $ sudo chown -h user6 file5, where file5 is a symbolic link to a file. The file ownership will not be modified, but the link to it in the current directory will have its owner updated.

For the recursive traversal case, see this example:

```
$ sudo chown -hR :newGroup Directory10
```

This command will recursively change the group of all files and links in `Directory10`, but no files located outside of `Directory10`.

The conclusion for any link-related uses of `chown` is always to double-check your commands, as using the wrong flag may lead to unexpected results. In addition, complex combinations have precedence rules, meaning some combinations may result in unexpected behavior.

- `chown` also allows multiple target directories or files – here is an example:

```
$ sudo chown -R user7 --from=john:administrators /home/SomeDir
Directory2 Directory3
```

Here is another example:

```
$ sudo chown user7 --from=john:administrators file1 file2 file3
```

- `$ sudo chown -Rv root /home/SomeDir`: The added v flag will enable verbose mode, which will report each change attempt. You can also add the `-v` option to single files. If you want to see only the changed files, use c instead of v.

- `$ sudo chown -hRv $USER /home/SomeDir`: This last example shows how we can use the environment variables to get the name of our user. This can be helpful in scripts. For me, the shell will replace `$USER` with `luke`.

## chgrp

`chgrp` is an alternative dedicated to group change. It is written by the authors of `chown`. Its primary use is to avoid the more complex `chown` syntax, in case you want to modify *only the group*. Its syntax is the same: `$ sudo chgrp newGroup FileOrDirectory`. Of all the `chown` options, it supports c, v, h, and R (with its extensions H, L, or P). Use the examples of `chown` for them; just skip the `user:` and colon parts when present.

## chmod

The last command in this section is `chmod`. With it, we change files and directory **permissions**, also known as **mode**, hence the name. It has the following format:

```
chmod [option] [permissions/mode] [path]
```

The most common options are as follows: -v to output verbose information, -f to suppress most error messages, -c to report verbosely only when a change is made, and -R to change files and directories recursively. There are three interesting long options:

- --preserve-root will explicitly fail attempts to change root directories permissions recursively.
- --no-preserve-root will explicitly accept attempts to change root directories permissions recursively. This is the *default behavior*.
- --reference=RFILE will *copy* the **permissions** settings *from a reference file* to the target path.

You can use chmod on your files and directories without sudo, but *you will need it for root-owned files*. Usually, a user is allowed to change the permissions of a file owned by another user only if the given file has enabled *write* **permission** for **others**. The second case when this is possible is when your user and the owner are *in the same* **group**, and *write permissions* for **group** members are *enabled*. The **root** can always change all permissions, no matter the file. If unsure of the result, you should add -v for full operation execution status, to revert your changes if you make a mistake.

We have covered the numeric representation of permissions in *Chapter 7*, specifically the *chmod, chown, and file permission settings* section. Here, I will cover in detail the letter options.

The permissions can be provided in the form [u/g/o/a] [+/-/=] [r/w/x/X/s/t]. This representation means that you can combine symbols from each square-bracket sequence. The resulting combinations might be, for example, u+x, o=r, or a-x. Here is the meaning of the letters:

- [u/g/o/a]: Are for *User*, *Group*, *Others*, and *All*, respectively.
- [+/-/=]: Here, + is to *add* the permission, - is to *remove* it, and = is to *set it exactly*.
- [r/w/x/X/s/t]: The first three are, respectively, the frequently used *read*, *write*, and *execute*. We will not look into X, s, and t usage here, as they are not necessary for daily usage. You can read more about them on the man page.

There are two exceptions from the mentioned equal sign format o=r. The first is when omitting a letter *after the equal sign*, =. Thus, o= means *remove all permissions* for the given users – in this case, for others. The second is when you copy permissions to one group from another – you can do this with any group on the two sides of the equal sign. Thus, g=o will mean that the *group* will copy the permission settings of *others*.

When you use chmod with -c or -v, you will get a report of the changed permissions both in numeric form and symbolic letter form, like this:

```
mode of 'File' changed from 0644 (rw-r--r--) to 0744 (rwxr--r--)
```

Thus, when using the mentioned information from *Chapter 7*, you can easily exercise and understand what some numeric forms mean.

The following are some example chmod uses:

- `$ chmod -c u+x /home/someFileOrDir`: Add execution permission to the user.

- `$ chmod -c o-w /home/someFileOrDir`: Remove write permission for others.

- `$ chmod -c g=rx /home/someFileOrDir`: Set all permissions for the group exactly and only to *read* and *execute*.

- `$ chmod -c -R u+rwx /home/someDirectory`: Enable all permissions for the user recursively, without changing any others.

- `$ chmod -c g=u /home/someFileOrDir`: This means the group copies the user's permissions.

- `$ chmod -c -R --reference=file1.txt /home/someDirectory`: Here, we take the permission string of `reference file1.txt` and set it recursively to `someDirectory` and its contents.

- `$ chmod -c 777 /home/someFileOrDir`: This sets all permissions to everyone; *please avoid using it*. Other famous numeric representations are 700 (everything for the owner only), 666 (no one executes), and 755 (the owner can do anything, but the group and others can only read and execute). Refer to *Chapter 7* for further details.

## Root user

As we have seen throughout this book, the **root** user has full access to the system and can perform any task. This account is intended for system administration and should *only be used when necessary*, as using it for everyday tasks is a direct security risk. The term **superuser** originated from Unix, where it was used to refer to the **root** user. As Unix evolved, the term became more used to refer to users with elevated privileges, regardless of their account name.

As we already know, regular users can use the `sudo` command to perform tasks requiring elevated privileges if they are in the `wheel` group. This allows them to temporarily gain administrative access to the system without logging in as the root user.

### su session

There is a possibility to open a root session directly in your user's terminal. This is done with the `su` command. It will ask you for the **root** password and change the default terminal's prefix string. In my case on Xfce (for the user `luke`), the prefix is the default: `[luke@atanas-machine test1]$ su`.

After executing `su`, it changed to `[atanas-machine test1]#` without the username before the @ sign. Be extremely careful – when in this session, your **home directory** *is not* /home; instead, it is /root. Navigating with `$ cd ~` will lead you directly into /root. Many **environment variables** will have *different values* – **$HOME** and **$USER** for starters. This is why `su` sessions *should be used on limited occasions and with a clear purpose*. To exit it, execute `exit`, and you will return to your user's terminal session.

### *Editing configuration files and getting sudo privileges*

Finally, we reach the goal of many users: making their user **root**-*powerful* and eliminating the tedious need to enter their password after sudo. Be warned – this *should never be done permanently* on systems that have *critical information*, *manage access to critical infrastructure*, and manage *critical networks and systems*. Providing a regular user with root privileges means effectively lowering one of the few strong barriers between malicious software and the capabilities of the root account to change anything on a system. With that in mind, on Manjaro (and most other distributions), to get sudo **privileges** for your account, you have to edit the /etc/sudoers file. You can cat it with sudo and access its manual with $ man sudoers. The other related file is /etc/sudo.conf, but it doesn't require a change in this case. You can access its manual with $ man sudo.conf.

By default, sudoers is set to be edited with **visudo**. This is because visudo checks the syntax of the sudoers file before writing changes, so making a mistake in this critical file is avoided. However, as we will not make serious changes, we will use nano to edit the sudoers file like this: $ sudo nano /etc/sudoers. After opening with nano, use the *PgDown* and *arrow* keys to navigate to its end. My user is luke, and for it I enter the line luke ALL=(ALL:ALL) NOPASSWD: ALL. Do the same for your user, then press *Ctrl+s* and *Ctrl+x*. Now exit the terminal and open it again. Try dumping on the terminal the /etc/sudoers file with sudo – you will not be asked for your password.

Some guides recommend doing this for the wheel group. This is effectively the same but valid for all wheel members. I strongly discourage it. To remove the sudo privileges, simply edit /etc/sudoers, add # to the additional line, and save the file.

## Summary

In this chapter, we started with an overview of **processes**, daemons, and how to inspect those running on our system. We then investigated **systemd**'s basic concepts and got deeper into **init** sequence analysis and **service** management with systemctl. After that, we briefly reviewed the virtual Linux consoles. We then explored the usage of journalctl and system logs, with detailed explanations and examples. Finally, we delved into the critical topic of user management, groups, and ownership settings to learn how to manage any user and file.

In the next chapter, we will continue our journey by learning how to optimize such a clean and effective distribution as Manjaro. We will also investigate how to use inxi and get specific system state reports when problems arise. We will finish the chapter with a short explanation of how to keep our home partition when reinstalling.

# 14

# System Cleanup, Troubleshooting, Defragmentation, and Reinstallation

In the previous chapter, we investigated processes, daemons, and systemd (diving deeper into the latter) and discussed the OS initialization process and service management. After briefly looking at Linux virtual consoles (TTYs), we learned how to investigate all system logs on Manjaro and checked a few tools related to them. In the last part, we dived into the basics of user management, groups, ownership, and root privileges.

This chapter continues our journey with file cleanup, from which any actively used system will benefit. We will then review hard disk partition defragmentation and when this is necessary on a Linux system. As *troubleshooting* is essential, we go on with the **inxi** tool, the best way to get system information. We will also look into an example of driver troubleshooting. We will close the chapter by learning how to keep our home partition during a Manjaro reinstallation.

The sections in this chapter are as follows:

- Cleaning unnecessary files, pacman, and caches
- Defragmentation – do we need it at all and how to do it
- inxi, troubleshooting, and mhwd
- Reinstalling and keeping our original /home partition

# Cleaning unnecessary files, pacman, and caches

Cleaning is an essential part of system maintenance. Manjaro is an optimized and highly trimmed OS that doesn't accumulate files without a reason. However, systems with low amounts of HDD space would benefit from cleanup. For this reason, we will review the systems on Manjaro that may need such checks, along with the inspection tools and ways of cleaning.

## Where do unnecessary files come from?

Manjaro is universal and supports a wide range of **hardware (HW)**, so it holds *thousands of packages*. As a **Rolling Release** distribution, it downloads updates frequently. In April 2023, by default, pacman checks for updates *every six hours* and keeps *three versions* for each package – two previous and the latest. If the latest version has some *bugs* or *incompatibility* with another package or running **software (SW)**, we can easily switch to a recent version. pacman will not consume more hard disk space with each new update. Once it reaches the three versions for frequently updated packages, it will stay at a certain amount of *consumed space*, not expected to grow much more. It is by design self-maintaining, limiting the number of kept packages.

> **Important note**
>
> I don't recommend occasional updates, as this leads to an accumulation of packages to be updated. For example, if I don't update some of my machines for over a month, I usually get between 1.5 and 2 GB of updates. This leads to longer download and installation times. I usually update at least once per week.

In contrast, **servers** are usually set up to never update automatically, so new packages are added only when necessary and with explicit action. They have thousands of users and work 24/7, and administrators update packages only when they are sure no service will be disrupted.

> **Important note**
>
> **Manjaro** is *not recommended* for **servers** due to its Rolling Release development model. This is officially stated in the Manjaro Forum by its creators. On the other hand, if we want to develop and use the latest SW and improvements, Manjaro is an excellent distribution, keeping us constantly up to date.

The other source of *extensive HDD space usage* comes from applications that keep a large amount of cached data. A classic example is the *browser cache* – many interactive websites use **JavaScript** and **Node.js** client libraries. Instead of downloading them each time, the browser gets them once until a newer version is available. In this way, websites are loaded much faster when opened. Of course, **cleanup** strategies that remove not recently used data and files are implemented.

The last source of excessive disk usage is too many and/or big packages and *user installations*. An example of such can be most **Steam** platform games downloads. The **Dota2** game needed 40 GB of space on one of my machines. Of course, if you are a gamer, you should consider this in advance and buy a machine with a larger HDD. You can also change your HDD or add one if you have an additional free slot.

Space problems are rare if your partitions have enough free capacity (e.g., over 100 GB for root). As mentioned in *Chapter 2*, Manjaro requires a minimum of *30 GB*. On one of my primary machines, after 6 months of extensive use and without installation of big applications, Xfce takes *40 GB* for root and *5 GB* for home. By comparison, a freshly installed KDE without additional software takes *11 GB* for root and *40 MB* for home.

## When do we have a problem?

Partitions with less than *a few percent free capacity* have **low performance**. Generally, having *at least 10%* free capacity and *at least 10 GB* free space is good for normal partition operation. If your home partition is not separate from your root partition, you should aim for an absolute minimum of *20 GB* of free space. A partition is like a book library – when quite full, adding new books from different genres and rearranging the shelves is hard. When you need lots of space, having an additional HDD for installations and big amounts of data is the best. As a result, we have the following main problematic cases:

- Limited free space of the root partition when the home directory is located in the root partition. This is the case with the *default Manjaro installation*.

- Limited free space of the root partition when the home directory *is on a separate partition*.

- Limited free space of the home partition when we need to keep a *larger amount* of personal data on it. In this case, backing up data we rarely need to an **external HDD** or a personal **NAS server** is the best option.

## How to analyze space usage

The preinstalled filesystem information terminal tool is df. In its listing, you must look for entries starting with /dev/sdXY, where X is a letter (a,b,c, and so on), and Y is a number. I prefer the duf command, which reports the status in a better-formatted table with colored text. Install it and run it without arguments. It will first report the local  devices we are interested in and then special devices, such as temporary filesystems. *Figure 14.1* presents an example of a Manjaro KDE installation I use for tests. Except if I did some big application installation, I never had problems with this one.

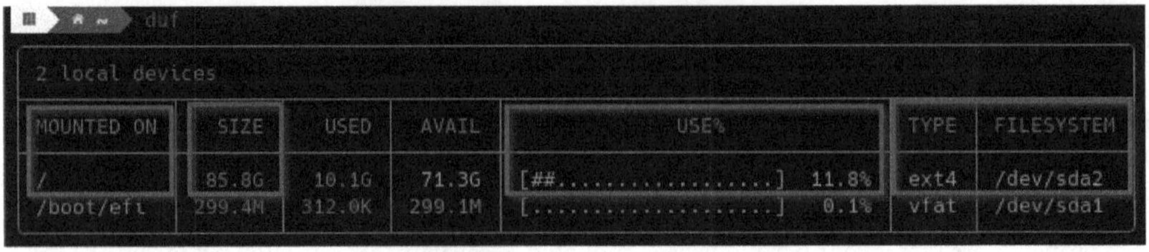

Figure 14.1 – The duf results for a fresh Manjaro KDE installation without a separate home partition

In *Figure 14.2*, we see another example from a 6-month-old Xfce installation. On it, the root partition is almost 84% full, and the separate home partition is mostly free. I will be happy to do some cleanup on it soon, as I prefer to be closer to the *at least 10 GB and 10 % free space* target.

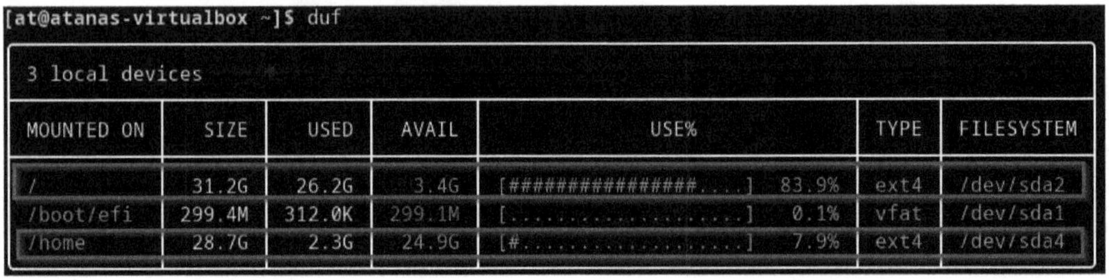

Figure 14.2 – The duf results for 6-month-old Xfce installation with a separate home partition

To investigate via GUI, you can always use **GParted** or **KDE Partition Manager**, presented in *Chapters 2* and *10*. In them, locate the entry named or labeled root, with **mount point** /; for home look for **mount point** /home. To *investigate visually* any partition or directory, use the GUI application **Filelight**. It has three inspection options – for root, home, or *a path of your choice*. Its visual representation with *pie charts* is comprehensive, and it can navigate subdirectories with double clicks and buttons.

## What can we do on Manjaro?

In this section, we will review the possible ways to perform cleanup on Manjaro.

### Cleaning the pacman cache

Before doing this, you should ensure your system is *stable* and all your applications work correctly. When you clean the cache, the older versions will not be available anymore, so you cannot downgrade packages or applications. That said, after all these years, I never had to downgrade regular packages and applications, except when a given application I installed was not explicitly ported for Arch or Manjaro.

We can use `pacman`, but with `pamac` is easier, as it has a more verbose set of commands. Both tools are reviewed in detail in *Chapter 8*. Here, we will look at a few specific `pamac` options. Remember that the `-d` short option stands for `--dry-run`, which only checks what will be done with the given command *without actually doing it*. Thus, remove `-d` for the following examples to execute them for real. The `-v` option is for verbose execution, providing details of what is or will be done. Use the `duf` command before and after the cleanup to check what amount of disk space you have:

- `$ pamac clean -v -d` – This command cleans the cache up to the number of versions set to be kept.

- `$ pamac clean -v -k X -d` – When you replace the X argument of the `-k` option with a number, this command will set the number of kept packages. Setting it to 0 will show you the total amount of kept packages with sizes.

- `$ pamac clean -v -k 1 -d` – On the extensively used installation, I got `To delete: 1414 files (2.9 GB)`. With `-k 0 -d`, I got `3374 files (5.3 GB)`.

- `$ pamac list -o` – The `-o` option stands for *orphaned* packages, so this command lists those *installed as dependencies* but *no longer required* by any installed package.

- `$ pamac remove -u -d <packageName>` – The `-u` option stands for *unneeded*. This option is used to force `pamac` to check and remove a single package only if it is not required from any other.

- `$ pamac remove -o -d` – This removes any *orphaned* packages. If you are unsure whether you need some of those and remove them all, you can easily reinstall only specific packages you need later.

Regarding the number of versions kept – you can change this as explained in *Chapter 8*, either by editing `/etc/pacman.conf` or changing it from the **Pamac GUI** application.

### Uninstalling any unnecessary application

When a PC has been used for years, we often install applications to test and often forget to uninstall them when they are no longer needed. The only recommendation here is to *remove such applications periodically*. It is best if you do it immediately after testing them. Be warned – on Linux, many applications are often required as **dependencies** for others, so consider removing only specific explicitly installed applications. If you want to remove applications installed by default on your system, check its **dependencies** with `pactree` or `$ pamac info <package>`.

### Following the Manjaro guidelines

The `https://wiki.manjaro.org/index.php/System_Maintenance` page provides several recommendations.

The first is about the `/home/user/.cache/` directory, which keeps *cached* content of different applications SW. As recommended in the guide, to generate a list of its contents with their sizes, execute the following:

```
$ du -sh ~/.cache/*
```

To automatically purge all `.cache` files that have not been accessed in 100 days, execute:

```
$ find ~/.cache/ -type f -atime +100 -delete
```

On my not-recently-cleaned Xfce installation, the only SW with significant data in `.cache` was Mozilla Firefox. It had 368 MB in `/home/user/.cache/mozilla`.

This guide also has a nice recommendation for cleaning up `journalctl` **logs**:

- `$ journalctl --disk-usage` – Gives the current log size summary. On my extensively used Xfce machine, I have 1.5 GB of systemd logs! I no longer need them and will delete most with one of the next two commands.

- `$ journalctl --vacuum-size=50M` – Removes all but the most recent entries by size.

- `$ journalctl --vacuum-time=2weeks` – Removes all but the most recent entries by time.

- To set a maximum size in MB for the journal, you can uncomment and edit the `SystemMaxUse=` line in `/etc/systemd/journald.conf`, adding, for example, `50M` after the equal sign for 50 MB maximum log size. This change is acceptable only if you don't need extensive **systemd** logs. For more information on `systemd` **logs**, read *Chapter 13*, the *journalctl and system logs* section.

Read the article mentioned at the beginning of this section for a few more hints.

### Using a cleaning tool

Among the many cleaning tools, notable mentions are **Stacer**, **BleachBit**, and **FSlint**. Of these three, only **BleachBit** is available in the official repositories; the others are available from **AUR**. Some users say any such tool is dangerous, as it might delete necessary files. They are partially correct, as some of them had issues in the past, but hardly anyone has had problems using **BleachBit** for basic cleanup in recent years. If you are unsure, you can use it at least for analysis. It has a simple GUI, shown in *Figure 14.3*:

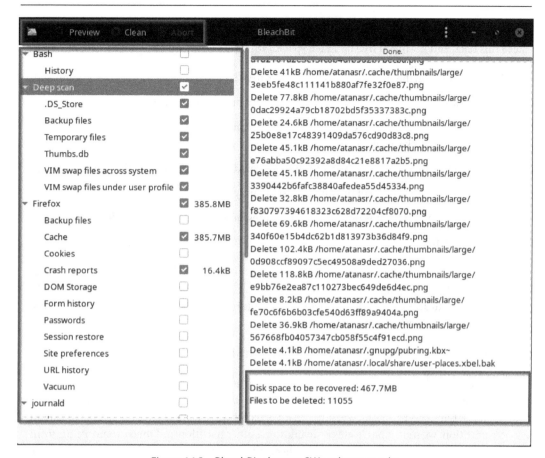

Figure 14.3 – BleachBit cleanup SW and scan results

On the top bar, from left to right, we have the menu icon and then the **Preview**, **Clean**, and **Abort** buttons. You can navigate to **Preferences** from the menu, but the default ones work perfectly. I only disabled the GUI's **Dark mode**.

The main left part of the screen provides several categories to clean. Currently, they are **Bash**, **Deep scan**, **Firefox**, **journald**, **System**, **Thumbnails**, and **X11**. Whenever you select a *slow-to-scan* category or subcategory, you will be warned. Select some or all categories on the left to scan your machine and press the **Preview** button. On a modern PC with a 4-core CPU and 8 GB of RAM, it took me less than 10 seconds to scan all. Then, you can deselect some of them and clean only what you need with the **Clean** button.

For the categories, we can consider the following recommendations. Some are pointless to clean, as we need this data, and it will not free up much space. Others are large and are always worth cleaning:

- **Bash** – Never clean this. Its history is small and we need it to operate with the `history` command.

- **Deep scan** – You can select the whole category, but on my machine, it was empty.

- **Firefox** – Clean only the *cache*, as it is the only big consumer. *Cookies* are pointless to clean as they are small and keep site preferences.

- **journald** – Don't select this; it is better to clean **journald logs**, as explained in the previous section.

- **System** – Cleaning the recent documents list is rarely necessary.

- **Thumbnails** and **X11** – It is pointless to clean these as they are typically small.

- To clean **Firefox**, you have to close it first.

If you want to clean the system directories (in my case, I had over *700 MB* of *localizations*), you must run **BleachBit** from the terminal with `$ sudo bleachbit &`. When you run it like this though, you will not be able to access the **Firefox** category for cleaning and analysis, as it is *a user space process*, and with `sudo` we run with `root` context.

### Extending the root partition if space is available

*Chapter 10* discussed the size change for `root` or other partitions but focused on *shrinking*. *Extending* is practically the same as we again change the partition size. It is equally dangerous due to changing the partition tables. On the other hand, when we do it correctly, there is no chance for issues. To extend the `root` partition, we must first delete or shrink some other partition and then extend `root` from a *Manjaro Live USB*. If you haven't done it yet, check all the information in *Chapter 10*, the *Storage management, partitions, and mounting* section.

### Moving home to a separate partition

With the default Manjaro installation, the `home` directory is placed in the same partition as `root`. A user's `home` directory contains a limited number of *hidden files and directories* necessary for system purposes. On a freshly installed Manjaro KDE, I have `.bash_logout`, `.bash_profile`, `.bashrc`, `.cache`, `.config`, `.local`, `.nanorc`, `.Xauthority`, `.Xclients`, `.xinitrc`, `.zcompdump`, `.zhistory`, and `.zshrc`. You can list them with `ls -la` or `exa -la`. These files and directories contain the configuration of different applications for the *current user*. With time, even if we use a personal **Network Access Storage** (**NAS**) for bigger files, the `home` directory will grow in size, at least due to the `.cache` directory.

As explained in *Chapter 10*, each partition has a **UUID** referenced in `/etc/fstab`. When installing with manual partitioning and a separate `home` partition, the OS creates an entry for it in `/etc/fstab`. With the default installation, the additional entry doesn't exist. If we have no second HDD, moving home will require first shrinking `root`, as explained in *Chapter 10* in the *Shrinking the /root partition* section. Once you have done this, follow the next instructions.

*If you add another HDD to your system, start here.*

Now that we have some unallocated space, we must create a partition, so open the partition manager of your choice. Create the partition from the *unallocated* space, adding the label home. If you work with **GParted**, for name, again use home. The next step is to mount it to some location. From the terminal, *identify* your partition with the listing from $ sudo fdisk -l (in my case, /dev/sda3) and *mount* it with the following:

```
$ sudo mount -w /dev/sda3 /mnt
```

Next, we have to copy all home contents to the new location:

```
$ sudo cp -pR /home/* /mnt
```

Some people might ask why not use the mv move program. With mv, we will have no backup, while with cp, we keep the original and can check the data before deleting the original. Just as general information, we could also use rsync instead of cp here (rsync is reviewed in detail in *Chapter 10*).

Our next step is to write in /etc/fstab stating that home is on another partition, and we want it *mounted* during **system boot**. First, execute $ sudo blkid to get the **UUID** of the new home partition. Its output will be like this:

```
/dev/sda3: LABEL="home" UUID="903cfe92-e09f-48e1-8b01-a6065ed8ead9"
BLOCK_SIZE="4096" TYPE="ext4" PARTLABEL="home" PARTUUID="1fc63a7f-
2e9e-4a08-af57-64efff6cad95"
```

You must copy the UUID value later, not PARTUUID, so be careful.

Now, execute $ sudo nano /etc/fstab, to edit the file. In it, at the end of the file, add the line UUID=903cfe92-e09f-48e1-8b01-a6065ed8ead9 /home ext4 defaults,noatime 0 2. Replace the UUID value with the one of your new home partition. Save the file with *Ctrl+s*, and exit nano with *Ctrl+x*.

The last step before rebooting is to *rename the old* home partition. We do this to keep it as a backup if some part of the process has issues. To do this, we must log out of our user session and log in as **root**. For **Xfce**, when you log out, you must select the **other** option instead of your user. For **KDE Plasma**, you will have a graphical button for **other**. For **GNOME**, you will have to click on the person icon to be able to select **Not listed?**. In all three cases, enter root as username and its password after this. Once you are logged in, open the terminal. As we are in a *root session*, remember that we will not need sudo for any command. Execute the following:

```
$ rename home homeOld /home
```

Now, reboot your machine. If everything is OK (which should be when following the described steps), your OS will load normally and no settings or history will be lost. In this case, open the terminal and delete the old home partition by executing:

```
$ sudo rm -rf /homeOld
```

If, by chance, you have made some mistake during the editing of your fstab file and the OS fails to boot, you have to:

1.  Log in as root.
2.  Delete the wrong /etc/fstab last line with $ sudo nano /etc/fstab.
3.  Rename homeOld to home with $ rename homeOld home /homeOld.
4.  Reboot.

If you cannot log in as root, you can do this from a root session from a Manjaro Live USB, although this means you have made a severe mistake during the previous steps.

## Defragmentation – do we need it at all and how to do it

**Fragmentation** is *scattering one file into tens of pieces* placed at non-consecutive locations on our **HDD**. In contrast, if we copy a *2 GB video clip* on a *new empty partition*, all its parts will be consecutively placed on an HDD.

Let's imagine another case – an *old partition with thousands of files*, to which we copied and from which we deleted thousands of files since its creation. If this has *only 3 GB free left*, and we copy the mentioned *2 GB video clip*, it will be fragmented on the **HDD**, as there is almost no space left. Fragmented files are like going to a public library where tens of genres are mixed. In this case, even if you have each book location, extracting scattered children's books from many shelves will be hard.

**Defragmentation** is rearranging the files' parts so that all are located consecutively in the memory. It requires *some free space*. To use an analogy, if you want to arrange a library with hundreds of thousands of books, you'd better have some extra space, or the task might be rigid and slow.

**Fragmentation** can happen when a given file grows, but no consecutive locations set is available for its parts. It can also happen if the filesystem doesn't handle frequent changes correctly. Many years ago, this was vital for Microsoft's NTFS filesystem.

This is rarely necessary on **ext4**, **btrfs**, **XFS**, and **ZFS**, as their file management growth strategies prevent fragmentation. It may be practically *necessary* after *one or more years of extensive partition usage at over 80% of its capacity*. Extensive usage means thousands of changes per week. The lower the amount of *free space*, the bigger the chance of **fragmentation**. That's why we said that, in general, a partition should work with *at least 10% or 10 GB* of free space. The file management and growth strategies work perfectly when free space is available.

With the aforementioned Linux filesystems, **fragmentation** is nearly impossible for a partition used as a rarely updated backup server, where data is mostly added but not frequently changed.

## How to analyze the fragmentation and perform defragmentation

Here, we will look explicitly at **ext4**, the standard for Manjaro and most Linux distributions. **E2fsprogs** is a package preinstalled on Manjaro that works with ext2, ext3, and ext4 filesystems. It has the `filefrag` tool for file fragmentation analysis and `e4defrag` for defragmenting **ext4**; check the tools' man pages for details. For more information on filesystems, check *Chapter 9*.

The `e4defrag` tool analyzes single files, whole directories, or whole partitions. It provides a nice summary with interpretation instructions. First, use `duf` to check your devices. For this example, I will analyze /dev/sda2, which has *31 GB total size*, is *82.3% full*, and is the `root` of one of my installations. It is 6 months old and I have done a lot of installations and experiments on it. Execute `e4defrag` with `-c` for the analysis:

```
$ sudo e4defrag -c /dev/sda2
```

The results for me were as follows:

```
<Fragmented files>                              now/best        size/ext
1. /var/log/wtmp                                13/1              4 KB
2. /root/.cache/mesa_shader_cache/index         3/1               4 KB
3. /var/log/samba/log.rpcd_classic              3/1               4 KB
4. /var/log/samba/log.rpcd_winreg               2/1               4 KB
5. /var/log/samba/log.samba-dcerpcd             2/1               4 KB
 Total/best extents                             561781/554243
 Average size per extent                        55 KB
 Fragmentation score                            1
 [0-30 no problem: 31-55 a little bit fragmented: 56- needs defrag]
 This device (/dev/sda2) does not need defragmentation.
```

As you can see, no files are fragmented. If you need to defragment a drive, use the same tool without `-c`, like this: `$ sudo e4defrag /dev/sda2`.

# inxi, troubleshooting, and mhwd

**inxi** is a *system analysis* Terminal tool that provides precise information on a system's CPU, RAM, disk space, network settings, HW and peripherals, installed SW, and more. Despite its broad capabilities, it is lightweight, and many Linux distributions include it by default. For the SW part, it can report information on the OS, kernel, some installed packages, and SW repositories. SW version numbers, dependencies, and installation locations may also be shown.

The complete list of topics `inxi` reports is as follows: System, Machine, Memory, PCI, Slots, CPU, Graphics, Audio, Network, Bluetooth, Logical, RAID, Drives, Partition (HDD and other drives), Swap, Unmounted (partitions), USB, Sensors, Repos (for repositories), Processes, and Info. We will explore the frequently used options, but to know them all, it is best to export the man page in HTML format with $ `man --html=firefox inxi` and search in it according to your needs.

We can combine its many options *in any order* (as long as some don't contradict) like this: `-Ab`. Any option requiring an argument should be put last or with a separate dash like this: `-A  -b`. Some of them require `sudo`; we *will be warned* if we have missed it. The following are the most frequently used options (I skip the single dash in front of the single-letter options):

- `A` – Report for *audio devices* HW, drivers, and SW versions.

- `b` – Brief general system report (lists CPU, machine name, graphics, network adapters, and drives).

- `B` – *Battery* information for *laptops*.

- `E` or `--bluetooth` – Shows **Bluetooth** devices.

- `C` – Detailed information for the **CPU**. Add `f` to also get a detailed **CPU** flags listing.

- `D` – Detailed **HDD/SSD** information.

- `G` – **Graphic cards** listing.

- `--edid` – Graphic cards data from the previous option, along with the **Extended Display Identification Data** (**edid**).

- `h` – The help menu.

- `i` – or `--ip` – Detailed network information for both **IPv4** and **IPv6** networks.

- `I` – General usage information for the system: number of processes, uptime, memory, and shell.

- `J` – Detailed **USB** device information.

- `o` – Lists *mounted* **HDD** partitions with *device name*, *size*, and *filesystem*. Add `l` to get the *label* as well.

- `p` – The same as `o`, but identifies the partitions by *mount point*. Add `l` to get the *label* as well.

- `j` – `swap` partition information. Add `l` to get the *label* as well.

- `L` – **Logical volumes** information if such are mounted. Such are, e.g., **LVM**, **LUKS**, and **bcache**.

- `m` – **RAM** information. When using this option, you must add `sudo` in front to get more **RAM slots** information.

- `M` – General machine information such as manufacturer, model, and motherboard model. Add `sudo` in front to get the motherboard serial number.

- `N` – Brief *network adapter* information.

- n – Detailed *network interface* information, including **MAC address** and connection speed.

- r – Repositories settings for *mirrors*.

- R – **RAID** data.

- s – Basic information from *hardware sensors* for *temperature* and *fan speeds*.

- --slots – **PCI** slots with type, speed, and status information, requires sudo.

- S – System information such as hostname, kernel, desktop environment, etc.

- F or --full – Full output for inxi. Includes *all Uppercase* single-letter options (except J and W) and adds --swap, s, and n. It does not show extra verbose output from options such as d, f, i, J, l, m, o, p, r, t, u, and x, unless you use those arguments in the command, such as inxi -Frmxx.

- z – A *filtering* option that adds security filters for **IP** addresses, serial numbers, **MAC**, location (w), and user home directory *name*. This way, when sharing results with someone for help, they see no *security-related details*.

- a or --admin – This option activates a much more detailed listing, which includes chip IDs and many other details.

- V or --verbosity – This option controls the level of provided details. It accepts a single parameter from 0 (the default) to 8 (the maximum). Thus, we will execute $ inxi -v 8 to get the most detailed complete system report. This option shall not be used with the b or F options.

## How to ask for help with inxi data

You can test all the listed inxi options to explore different parts of your system. Say you are looking for a new driver or for information on which driver is currently active on your system for audio, USB, or Bluetooth. Call inxi with the corresponding option, and you will have the information.

When we ask for help on the Manjaro Forum, we are often asked to provide inxi *results in advance*, as without them, adequate help cannot be provided. When you need help, please follow these rules:

1. First, *always search for the given issue* in the Manjaro Forum, then use a web search. Try different terms – for example, Manjaro Linux HP Probook Wireless driver not present, Manjaro Linux HP Probook Wireless not present, Manjaro Linux HP laptop Realtek Wireless not working, etc. When applicable, use filtering options – for example, on Google, we can look for results from the last year or a specific period. If a solution exists, new posts are redundant.

2.  When you find information, make sure to *read the given post thoroughly*. Ask additional questions at the end *only when you know nobody else has asked them*. If someone has already asked your question, it is perfectly OK to post a short sentence such as: `Hello XXX, I have the same issue and tried the provided steps but couldn't solve it.` If the topic has *not been updated recently*, you can add a *polite request* such as: `Has anyone found something more on this issue recently?`

3.  When you find a Manjaro Forum post that *solves your issue*, it is nice to *like it with the heart icon*. If you *can add something more to the topic*, post a reply stating, for example, `Hey John, thanks a lot for your hints. This solved the issue for me. I also found out that….`

4.  If you cannot solve your problem after all your efforts, you are *welcome to make a new post* asking for help. In this case, you must read the following post to get help in the best possible way: `https://forum.manjaro.org/t/howto-request-support/91463`. You should also read this one on how to provide system information: `https://forum.manjaro.org/t/howto-provide-system-information/874`. In particular, the commands you have to execute are:

    *   `$ export LANG=C` – This command will explicitly set the locale for the current terminal session to English. It is always required when your locale is in another language but you post in the English part of the Forum. The command has no output on the terminal.

    *   `$ inxi --admin --verbosity=7 --filter --width` – This provides the detailed system report you need to add to your post, necessary to the advanced users to help you. It is essential to put the Terminal copied reports enclosed either in `[code] Terminal tens of lines [/code]` or in `--- Terminal tens of lines ---`. As the Manjaro Forum gives you a preview, you will see how this puts the tens of lines in a nice scrollable sub-window. This helps not to overload your post and is **required**.

If you need to check something for yourself, try the `$ inxi -Fa` short form. It will activate all main options with the F flag and explicitly force them to provide more details. In case you have missed it, to get *the most extensive complete report*, call `$ inxi -v 8`.

## A troubleshooting example for wireless not working on a new laptop

This is the original way I resolved my issue. However, in the next section, we briefly review the **mhwd** tool, which can also help in such situations. If you experience issues, check this solution, and also read the next section before dealing with any problems you may face. Combining both approaches may help quickly understand the drivers or modules causing problems.

In my company, in March 2023, we bought a new *HP Probook 440 G9* laptop. When installed with Manjaro (tested with **KDE** and **Xfce**), we *had no wireless internet settings, icons, or menus*. The **Ethernet** (i.e., cable network) worked *flawlessly*. Wireless worked without issues on a *G8* version of the same laptop bought in September 2022. I searched the Forum and the internet, but *the tens of results were not helpful*. Then, I visited the *HP product page* and understood that the possible brands for the wireless adapter on the *G9* were either **Realtek** or **Intel**.

In the following listings, I explicitly replace the non-relevant parts of the output with *ellipsis*. A **wireless adapter** is usually connected via a **PCI** bus, so I executed $ `lspci -k`. From all reported devices, I got the following output for the working Ethernet network controller:

```
Ethernet controller: Realtek Semiconductor…. PCI Express Gigabit
Ethernet Controller (rev15)…
Subsystem: Hewlett-Packard Company device 8a9e
Kernel driver in use: r8169
Kernel modules: r8169
```

*I also got another output for a Realtek device*:

```
Network controller: Realtek Semiconductor Co., Ltd. Device b852
Subsystem: Hewllet-Packard Company device 8a9e
```

There was no report of a *working kernel driver related to it*, so this should have been our missing device driver.

In the $ `inxi -v8` output, I had the following for the non-working device:

```
Network:
    Device-1: Realtek vendor:… driver: N/A speed: Unknown…
    … gen: 6…
    chip-ID: 10ec:b852 class-ID:…
```

The conclusion was I must look for a *wireless driver* for *Realtek Wi-Fi* with `chip ID 10ec:b852`. The first part of this ID is a generic *vendor-related ID*, while the second is the **chip ID** itself. I searched on Google for `Manjaro wireless driver Realtek b852` and got the following results:

- `https://forum.manjaro.org/t/wifi-adapter-not-recognized/124742/4`

- `https://forum.manjaro.org/t/thinkpad-e15-wifi-not-working-on-fresh-install-realtek/110835`

- `https://forum.manjaro.org/t/problem-is-no-wifi-adapter-found/125090`

They all had solutions and pointed out I must install the `rtl8852be-dkms-git` AUR package. Some of the posts stated that I also need the *kernel headers* for my kernel version. I activated the **AUR** `pamac` option, and with `inxi` (also possible with `neofetch`), I identified my kernel as version *6.1.26*. The older solution from the Forum used the `linux515-headers` package, which is for kernel version *5.15*.

However, there is an easier way. We can use the `$ mhwd-kernel -li` command, which we will review in detail in *Chapter 16*. For now, know that it will show you the currently running kernel and other installed versions. It reported the currently running kernel package as `linux61`, and the corresponding headers were suffixed with `-headers`, i.e., the `linux61-headers` package. We can also use `grep` on a `pamac` search like this: `$ pamac search headers | grep -C10 -I linux6`.

With `$ pamac search 8852`, I identified the currently available corresponding **Realtek** driver packages from **AUR**: `rtw89-dkms-git` and `rtw89bt-dkms-git`. The second one was explicitly for **Bluetooth** devices, so I chose the first one.

Finally, I installed the following:

```
$ pamac install linux61-headers
$ pamac install rtw89-dkms-git
```

After rebooting, *I got options for connecting to a wireless network* when right-clicking on the Ethernet network icon (on the bottom menu bar, tested with Xfce and KDE Plasma). I liked the related posts that had helped me. This is how we troubleshoot such a problem.

## The mhwd tool

We have mentioned the **mhwd** tool. Its name stands for **Manjaro Hardware Detection**, and it is automatically started only during installation. However, we can run it manually later to get information. As of September 2023, `mhwd` doesn't provide a man page, only a `-h` help menu, sufficient for basic work. Its most important options are:

- `$ mhwd -l` – Lists available device configurations/drivers for our system. It usually shows installed and potentially working drivers. Add `-d` for full details.

- `$ mhwd -la` – Lists all available device configurations/drivers, even if not appropriate for our system.

- `$ mhwd -lh` – Lists all devices' HW information, including BUS, CLASS, VENDOR, and DEVICE IDs. Add `-d` for full details. For the wireless issue case, I got the following in *September 2023*:

```
TYPE                BUS            CLASS VENDOR DEVICE CONFIGS
Network Controller  0000:03:00.0   0280  10ec   b852    0
Network Controller  0000:04:00.0   0200  10ec   8168    0
```

The second was my working Ethernet adapter, while the first was my *non-working* **wireless adapter** with device **ID** b852. As there were 0 CONFIGS, nothing could be installed. In this case, we still must follow the previous section's solution. The good thing about **mhwd** is that it provides an excellent HW summary with the -lh option, which can hint at what is missing. You get many more details different from the inxi detailed output when you add -d to it.

This is the official **mhwd** page, which shows a few more examples of the command: https://wiki.manjaro.org/index.php/Manjaro_Hardware_Detection_Overview. It is worth checking out.

**mhwd** has a Linux kernel addition, mhwd-kernel, which we review in *Chapter 16*.

## mhwd for graphics cards

For video adapters, mhwd can be an irreplaceable tool. The Manjaro team has made a great 10-page guide that I will not duplicate. Check it out here:

https://wiki.manjaro.org/index.php/Configure_Graphics_Cards

## A troubleshooting point on Secure Boot and monitors

Here, I have a short story to tell. A friend installed Debian on a laptop with an NVIDIA graphics card. He tried to make the extra monitor output work for a week and then gave up. Several months after this, another friend with Ubuntu had the same issue and discovered that the second output started working normally after disabling **Secure Boot** from BIOS (explained in *Chapter 2*). Even though the two laptops were from different brands, both had issues connecting an external display. The problems are often not in the distribution but in the Linux drivers for a specific device (in this case, NVIDIA graphics card drivers). There are two conclusions:

- First, if you are using any Linux distribution on a given machine, it is obligatory to disable Secure Boot.

- Second, when facing a problem for which nobody can help you, try to find information on whether other people with the same laptop, graphics card, or other HW have the same issue on another distribution. Many times, this helps.

To conclude, Linux is magnificent and ubiquitous. There are solutions for most problems, but they sometimes require a lot of digging. For others, we have to wait. Linux and all its distributions are constantly in development, so we at least know that most of the time, a solution will be developed – and, as always, for free.

# Reinstalling and keeping your home partition

Whether you have your Manjaro installation with a separate home partition or not, the most secure way to keep your data between reinstallations is to *save it on a backup drive*. Before reinstalling, update your backup, and after this, restore it as it was. There is only one potential issue – if your old and new users have different **UIDs** and/or usernames, you might need to use chown after restoring to avoid potential ownership issues. (We reviewed chown's usage in *Chapters 7, 10*, and *13*.) However, if they have the *same name* and were the first user to be registered for the OS during installation, both will have a **UID** 1000, so there will be no issue.

You must have your home directory on a separate partition to avoid the *backup* and *restore* steps. If this is not your case, first follow the instructions in the earlier subsection of this chapter, *Moving home to a separate partition*.

During installation, Manjaro copies a set of *hidden files* to the home partition. If you don't remember, *hidden filenames* start with *a dot*. They hold personal settings for your **shell**, a few other *applications*, and your *graphical environment*. Usually, personal images, projects, and other personal data are more important. In this case, simply start the installation following the steps from *Chapter 2*, in the *Manual installation for EFI-based computers* section. Here is a summary of them, pointing out how you *should not format* the home partition:

1. Restart and boot from the Manjaro Live USB stick.
2. Choose **Manual** installation.
3. Select the old root and set its *mount point* to root, designated by a single slash /. *Select the* **Format partition** *option.*
4. Select your old home, *mount point* /home, **Format partition** should *NOT be selected*.
5. Select your boot partition, *mount point* /boot/efi.
6. When you restart, your old data will be in your home directory.

Although the easiest, why is this not considered the best option, and instead, having a backup is *always strongly advised*? If you make even a single mistake and corrupt your partitions, or Calamares has a bug triggered precisely in your situation, without a **backup**, your data is potentially gone for good. Some recovery tools exist for **ext4**-formatted partitions, but if your **HDD GPT** table is ruined, recovery *is not guaranteed*. Thus, it is essential to always have a backup of your important data on an *external HDD*, personal **NAS**, or some **cloud storage** service.

# Summary

In this chapter, we first reviewed unnecessary files and data *system cleanup* in detail. We then reviewed **fragmentation**, how to analyze it, and when to defragment an **ext4**-formatted drive. Our next stop was inxi, troubleshooting, and mhwd, providing two practical examples for issues. Finally, we saw how to keep our home partition data when reinstalling.

In the next chapter, we will continue with Shell scripts and automation. We will start with the basics of BASH shell script writing. We will then continue with time-based execution of programs and scripts. In the last part, we will see how to trigger actions based on systemd timers and specific OS initialization events.

# 15

# Shell Scripts and Automation

In the previous chapter, we looked in detail at filesystem cleanup, exploring the pacman cache, uninstallation, and cleaning tool approaches. We then briefly covered the topic of defragmentation. After that, we moved on to system information extraction with **inxi**, along with a practical example of troubleshooting. We finished the chapter with easy instructions on reinstalling and keeping our original home partition.

In this chapter, we will continue with the basics of **BASH shell scripts**. They can execute any commands automatically, including those for system administration. We will explore their syntax, basic rules, and where to get more information and examples. We will then review the setup of *periodic and calendar-based execution* of commands and scripts, based on **cron** jobs and **systemd timers**. We will close the chapter by revealing how systemd timers can execute tasks related to system events for us.

The sections in this chapter are as follows:

- The basics of BASH shell scripts
- Time-based execution of programs and scripts
- Configuring systemd timer triggers related to system initialization via monotonic timers

## The basics of BASH shell scripts

In this section, you will see a lot of information cited from the original **BASH** documentation, available as a man or info pages as well as online at `https://www.gnu.org/software/bash/manual/html_node/index.html`. It is excellent but hard to understand initially and not rich with examples. For examples, I find the site `https://linuxhint.com/` one of the richest sources for all possible BASH options and functionalities. Our purpose here is to provide a crash course on the most important parts of the official GNU documentation. You will learn how to write a basic script and where to look for further information and examples.

As BASH is a scripting language, gaining a deep understanding of it would require a whole book. Hence, at the end of this section, I have recommended two BASH books from Packt Publishing, which provide a much deeper explanation and examples.

## What a shell script is and how to run it

On Linux, the shell is the command-line interpreter and interface that allows you to interact with the OS by executing commands. It provides a way to run programs, manage files, and perform various system operations. As you already know, each command is a program itself. We reviewed shells in *Chapter 7*, which you must ensure you have read before reading this chapter. **BASH** has been one of the most widespread scripting languages for decades, so we will use it for out scripts here. **Zsh** is another popular shell lately, the default one on **Kali Linux** and **macOS**. It is partially compatible with many **BASH** basic features, but not all, and different for the advanced ones. That's why there is a way to choose the shell interpreter for your scripts, which we will explain further in this and the following sections.

It is important to note that the default Terminal interpreter *may differ* from the shell script interpreter. On most Linux distributions, there will be a preset environment variable, SHELL, which is *a link to the interpreter called for your scripts*. As of *May 2023*, on all official Manjaro flavors, it is currently set to /bin/bash.

By default, ~:zsh-Konsole is written on top of Konsole on Manjaro KDE. This means that **Zsh** will interpret our commands. The easiest way to check our Terminal interpreter is to write some gibberish (such as adfjkh) and press *Enter*. The interpreter will not recognize the command and print an error, *starting the line with its name* like this:

```
zsh: command not found: adfjkh
```

Thus, although *the script interpreter* is **BASH**, the *Terminal interpreter* on Manjaro KDE is **Zsh**. The same is the case for Manjaro **GNOME**. For **Xfce**, the interpreter is **BASH** for both the Terminal and executed scripts.

A **shell script** is a text file that *has the execute flag set*. Thus, if our script file is named MyScript.sh and has the flag set *for the current user or one of their groups*, it will be executed when we call it *with a dot and forward slash* like this: $ ./MyScript.sh. This practically forces the Terminal to start *a new session* with the interpreter defined in the script itself (more on this in the next section). This new session will block the Terminal, so we cannot perform any task until it finishes. We will potentially only see some messages if the script reports something to the calling Terminal; otherwise, we must wait for it to finish. The new session *gets the environment variables values from the current Terminal session. It can use them*, but as a child session, it *cannot change* their values permanently for the calling Terminal. This means that the usage of export and unset in the script will have no effect *after the script execution has finished*. Stopping a started script before it finishes is done by hitting *Ctrl+C one or more times*; however, if it performs certain administrative tasks and we leave them partially done, we will not know the current system state.

Be careful with who *owns* the script file and what *permissions* are assigned to it. If the owner is **root**, and the execute permission for the *group* and/or *others* is not set, trying to execute the script *as our user* will fail. If necessary, you can use chown and chmod (as explained in *Chapters 7* and *13*) to change the *permissions* and *ownership* of your script.

Another way to execute a script is to pass it directly as *an argument* to a shell application – $ /bin/bash MyScript.sh. However, as the script's first line often contains a definition for the interpreter, this is typically done in special cases for scripts without a shell interpreter definition.

Our last option is to source our script with the source command. However, if you have the exit command at the end of your script, this will close the current Terminal session. Consider this when using source. I will not review source here. It is a BASH function, and you can read more about it here: https://www.tutorialspoint.com/linux-source-command.

## A simple script example and its contents

Each line in a script can have any valid for the interpreter command. However, only for the first line, if it starts *with a hash and an exclamation mark*, #!, followed by a path to a shell interpreter, it is called a **shebang** or a **hashbang**.

For **BASH** the **shebang** is #!/bin/bash. As the /bin directory is a symlink to /usr/bin, many scripts call the env command to refer to the BASH location with this **shebang** – #!/usr/bin/env bash. On Manjaro, both are valid.

The primary use of a shell script is to *execute a set of standard commands*. Instead of typing them each time we have to execute them, we create a text file with nano, **Kate**, or another text editor, and we write in it the following script:

```
#!/bin/bash
# This line is a script comment example
echo Test script executed by:
echo $USER # prints the current username
echo Shell interpreter:
echo $SHELL # prints the SHELL environment variable value
# In the next echo call we use quotes as
# otherwise the asterisk is considered a script operator
echo "**********************************************"
echo Environment Variables for $USER:
env
echo "**********************************************"
echo Entering directory Downloads
cd /home/luke/Downloads
echo Listing Downloads directory contents
exa -lah # example for command with options
```

Each new line in the script marks a command's end. We can write every command we have used in the book so far.

An excellent example is to take all **ufw** settings presented in *Chapter 12*, in the *A standard, good, default configuration* section. As ufw needs sudo, create the script with *owner* root via $ sudo nano ufwConfigScript.sh, or change the owner to root after writing it. When copying and modifying the commands as needed, you should remove all references to sudo from them, as calling a command with sudo requires you to provide your password each time. Otherwise, the script will not be able to work without your interaction. When ready, change its permissions with $ sudo chmod u+x, then execute it with sudo: $ sudo ./ufwConfigScript.sh.

Before going any further, remember these two hints. First, a good practice is to always name your scripts meaningfully so that you know what you created them for when you return to them. Second, if you make a mistake in a script so that it blocks the Terminal and doesn't return, usually hitting *Ctrl+C* once or several times breaks its execution. As mentioned earlier, interrupted scripts may leave our system in an unknown state, so be careful.

## Considerations and limitations for script commands, comments, and syntax

There are a few general considerations for the commands called in a BASH script:

- The commands used in a script should be *present on the system*.

- The command **syntax** inside the script should be *correct*. An example in the previous section used *quotes* for echo to output a series of asterisk symbols, serving as visual separators of the env command output. If your script contains incorrect commands, don't be surprised by unexpected results; instead, look for the correct usage. Testing the commands on the Terminal in advance helps.

- The script accepts the values of the *current environment variables*. It can echo them or use their values as appropriate.

- The environment variables values *for the* **current user** *and* **root** *are mostly different*. Whether we execute the script with or without sudo can result in different results if *environment variables* are referred to in the script.

- If you do **root**- or **user**-*specific actions* inside a script, you have to consider that the users of the script shall be informed of this in advance. In general, the usage of sudo inside a script is strongly *discouraged*; instead, it is preferable to call the script itself with sudo to execute it in a **root** context.

- When *setting an environment variable value inside a script*, its value remains *local for the time of the script execution only*.

- **BASH** has some basic commands integrated to not start extra processes for simple commands such as `cd`. As the script is executed in a separate environment, after it finishes, we will continue to be on the Terminal in the same directory we were in before starting the script.

  All BASH built-in commands are described here: `https://www.gnu.org/software/bash/manual/html_node/Shell-Builtin-Commands.html`.

- No matter whether at the beginning or at some point in the middle of a line, when the BASH interpreter sees the *hash symbol*, #, it and all the characters until the end of the line are considered a comment.

- Single and double quotes have the same general purpose – they say to BASH, *"Process this string as a whole piece."* However, there are key differences between `'some string'` and `"another string"`.

  **With single quotes** (e.g., `'some string'`), the string inside will be taken as it is. No other variables can be referenced (expanded) in it. In single quotes, no commands can be substituted. The special symbols $, ', ", \, *, and @ are always considered literal symbols (not special characters). This means that if we execute `echo 'someWord $1! \ " * @'`, the same string provided in the single quotes will be returned on the terminal.

  **With double quotes** (e.g., `"other string"`), when a reference to an environment variable is used, such as `"string with an inserted value from $SHELL"`, the $SHELL string will be substituted with its value. The special symbols $, ', ", \, *, and @ have additional functionality. Their full list is split into several sections in the GNU BASH manual. The main page is available at `https://www.gnu.org/software/bash/manual/html_node/Double-Quotes.html`. We will not review all of their uses in detail here and only mention a few special cases in the following subsections.

## Variables, arguments, and arithmetic calculations

Apart from the **environment variables**, in a BASH script, we can have *user-defined variables*. They are defined like this – `VarName=SomeStringOrNumber`. A variable name can *start* with letters or underscores, *contain* numerals, and is *case-sensitive*. To add spaces in its string value, you must *enclose the string in quotations* like this – `Variable_1='Some string with spaces'`. It is crucial never to name variables and functions with any of the BASH-reserved keywords listed here: `https://www.gnu.org/software/bash/manual/html_node/Reserved-Words.html`.

In the same way that we reference environment variables, we can access the *user-defined script variables* with a dollar sign, $ in front – `echo $VarName`. We can also use the full syntax to reference a variable, which has additional curly braces around it – `echo ${VarName}`.

A variable can get the value of another variable – `Var2=$Var1`. You can also use it inside a longer string, obligatory enclosed in *double quotes*, in which case, the curly braces are also obligatory – `echo "The value of Var2 is ${Var2}"`. We will see a practical example later.

The same general rules for variables apply in the Terminal for **BASH** and **Zsh**. Simply type `VarName=SomeStringOrNumber` and press *Enter*, and this variable will be valid for the current session. When we close the Terminal, we lose all user-defined variables.

**Arguments** of a program are the strings we call it with. In the case of `$ sudo grep -E -w "^.$"` `-i --color -A 2 -n -r /etc`, every string after `grep` separated by spaces is an **argument**. A BASH script can take arguments the same way we do so with other commands. We access the first 11 arguments passed to a script with `$1`, `$2`, ..., `$9`, `${10}`, and `${11}`. An example is the following script, which will create three files with names of your choice:

```
File: createUpTo3.sh:
#!/bin/bash
ext="my.config" # here we create the variable ext
touch $1$ext  # these commands will take the given input
touch $2$ext  # arguments, add to them the value of $ext, and
touch $3$ext  # create three files with corresponding names
```

When we call the script from the command line with the following arguments – `$ ./createUpTo3.sh` `file1 test2 test3`, it will create three files named `file1my.config`, `test2my.config`, and so on. Of course, you need to provide the correct number of arguments, so checking the number of arguments in advance with an `if` statement (explained later) is recommended. The number of arguments passed to the script can be read from the built-in variable `$#`. In the `createUpTo3.sh` example, it should return 3 to allow the execution of the script. Command-line arguments accessed via `$` from within the script are called **positional parameters** in the BASH manual.

Although variables are strings (and not numbers), BASH can work with them as integer numbers – that is, natural numbers, 0, and negative numbers. If a variable string is a numerical value, you can add, subtract, divide, multiply, and do many regular C/C++ language logical and conditional operations with them. To do this, you have to use the *double braces* operator `(())` like this – `((Var1+Var2))`. Then, you can echo this value or assign it to another variable. Just remember that **BASH** only supports **integer arithmetic**. There is one thing to note about **division** – when dividing, for example, 7 by 3, you will get a result of 2, as this is the whole number part of the result of the division operation. An interesting example is `echo $(($Var1+$Var2-$Var3*4))`. Here is one more – `newVar=$(($Var1/` `($Var3*4)))`. If any of the variables in such operations are not a pure numeric value – that is, they contain underscores and letters – *their value will be ignored* in the arithmetic operations. Be careful with such errors, as they may lead to unexpected results.

In the case of division combined with multiplication and other operations, it is better to use braces, as in the last example, to avoid potential issues with the execution order of mathematical operations.

The full list of supported operators is described in this part of the BASH manual:

`https://www.gnu.org/software/bash/manual/html_node/Shell-Arithmetic.html`

For more examples of arithmetic operations, you can check the first part of these 74 operator examples: `https://linuxhint.com/bash_operator_examples/`. In some of them, the `expr` command is used, and for `expr`, spaces are required between the operators and the variables.

## Writing multiple commands on one line

To write a sequence of commands and constructs on a single line, we separate them with a semicolon `;` like this:

```
ls -l ; echo $(($Var1+$Var2-$Var3*4)) ; exa -lah
```

The other way is to use the logical AND operator, `&&`, which will ensure *the second command will be executed only if the first doesn't return an* **error**. Thus, we can combine two or more commands on a single line like this:

```
exa -lah && pwd && echo Test
```

For each new command, we add `&&` and the call for the command.

## Decision-making, file testing, arithmetic, and logical operators

A decision is a point in a script where we choose one or another action based on some parameter or variable value. A nice example of the tens of different decisions is the variety of `ls` results, modified with different combinations of its options (arguments).

To write **decisions**, we can use an `if [ testCommands ]; then … fi` construct, where `fi` marks the end of the `if` statement. The `then` clause contains the commands to execute. An `if` statement can contain multiple *optional* `elif` statements and *one optional* `else`.

*The spaces around the square brackets are crucial.* You can skip the space after the closing square bracket, `]`, only if it's followed by a `;` character and the `then` keyword. It is also important to know that the *opening square bracket*, `[`, is equivalent to the `test` BASH Bourne Shell builtin, as described at `https://www.gnu.org/software/bash/manual/html_node/Bourne-Shell-Builtins.html`. In other words, with *square brackets*, we tell **BASH**, "*I want to use* `test`."

The official name for `testCommands` is *Conditional Expressions*.

Here is a metacode example:

```
if [ testCommands ]; then  # a semicolon is required here
    execute_if_true_1

    …
    execute_if_true_1
elif [ testCommands ] # a semicolon is not needed, then is on a new
line:
then
```

```
    execute_if_true_2
else  # if none of the previous is true execute:
    commandsToExecute3
fi
```

The testCommands expressions are constructs that will return *true* or *false*. We will inspect some of their most common variants later in the section. If testCommands return *true*, the block of commands after it is *executed*, and any further elif or else sections are *skipped*. If the testCommands expressions return *false*, BASH looks for the next testCommands to check (if sections with testCommands are present).

As a result, if there is more than one section with testCommands, *only one of the several sets* of commands will be executed.

There can be multiple elif test and commands sections, but it is recommended not to have more than a few of them. Usually, you would see up to four elif statements. If we need more than a few elif cases, we should use a case statement, which I will not review here.

There can be multiple commands inside an if, elif, or else block, each *on a separate line*.

Here is an example of using the if construct with comparison operators for **strings**. Imagine we are writing a script to search for files in a directory with the -s option, search in a provided text file with the -t option, or print an error if no arguments are provided. In this case, we can write the following:

```
if [ $1 == "-s" ]; then
  search_In_Directory_Commands
elif [ $1 == "-t" ]; then
  search_In_Provided_File_Commands
else
  echo  Please provide options and arguments! Use -h for help.
fi
```

There are many *Conditional Expression* operators (testCommands), described at https://www.gnu.org/software/bash/manual/html_node/Bash-Conditional-Expressions.html.

The *most essential Conditional Expressions* for **strings** are the following:

- string1 == string2: *true* if the two strings are *the same*
- string1 != string2: *true* if the two strings *differ*
- -z string1: *true* if the given string is *empty*
- string1: *true* if the given string is *not empty*
- -n string1: *true* if the given string is *not empty* (just like the previous one!)

For **files**, we have a separate set of testCommands expressions, which are helpful, as BASH scripts operate directly on files inside the filesystem structure. The most important testCommands for them are:

- -e or the equivalent -a: *true* if the file *exists*

- -d: *true* if the file *exists* and is a *directory*

- -f: *true* if the file *exists* and is *a regular file* (and not any other **Unix** filetype as defined in *Chapter 9*, in the *Linux filesystem basics* section)

- -h: *true* if the file *exists* and is a *symbolic link*

- -r: *true* if the file *exists* and is *readable*

- -w: *true* if the file *exists* and is *writable*

- -x: *true* if the file *exists* and is *executable*

- -N: *true* if the file *exists* and has been *modified* since it was last read

The following is an example of how to use these, taken from example 52 at https://linuxhint.com/bash_operator_examples/#o52:

```
filename=$1
if [ -e $filename ]
then
  echo "File or Folder exists."
else
  echo "File or Folder does not exist."
fi
```

More examples are available on the same web page, numbered 52 to 74.

For **numeric** values, we can carry out a regular comparison operation on any two numeric variables with the -eq, -ne, -lt, -le, -gt, and -ge options. However, the C-language operators for these operations, such as && and ||, are much more convenient. They are all described in the *Shell Arithmetic* section of the GNU BASH manual mentioned earlier. To ask BASH to handle testCommands as a *Shell Arithmetic expression*, we must use the *double square brackets* form [[ testCommands ]]. This form also allows us to combine several testCommands expressions with the logical operators ||, &&, ==, !=, and !.

The following is the basic set of tests for numbers:

- [[ number1 == number2 ]]: *true* if the two numbers are *equal*. We can also write the alternative [ number1 -eq number2 ]. For the rest of the operators, I will skip the brackets and just mention the letters alternative.

- number1 != number2: *true* if the two numbers are *not equal*. The alternative is -ne.

- `number1 < number2`: *true* if the first number is smaller than the second. The alternative is `-lt`.

- `number1 <= number2`: *true* if the first number is smaller than or equal to the second. The alternative is `-le`.

- `number1 > number2`: *true* if the first number is greater than the second. The alternative is `-gt`.

- `number1 >= number2`: *true* if the first number is greater than or equal to the second. The alternative is `-ge`.

As mentioned, the *double brackets* allow us to *combine* `testCommands`. Here is how to test whether two numbers are *equal* AND two others are *different*:

```
if [[ num1 == num2 && num3 != num4 ]]; then …
```

The main logical operators for `[[ … ]]` test combinations are defined as follows:

- `||` – logical OR: *true* if at least one of the left- and right-side expressions is *true*

- `&&` – logical AND: *true* only if both expressions on the left and right sides are *true*

- `==` – logical `equality` check: *true* if both expressions on the left and right side have *the same* value (both are *true* or both are *false*)

- `!=` – logical `inequality` check: *true* if both expressions on the left and right side have different values (one is *true* and the other is *false*)

- `!` – logical `inversion` operator: Switches the result value of its operand from *true* to *false* and vice versa

When combining several operators, special rules define the sequence in which they are applied. These are called **precedence** rules and are exactly as in the C programming language. All possible Shell Arithmetic expressions are listed in the GNU BASH manual in order of decreasing precedence: `https://www.gnu.org/software/bash/manual/html_node/Shell-Arithmetic.html`.

When we use braces `()`, we can define the precedence ourselves. The execution of such complex expressions will start with the innermost braces and continue to the outermost ones. Say we have the following expression:

```
if [[ (num1 == num2 || num3 != num4) && (num3 == num6 || num5 != num4)
]]; then …
```

It will first check the `==` and `!=` inside the `()` braces of the first part. Then, despite `&&` having higher precedence than `||`, it will execute the `||`. Then, it will work on the second `()` braces and execute `&&` at the end. For a complete precedence listing, check out the *Shell-Arithmetic* link.

## Command substitution

The GNU BASH manual states, *"Command substitution allows the output of a command to replace the command itself."* It occurs when a command is enclosed in braces like this – `$(command)`.

This helps to use the results of a command as a variable or directly as a parameter in an expression. Here is a simple example with the `date` command:

```
currentDate=$(date)
echo $currentDate
```

## Other constructs and where to learn more

There are many other valuable constructs and functionalities in BASH. To learn more, check out the GNU BASH **manual**, particularly the sections on `case`, **loops**, **aliases**, and **functions**. It also includes a few nice indexes here: `https://www.gnu.org/software/bash/manual/html_node/Indexes.html`. The site `https://linuxhint.com/category/bash-programming/` offers thousands of examples, and you can use BASH-related keywords directly in its search to find examples of a given construct or topic.

If you prefer a systematic learning approach, I highly recommend two Packt Publishing books:

- *Learning Linux Shell Scripting – Second Edition, by Ganesh Sanjiv Naik*: `https://www.packtpub.com/product/learning-linux-shell-scripting-second-edition/9781788993197`.

  This book is a good manual, with detailed explanations and examples on each topic. Take a look at the summary of its contents from the preceding link.

- *Bash Cookbook, by Ron Brash and Ganesh Sanjiv Naik*: `https://www.packtpub.com/product/bash-cookbook/9781788629362`.

  This book is a collection of over 70 recipes for practical use cases. It starts with a 30-page BASH introduction and the rest of the book consists only of practical examples. Going through the summary of its contents will give you an idea of what you will learn from the recipes.

Apart from the preceding books, these three summary sheets can be beneficial once you know the basics: `https://mywiki.wooledge.org/BashSheet`, `https://devhints.io/bash`, and `https://quickref.me/bash.html`.

# Time-based execution of programs and scripts

There are two main ways to execute a script at *a specific time*, which we will review in this section. The first is based on a **cron** package, and the second is based on **systemd** timers. The systemd and process basics described in *Chapter 13* are essential in both cases. While using **cron** is simpler, it is started at a certain moment during the OS initialization sequence, so cron jobs can only be triggered after this.

*A specific time* means at an exact calendar time, such as *10.12.2023 at 23.45 UTC*, supported by *both cron and systemd*.

The other way is based on timers, offered *only by* **systemd**. A timer is a hardware or software module that can trigger events based on an elapsed time period – for example, every 35 seconds or every 75th hour. These are often called **monotonic** timers and are configurable; we can set their activation period to custom values.

The initial release of **cron** was for **Unix** in *May 1975*. Multiple posts on the internet state that it is obsolete and that **systemd**-based distributions are strongly advised to use **systemd timers**, as they provide more options and flexibility. While this is true, **cron** is not obsolete as the two latest versions of **POSIX** (from *2008* and *2017*) include a section on `crontab`, which is *a system utility to start periodic jobs*: `https://pubs.opengroup.org/onlinepubs/9699919799/utilities/crontab.html`. Nowadays, the original implementation has been replaced by different ones, listed on this Arch Linux page: `https://wiki.archlinux.org/title/Cron`. Considering that the `crontab` command is part of the latest **POSIX** standard and that *non-systemd-based Linux distributions* benefit from it, it continues to be actively used.

On Manjaro Xfce and KDE Plasma, we have the core repository package **cronie** pre-installed. It includes the **crond** daemon, the `crontab` command, a `cronie.service` systemd file, and the `anacron` program for systems that don't operate 24/7. **cronie** is actively developed, with two releases in 2022 and one *in development as of September 2023*. From the cronie alternatives, the latest release of **fcron** is old – from 2016, **bcron**'s is from 2015, and **dcron**'s is from 2011.

On the **GNOME** flavor, we must install **cronie** manually. There are some concerns about the security of the **cron** implementations. For **cronie** in particular, it is officially stated that it supports security enhancements based on **PAM** and **SELinux**.

Before we go on, it is essential to state that, here, we are discussing the triggering of relatively short tasks compared to the periods between which they will be executed. This means, for example, executing some specific programs or commands that will finish after a few seconds, minutes, or hours. The next time we would trigger their execution is after they have finished the previous execution. To make up the difference, an enabled daemon, such as a Bluetooth one, will keep working on standby when the Bluetooth service and function are enabled. It will constantly be waiting for a Bluetooth device to connect.

## Executing periodic tasks with cronie

Executing tasks with **cronie** is simple and can happen at intervals with *an accuracy of minutes.* The manuals from the package are for `cron` and `crond` (the same), `cronnext`, `crontab`, `"crontab(5)"`, `anacron`, and `anacrontab`. This means you must call `man` with them, and not with the package name.

In the Manjaro releases from *May 2023*, all users can add `crontab` jobs based on the presence of a single empty file – `/etc/cron.deny`. If a user is added to it, they will not be allowed to use `crontab`. For more information, check out the `crontab` man page.

To set up cron jobs, we call the `$ crontab -e` command, which opens the cron configuration via our default Terminal editor. My default editor is set to `nano` in the `/etc/environment` file by adding the line `EDITOR=/usr/bin/nano`.

We can also write the configuration in a file and provide it to `crontab`, which I prefer. Having everything in a separate file means I can transfer it between machines, keep different versions, and so on. The `crontab` command has multiple configurations, which we will go through here. Calling `crontab` without arguments will print its help.

My example is based on a text file in my home directory named `crontab.config`.

**cronie** executes scheduled commands via the old Bourne Shell (`/bin/sh`). This can be changed via the supported `SHELL` variable at the beginning of the config file with the following line:

```
SHELL=/bin/bash
```

**cronie** extracts the current user's *name* and *home directory* values from the `/etc/passwd` file and assigns them to the `$LOGNAME` and `$HOME` environment variables. As `$LOGNAME` can't be overridden, cron jobs scheduled for a given user can be executed only from their username. `$HOME` can be overridden, but this is applicable only in special cases.

Here is a summary of the job scheduling rules syntax:

- The rules consist of five fields with time *units*, followed by a shell command in this order:

  ```
  min hr day_of_month month day_of_week shell_command
  ```

  All are separated with a *single space*.

- Each *unit* can also be replaced by an asterisk `*`, which means all its values will trigger the command execution.

- Here is a simple rule example:

```
15 12 7 JAN * /home/luke/script.sh
```

This script will be executed on January 7 at 12:15, no matter the weekday.

- The units can have the values defined in *Table 15.1*:

| Time unit | Allowed values |
|---|---|
| **Minute** | 0-59 |
| **Hour** | 0-23 |
| **Day of month** | 1-31 |
| **Month** | 1-12, or the first three letters of the month name, case-insensitive (JAN, jan, Mar, etc.) |
| **Day of week** | 0-7, where 0 and 7 both mean Sunday, or again, the first three letters of their names, case-insensitive |

Table 15.1 – The crontab time units and their allowed values

- Each unit can be specified by a *single value*, a *comma-separated list* of values, or a *range*. *Lists* and *ranges* can be combined. For example, let's say the **hour** is set to 10:

  - If we have 10,20,30 for the minutes, this *list* will execute on the selected day and date *three times* – at 10:10, 10:20, and 10:30

  - If we have 10-20 for the minutes, this *range* will trigger the execution *every minute* from 10:10 to 10:20 – that is, 11 times

  - If we have 10-20,30,40 for the minutes, this variant will trigger the execution for the previous *range* at every minute from 10:10 to 10:20 – that is, 11 times – and then *twice more* at 10:30 and 10:40

- Here are a few other examples:

  - 45 15 2 12 * command: This format will execute the command at 15:45 on *December 2*, no matter the weekday

  - * * * * * date >> /home/luke/dateLog.log: This will execute the command every day *at each minute* and append the result to dateLog.log

  - * * * * 7 date >> /home/luke/dateLog.log: This will execute the command *at each minute*, but only on *Sunday*

You can find a few more examples in the "crontab(5)" man page. In it, there are also explanations for using the tilde sign ~, which is used for the randomization of a time unit value chosen from a list. It also explains the usage of the *forward slash division sign* / – to choose only some unit values at equal intervals when using a list.

The crontab.config file I used to test the commands for this chapter looks like this:

```
SHELL=/bin/bash
* * * * * date >> ~/DateReport.log
10,40 8-12 * * * /home/luke/someScript.sh > ScriptReport.log
5,10,15,20,30,40,50 * * * * echo $USER >> ~/UserReport.log
```

With the file ready, we must ensure cronie.service is enabled. Call $ systemctl status cronie.servce to see the results. In this command status, you shall see the following on the second and third lines:

```
cronie.service - Periodic Command Scheduler
   Loaded: loaded (/usr/lib/systemd/system/cronie.service; enabled;
preset: disabled)
   Active: active (running) since Thu 2023-06-01 05:30:28 EEST; 1 day
10h ago
```

We need to have enabled just before the preset keyword. preset: disabled means that default systemd settings would never be applied to this service. On the third line you can see active. If **cronie** is not activated, use $ sudo systemctl enable --now cronie.service.

Now, call $ crontab crontab.config. You will be warned if you have format errors. This will add the configuration for the **current user**. Adding sudo in front will activate the given tasks for the **superuser**. crontab will warn you if you have some errors in the config file.

To list the **current user**'s **cronie** tasks, call $ crontab -l; add sudo in front to list them for the **root** user. Only root is allowed to set **cronie** jobs for other users, and for this, you should use the following syntax:

```
$ sudo crontab -u someUserName crontab.config
```

Whenever you change the crontab.config file, call crontab again with it to apply the changes. To read the detailed rules and see more examples, call $ man crontab and $ man "crontab(5)".

It is *essential* that when you schedule cron jobs, whether for **root** or a **regular user**, even if nobody logs in, *the jobs will be executed if the PC runs*. This is because the crond **daemon** is *initialized at the end of the OS load* and starts working before logging in to any account (yours, root, or someone else's). You can check it with systemd-analyze. The difference between user and root accounts is in the context (different services, file access, and privileges) and the corresponding account's basic environment variables, set earlier during the OS init.

**cronie** can also send emails if a mail service is set up. One of the most common recommendations is to install and set up the **msmtp** package, along with **msmtp-mta** and **mailx**. You will have to add the following line to the top of your crontab config file:

```
MAILTO="someEmail@someServer.com"
```

To set up the mail service, follow this guide: `https://wiki.archlinux.org/title/Msmtp`. Remember that depending on your email provider, you might need additional verification, authentication, or application-enabling steps. For example, **Gmail** uses *special verification for applications* and requires a two-step verification to allow third-party applications to send mail from their accounts. I will not go into detail here.

The last note in this section is on **anacron**, which I will not review, but it is nice to know what it is used for. It also executes jobs periodically, but instead of using exact time schedules, it uses periods of job executions. If your system has been turned off for some time, when started, any past due **anacron** jobs will be executed. This is useful for tasks not required to run periodically at a specific time or date, such as system maintenance. If you want to learn more, read the **anacron** man page.

## Executing periodic tasks with systemd

As mentioned earlier, you must have read *Chapter 13* before reading this section, as it covers all systemd basics. To start with, let's examine the pros of **systemd timers** in comparison to **cron** implementations:

- You don't need to install any packages since systemd is already installed on any Manjaro flavor.

- With the systemd `systemctl` command, we can call, activate, or stop any configured process.

- All the standard output goes to `journalctl` by default, so logging is integrated.

- systemd supports *restarting* any skipped or unsuccessful tasks upon next boot.

- Setting up *randomized delays* is easy – it supports not only **calendar**-based *events*, but also events related to systemd targets and services, user logins, the completion of a task or target, and other OS-related events.

- *Debugging* is easier as systemd services *can be triggered independently*, without waiting for scheduled events.

- Time zone handling is provided.

- Each task can be set up to operate in a particular environment.

- The provided accuracy can be configured from minutes to microseconds.

- Although a bit complex, setting up a script to be executed periodically (or at a system event) is easy for any simple task. We will take a look at such an example in this section.

Now, let's examine the cons:

- To use the complete functionality, you need to understand the basics of systemd unit files (in particular service and timer units).

- You need a service and a timer unit file for any scheduled job.

- The default timer accuracy is one minute but *can be trimmed to microseconds*. However, on a general-purpose PC, whenever you overload it, it may delay the execution of other tasks, including the systemd timers. Absolute accuracy cannot be guaranteed. For any systemd-timer-triggered task, we should know that unspecified tolerances of the trigger times are possible, and there are explanations for this in the man pages. This is also valid for **cron** jobs, as it is a general characteristic for any general-purpose OS like Linux. A Real Time kernel and special software should be used for special time-critical tasks. In any case, the offered systemd accuracy is *much better and configurable* in comparison with **cronie**.

- There is no integrated ability to send emails; this action requires creating your own notifications script. Despite this, there are some hints in *Section 6.3, MAILTO*, at `https://wiki.archlinux.org/title/systemd/Timers`.

Before we continue, as timers and services are systemd units, remember that the general manual for them (together with a few examples) is available via `$ man --html=firefox systemd.unit`.

Now, let's examine what we should write in the service and timer *unit files* to trigger the execution of a command, script, or program. I will use the names `custom_script.service` and `custom_script.timer`; name yours according to the task they will be used for. The only rule is that they should have the same name, *differing only in the extension after the dot*. This way, systemd will *automatically activate the service* when the given timer elapses (i.e., is triggered). If the names differ, we can connect a timer to a random service unit to be activated when it hits a configured time point via the `Unit` timer file definition. We will see this later in this section.

There is a difference between *system-* and *user-level* systemd timers/services. The *system-level* ones are initialized and available earlier in the boot process and *may work even without any user being logged in*. The *user-level* services can be triggered after a given user *has logged in*. Systemd supports a *separate additional systemd instance* for each logged-in user, as we learned in *Chapter 13*.

If we want to create a *system-level service*, we should put the `.service` and `.timer` files in `/etc/systemd/system/`. For a *user-specific service*, they should be placed in `/etc/systemd/user`.

Here, we will create a *system-level* timer-triggered service. As a result, the script should either be owned by root (the better approach), or if the user owns it, the *execute permissions* for *others* must be enabled.

Create the service file with the following command:

```
$ sudo nano /etc/systemd/system/custom_script.service
```

In it, enter the following settings, skipping the commented lines if you wish and editing the corresponding paths and commands:

```
[Unit]
Description=Perform custom task or command

[Service]
Type=simple
ExecStart=/home/myUser/myScript.sh
# Alternatively, for a command (optionally with arguments):
# ExecStart=/usr/bin/command -opt1 SomeValue
# Optionally, you can as well redirect the stdout via:
# StandardOutput=file:/home/myUser/myService.log
# The stdout redirection is essential if your program or
# script would generate an extensive amount of logs.

[Install]
# Required optionally
# If you want this service to be loaded for a specific user or
# system target, you specify it here. The example is for a
# system target
# WantedBy=multi-user.target
```

Type=simple is the default when ExecStart is specified; still, it is better to write it explicitly. simple means that the service will be considered *successfully started immediately when the command is executed* (even if the service's binary cannot be called successfully). To read more about all possible types and systemd services in general, check out $ man systemd.service.

With ExecStart we define what needs to be executed. You can also provide a command with parameters. If, for some reason, you want to set a custom shell interpreter for your command explicitly, you can use the following form:

```
ExecStart=/bin/bash -c '/path/to/command -opt1 value1 -opt2 …'
```

With -c, we tell bash that the string enclosed in quotes has to be executed as a whole command. However, this is not necessary for a BASH script with a proper **shebang** at the beginning, as it will be automatically executed by it.

Keep in mind that ExecStart *cannot expand environment variables and paths*! An exception is replacing a user's home directory with %h, but I would not recommend using it. Using *absolute (full)* **paths** is more *evident* and strongly recommended. In addition, executing a BASH script with user-specific environment variables and paths (such as using ~ to designate /home) via a service may lead to issues if executed by the system instance of systemd. In such cases, you must make the service and timer user-owned. Again, the better option is using *absolute* **paths** and *values*.

Regarding `WantedBy`, you can list the different targets for the system via `$ systemctl list-units --type=target`. On Manjaro, for system-level services, there are two frequently used targets. The first is `multi-user.target`, which designates the OS loaded without a graphical environment but with the ability for any user to log in via a Terminal (in text mode). The other is with a graphical environment loaded, named `graphical.target`.

To list the *user targets*, use `$ systemctl --user list-units --type=target`. You will most frequently use `default.target`. To see its dependencies, call `$ systemctl --user list-dependencies default.target`.

When ready, press *Ctrl+S* and *Ctrl+X* to save and close the file.

Now, create a systemd timer `Unit` file by calling:

```
$ sudo nano /etc/systemd/system/custom_script.timer
```

In it, enter the following settings for a calendar-triggered event, specifying the date trigger as you need it. Skip the commented lines if you want:

```
[Unit]
Description=Call the custom script or command
Requires=custom_script.service

[Timer]
# The next is not required if the timer and service unit names
# coincide. This is typically used to override the default
# service to be triggered.
Unit=custom_script.service
# Execute the task if it missed a run due to PC being off and
# when OnCalendar triggers the given Timer
Persistent=true
# AccuracySec defaults to 1min. It can be 1us, 50ms or other
# similar value. 1us means the best accuracy according to man.
AccuracySec=1s
OnCalendar=*-*-* *:*:00
# on each second: OnCalendar=*-*-* *:*:*
# on each minute: OnCalendar=*-*-* *:*:00
# each hour: OnCalendar=*-*-* *:00:00
# each day at 10:15 : OnCalendar=*-*-* 10:15:00
# each Month on the 3rd at 12:14:22 :
# OnCalendar=*-*-3 12:14:22
# each year on a certain date: OnCalendar=*-02-14 12:00:00
# each year on a certain date only if it is Friday:
# OnCalendar=Fri *-Feb-14 12:00:00
# each Friday: OnCalendar=Fri *-*-* 12:00:00
```

```
# each Friday, every minute from 12 to 12:59:
# OnCalendar=Fri *-*-* 12:*:00
# each workday, at 12: OnCalendar=Mon..Fri *-*-* 12:00:00

# You can also use the verbose options minutely, hourly,
# daily, monthly, weekly, yearly, quarterly and semiannually.
# They refer to the beginning of this period, i.e., 00:00
# minutes/hours/etc...
# This means that OnCalendar=monthly will trigger at 00:00 on
# the 1st day of the month.
[Install]
WantedBy=timers.target
```

The `Requires` option points to the service – that is, it describes the dependency. `Unit=custom_ script.service` defines *which service unit* will be *triggered* by this timer configuration. `Persistent=true` is as explained in the comment.

`AccuracySec` is essential for services triggered relative to other events, which we will review in the next section. For the `OnCalendar` events, `1s` accuracy is enough, as they are anyway scheduled up to seconds. Remembering that you *can combine several triggers for the same timer unit* is crucial. For example, these could be several `OnCalendar` time points with `OnBootSec` or `OnActiveSec`.

The last line specifies that this timer should be configured during the *systemd timers initialization* (which is the standard way). As on Manjaro such a target is available both for the system and the user systemd contexts, there are no special related considerations for `WantedBy` (unlike for the service file).

For details on the time and date formats used for systemd timers and other related configurations and commands, read `$ man --html=firefox systemd.time` and `$ man --html=firefox systemd.timer`. Saving the exported HTML files is a good idea (to access them later without calling man in the Terminal each time).

Creating the two files in the `/etc/systemd/system/` directory is enough for systemd to work with the service and timer.

Our next step is to *enable* the timer and understand the *start* and execution. As mentioned on the `systemd.timer` man page, the service should be *stopped*. If it is running when the timer event hits, systemd *will not reactivate* it.

The `systemd.timer` man page also explains `DefaultDependencies`: "Timer units will automatically have dependencies of type `Requires=` and `After=` on sysinit.target, a dependency of type `Before=` on timers.target, as well as `Conflicts=` and `Before=` on shutdown. target to ensure that they are stopped cleanly prior to system shutdown. Only timer units involved with early boot or late system shutdown should disable the `DefaultDependencies=` option."

Regarding timer units triggered by an `OnCalendar` event, the manual explains that by default: "Timer units with at least one `OnCalendar=` directive acquire a pair of additional `After=` dependencies on `time-set.target` and `time-sync.target`, in order to avoid being started before the system clock has been correctly set. See `systemd.special(7)` for details on these two targets."

This means we should *enable only the timer* (without the service) to enable our timer-based activities. We do this with the following:

```
$ systemctl enable [option1] [option2] timerName.timer
```

The possible `[option1]` values are as follows:

- `--system`: The *default*, which means that if we write no string for `[option1]`, `--system` is implied. These are units that will usually be initialized during OS loading, without the requirement of a user to log in.

- `--user`: With this value, we enable a unit *especially for the current user*. As mentioned on the systemd man page, when we add this option, we tell `systemctl` we want to manage the *current user systemd instance*. When it is *missing*, we default to the *system-level systemd instance*.

- `--runtime`: This value enables the unit *temporarily, only for the current OS run*. After shutdown or reboot, the applied settings will be lost.

- `--global`: This option will enable the unit *for all future logins of all users on the system*. This option requires the execution of `$ systemctl --user daemon-reload` after it if we want the given timer to be activated in the current login sessions of users. I recommend you read all the relevant systemd manuals thoroughly before using it. I have described it here only for completeness.

`[option2]` can only have the `--now` value. If provided, it will additionally start the given unit immediately. The `disable` command has the same options. `enable` and `disable` do not `start`/`stop` a unit but *put it in a state of readiness*. To make the unit operational, you must explicitly use `--now`. If you skip `--now`, you must reboot or start the service separately via `$ systemctl start timerName.timer`.

Finally, for our example, we have to execute the following:

```
$ sudo systemctl enable --now custom_script.timer
```

To conclude the section, the following are a few other useful commands to debug potential problems and one to disable the timer. The next one checks a system timer status. Add `--user` before `status` if it is a user-based one:

```
$ systemctl status custom_script.timer
```

To check the status of all timers (which will include yours), execute:

```
$ systemctl status *timer
```

If your timer is enabled but you don't see your script results (in my case I've used a simple script printing the date in a file), you can check your **service status**. It will contain *the **journalctl** entries for the last execution* so that you don't have to run it separately to see them. Execute the following:

```
$ systemctl status custom_script.service
```

In my case, during one of the tests of the preceding files, I used *wrong script path*, so I got status `failed`, as shown in *Figure 15.1*:

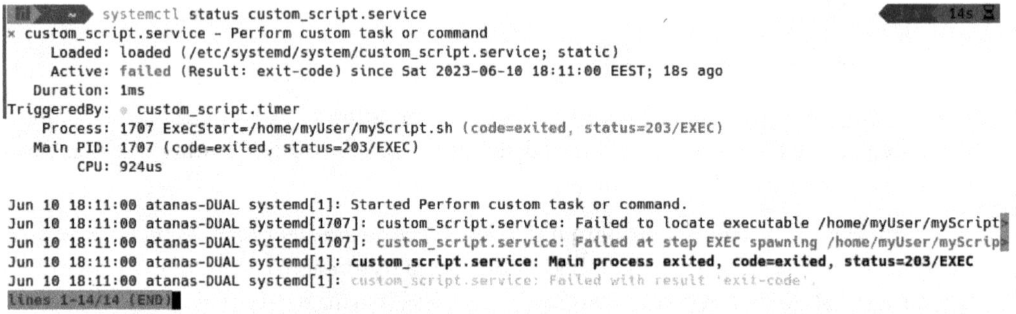

```
        systemctl status custom_script.service                                    14s
× custom_script.service - Perform custom task or command
     Loaded: loaded (/etc/systemd/system/custom_script.service; static)
     Active: failed (Result: exit-code) since Sat 2023-06-10 18:11:00 EEST; 18s ago
   Duration: 1ms
TriggeredBy: ● custom_script.timer
    Process: 1707 ExecStart=/home/myUser/myScript.sh (code=exited, status=203/EXEC)
   Main PID: 1707 (code=exited, status=203/EXEC)
        CPU: 924us

Jun 10 18:11:00 atanas-DUAL systemd[1]: Started Perform custom task or command.
Jun 10 18:11:00 atanas-DUAL systemd[1707]: custom_script.service: Failed to locate executable /home/myUser/myScript
Jun 10 18:11:00 atanas-DUAL systemd[1707]: custom_script.service: Failed at step EXEC spawning /home/myUser/myScrip
Jun 10 18:11:00 atanas-DUAL systemd[1]: custom_script.service: Main process exited, code=exited, status=203/EXEC
Jun 10 18:11:00 atanas-DUAL systemd[1]: custom_script.service: Failed with result 'exit-code'.
Lines 1-14/14 (END)
```

Figure 15.1 – A systemd system service failed status

*Figure 15.2* gives a readable form of the error messages:

```
Jun 10 18:11:00 atanas-DUAL systemd[1]: Started Perform custom
task or command.
Jun 10 18:11:00 atanas-DUAL systemd[1707]:
custom_script.service: Failed to locate executable
/home/myUser/myScript.sh: No such file or directory
Jun 10 18:11:00 atanas-DUAL systemd[1707]:
custom_script.service: Failed at step EXEC spawning
/home/myUser/myScript.sh: No such file or directory
Jun 10 18:11:00 atanas-DUAL systemd[1]: custom_script.service:
Main process exited, code=exited, status=203/EXEC
Jun 10 18:11:00 atanas-DUAL systemd[1]: custom_script.service:
Failed with result 'exit-code'.
```

Figure 15.2 – A systemd system service journalctl extract

I edited the file but forgot to add a **shebang** and got another failure – again, visible by the service status. After adding the shebang, the script worked like a charm. I have included this to show you *we all make mistakes; knowing how to debug helps us to fix any issues.*

To permanently stop and disable the service and timer, execute the following:

```
$ systemctl disable --now custom_script.timer
```

Add `--user` before `status` if it is a user-based one.

# Configuring systemd timer triggers related to system initialization via monotonic timers

Sometimes, we need a script or command to be executed earlier in the initialization process, at an exact time point after initialization, or at predefined intervals of execution. For this, we use **monotonic** timers. The five common monotonic triggers are listed in *Table 15.2* as defined in the `systemd.timer` man page, with simplified explanations:

| Trigger | Meaning |
| --- | --- |
| `OnActiveSec=` | Sets a timer based on when the timer itself was last activated. It triggers a service at equal intervals – for example, every 400 minutes. |
| `OnBootSec=` | Sets a timer based on when the computer started – for example if we want to start a service precisely 27 seconds after this. This trigger is typically used for system-level tasks not related to user login. |
| `OnStartupSec=` | Sets a timer based on when a given user service manager starts running. This is useful for user-specific tasks relative to the moment a given user has logged in. |
| `OnUnitActiveSec=` | Sets a timer based on when a specific unit (e.g., a service) was last activated. |
| `OnUnitInactiveSec=` | Sets a timer based on when a specific unit (e.g., a service) was last deactivated. |

Table 15.2 – A list of event-based triggers of systemd timers

It is good that we can combine the triggers as we want, including *with* `OnCalendar`. If there are multiple triggers, the given timer will trigger its target each time any of the events happen.

`AccuracySec=` plays a critical role here, as we can schedule systemd timers with *accuracy up to a microsecond*. Thus, if you use any of the preceding triggers with short time periods, set the accuracy correspondingly. For example, if we want a script to be executed every 10 milliseconds, we should set the accuracy to units at least 100 times smaller, which is 100 microseconds.

The time units each of these definitions and the `AccuracySec=` setting can accept are as defined by $ man systemd.time:

```
usec, us, µs
msec, ms
seconds, second, sec, s
minutes, minute, min, m
```

```
hours, hour, hr, h
days, day, d
weeks, week, w
months, month, M (defined as 30.44 days)
years, year, y (defined as 365.25 days)
```

You can use the abbreviated forms, considering that the first two rows are for microseconds and *milliseconds*. Here are a few examples:

- `OnBootSec=2s`: Sets a single execution of a system task 2 seconds after the machine boot. Of course, we have to be sure that any called or related program will work correctly that early in the boot process. This can be an issue if the complete initialization of our system takes 15 seconds.

- `OnStartupSec=15m`: Sets a single *user-specific* task execution 15 minutes after the user logs in.

- `OnActiveSec=17300ms`: Sets a *repeated* task execution every 17.3 seconds after its last activation, triggered by the timer elapsing itself. This one is frequently combined with `OnBootSec` or `OnStartupSec`, so once the given task has been triggered, it is then started at equal intervals.

- `OnUnitActiveSec=200s`: Sets task execution 200 seconds after the unit activated by the timer was last activated. The difference with the previous one is that the trigger here is not the timer itself but the unit's activation. Thus, the timer will start if the target service is manually activated. Suppose the given service is manually activated at the 100th second of the last activation. In that case, the timer will start counting again from 0 (to trigger the subsequent execution after 200 seconds).

- `OnUnitInactiveSec=200s`: This trigger is exactly like `OnUnitActiveSec` but starts counting when the given unit is deactivated. *For services of* `Type`=**simple** this means practically *the same result as activating based on* `OnActiveSec`. In general, `OnUnitInactiveSec` and `OnUnitActiveSec` are used in exceptional cases for *longer programs*, with a *service type different* from `oneshot` and `simple`.

As mentioned earlier, it is nice that we can combine any of the preceding triggers with `OnCalendar`. Let's say we want to receive a report from a remote machine periodically. Say we expect the report at 00:00 every day. We also want to get the same report at boot and two hours after a user logs in. In this case, the `[Timer]` section of the timer file would contain the following:

```
OnCalendar=*-*-* *:00:00
OnBootSec=10s
OnStartupSec=2hr
```

To conclude the chapter, note that the HTML conversion of this man page is not perfect, losing the formatting of the main table. This is the only man page with which I've spotted such a problem. Reading it from a separate terminal is recommended. There are a few more triggers in the `systemd.timer` man page. Check it out. It's obligatory if you work with timers.

# Summary

In this chapter, we reviewed the basics of **BASH shell scripts**. We learned about their primary constructs, such as basic conditional expressions, variables, arithmetic and file operators. We also saw a few simple examples. We continued with **cron** and systemd **timers,** covering how to set up periodic jobs with both. We finished the chapter with the triggering of systemd services based on timers related to system events.

In the final chapter, we will unveil the basics of the Linux kernel, the parent of all Linux distributions. We will start by explaining what a kernel is and what it does. We will then cover essential points of managing memory and resources, explaining them with simplistic metaphors. We will also review the kernel releases, how to change or upgrade the kernel on Manjaro, and the corresponding considerations. We will finish our journey by explaining the Linux Real Time kernel, a special modification available for over a decade.

# 16

# Linux Kernel Basics and Switching

In the previous chapter, we reviewed the basics of **BASH shell scripts**. We went through what a shell script is, how to execute one, what the differences are when starting it from root or user context, and how to write basic scripts. We then moved on to task automation based on calendar and timer events. We reviewed **cron** in detail, focusing on the latest and best implementation available on Manjaro, and looked at **systemd** timers as an alternative.

In this last chapter, we will end our journey by explaining the Linux kernel's complicated functions in a simple way. We will first go through its basic features, revealing how it works, what it provides to the **operating system** (**OS**), and how we can inspect the currently available kernel modules. We will then look at loading and unloading kernel modules and how to switch the currently working kernel version on Manjaro. Our final stop will be a brief explanation of the Linux **real-time** (**RT**) kernel version.

The sections included in this chapter are as follows:

- Basic Linux kernel characteristics for beginners
- The kernel from service, driver, and process perspectives
- Kernel releases and development
- Upgrading/changing the kernel release on Manjaro
- What is the Linux RT kernel?

## Basic Linux kernel characteristics

An OS kernel is the primary **software** (**SW**), managing the **hardware** (**HW**) resources and providing them in a unified way to any user application. Whenever a program needs to access a USB stick, the internet, or open or save a file, it accesses OS SW interfaces. These interfaces are called **Application Programming Interfaces** (**APIs**). In this context, a **unified** way means the kernel API is *the same*, whether our PC is ASUS, Dell, HP, Lenovo, or another brand, and no matter the HW differences, such as the amount of **Random Access Memory** (**RAM**), the **Central Processing Unit** (**CPU**) or **processor**), or a different audio card.

Before going further, we have to define the terms **task** and **process**. They have strict definitions in the kernel development world. Despite this, many people use them as synonyms, as from a general perspective, the kernel's job is to allocate the resources for a given process, no matter how many tasks it will perform. I will use them in a simplistic way in my explanations – a **task** is a single problem to be solved, while a **process** can work on a (un)limited series of tasks.

The kernel manages the allocation of **processing time** and **RAM** so that hundreds of applications work *simultaneously* with the same resources. A single CPU core can only work on one task at a given moment. Even if we have a 16-core CPU, the maximum number of processes that can run simultaneously is 16. However, we see *hundreds of processes* via $ ps -1A or htop. To use a metaphor, imagine the computer as a big supermarket where each customer is a task or process. When entering the supermarket, the customer can take a basket or a small, big, or huge trolley (like different amounts of **RAM**). A big family may even take three big trolleys (just like *a parent process with a few child processes*, such as a browser). The store provides enough for everyone, and employees return the trolleys to the trolley park so that new customers can use them. Whenever a product on a shelf finishes, a supermarket employee goes to the storage to bring out more. The shelves represent HW resources such as HDD, network, audio, and video. **Filesystems** with *user-accessible* **files** are also available to users as products on the shelves.

Ultimately, the store has 16 counters (like 16 CPU cores), and people form queues at each to have their products *processed*. If a counter closes for a while, an employee informs customers not to put more products on its belt line and redirects them to other counters. If 10 counters work but most people go only on a few of them, an employee announces it via the store audio system, and customers are redirected to open counters with their trolleys (RAM resources).

While this is a simplified analogy, it represents the hundreds of responsibilities an OS kernel handles. A single customer (or process) doesn't care how many trolleys there are in total, which chain the store belongs to, and how many customers it can handle per minute (system capacity). The interface (API) is always the same, and most people (no matter the country) know how to use it. This is the power of a **unified API**.

Linux is famous for being one of the most effective kernels ever. In addition, it is the only *completely* **configurable kernel**. We can scale the same system, from *a small store* with one counter to a *big supermarket* with tens of counters. In technological terms, Linux can operate on tiny battery-powered microdevices, embedded devices, small robots, big robots, old and weak PCs, new and powerful PCs, smart TVs, big servers, and supercomputers.

*Figure 16.1* presents this concept with multiple applications on top, one of which has several child processes. Each user application accesses the resources it needs via the kernel API. Kernel space is explained later in the chapter.

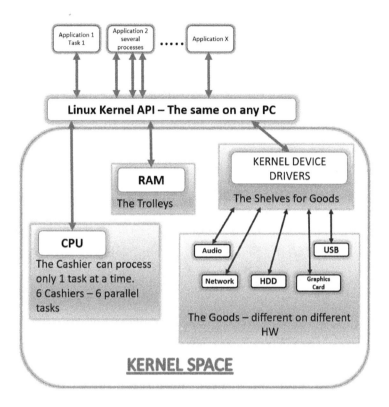

Figure 16.1 – Basic resources managed and provided by the kernel

## Memory types

Computers have three types of memory:

- **Persistent storage**: When we shut down the PC, its contents remain. These are **HDDs**, **SSDs**, and **flash cards** (such as memory sticks and micro SD cards). They are relatively slow and contain files accessible by the kernel and the user. These are also like shelves with goods a customer may take to put in their trolley (RAM) – in other words, user-accessible files. The warehouse is all the files, *accessible only by root*; regular customers don't have access to it, except special ones with special permission (with `sudo`).

- **Dynamic storage or RAM**: This needs power to keep data – that is, it keeps content only while the PC is powered on. It is managed entirely by the kernel; user applications can use parts of it after the kernel provides those resources. This is also called operational memory, where we store intermediate results while working (analogous to customers' trolleys). The kernel-reserved RAM is the employee trolleys to load shelves. They are special and never provided to customers.

- **CPU cache memory**: In the supermarket metaphor, this is the *conveyor belt* for the goods a customer transfers to the cashier to be processed. The belt is **short** and only operates with the current buyer goods – that is, *the current process* **data**. The goods on the belt change depending on the customer. This is super-fast memory located inside the CPU core, again managed by the kernel.

*Figure 16.2* presents the three types of memory:

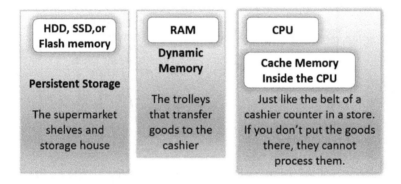

Figure 16.2 – Types of memory on a computer

## Kernel space, isolation, and user space

As we said, **RAM** is the **operational memory**. The kernel manages it all, and part of it, just like the management room in a store, is never accessible to regular users, nor do they know where it is. This is what we call **kernel space** and why we also say it is **isolated**. The kernel space has systems to monitor tasks/processes (buyers), RAM allocation (trolleys), filesystem files, kernel modules (shelved goods), and CPU time (counter load). Each user is provided with the RAM they need, and this is the only portion of RAM they know about. They know other processes are around but cannot access their RAM, which we call **process isolation**.

On a PC, the HW device that manages the individual RAM spaces and monitors **isolation** is called a **memory management unit** (**MMU**), again under kernel management. The user-provided RAM is called virtual memory, and from the user's perspective, it starts from address 0 (empty trolley) up to the maximum of the provided RAM chunk. In our supermarket example, the supermarket has a number and tracking device on each trolley, no matter its size, and at each moment, the store management (that is, the kernel via the MMU) knows how many trolleys are in use and which trolley was given to which customer. As Linux is super flexible, at any moment when a user needs more RAM (trolleys), a special service provides as much as requested and takes care of managing them.

From the **user's** *perspective*, they are like a perfectly cared-for customer. They access services through the API and ask the kernel to open files, be given an interface to send graphics to the monitor, be provided with packets from the network, and so on. When they have the resources in their RAM chunk (trolley), they can do whatever they want with them. The system is so flexible that it can serve any request type, just like how a store can serve kids, regular people, enterprise customers, and older people with special needs.

Now, if all essential services come from the kernel, what is the **user space**? For starters, this is the **RAM** chunk *provided to a* **user application**. In addition, the user can access user-owned files and some **tools** in **user space**; they don't need the kernel to access and operate with them. The most notable is **libc** (or **glibc**) – the **GNU standard C library**; another example is **Python** and its packages. If a user process doesn't access some special HW resources (such as audio or network), once the kernel provides it with RAM and a PID, it only periodically gives the CPU time to do its work. From this perspective, the supermarket metaphor is not ideal, as in it, the CPU is the counter, and we don't use the CPU once but all the time. Still, it is reasonable to imagine the general structure. After all explanations, it is essential to emphasize that **systemd** is also a user space process.

We see **user** and **kernel space** processes when we list all processes with $ `ps -lA` (or `htop` with no hidden data). As you already know, each process has a parent. On Manjaro, PID `1` is **systemd** (sometimes depicted as `/sbin/init`), and PID `2` is **kthreadd** (the kernel thread daemon, operating in kernel space). They have a **Parent PID (PPID)** `0` and are started from the kernel. Any process with **PPID 1** (started from **systemd**) is a *user space* process, while any with PPID 2 is started by **kthreadd** and is a *kernel space* process. With the `htop` *Shift+K* key combination, we hide/show kernel **threads**, and with *Shift+H*, we hide/show user **threads**. To clarify, a **thread** is a "lightweight process"; one process usually may have many threads.

## Architecture

In terms of architecture, kernels are either **monolithic** or **microkernels**. Linux is a **monolithic** kernel.

A **microkernel**, to use another metaphor, is like a bazaar – each stand manages itself, so the kernel is limited to only providing basic services. Each additional service is a separate user-space module. From a generic perspective, this kind of kernel only provides memory management, CPU management, and **Inter-Process Communication (IPC)**.

**Monolithic** kernels are like the great supermarket example – all services are centralized. Due to this, its management is much more complicated and extensive but also *optimized*, *effective*, and *scalable*. From a technological perspective, Linux is also **monolithic**, as *all its resources reside in one centralized part* of **RAM**.

Despite being **monolithic**, Linux can easily add additional services and functionalities, as it is also **modular**. The kernel **modules** are **built in** and **dynamically loadable**. Changing built-in modules requires re-compilation and changing the kernel. The dynamically loadable modules can be *changed while the kernel is running*. This versatility increases the availibility of Linux applications to thousands of different systems.

In addition to the previous characteristics, the Linux kernel is highly **configurable**. This, however, can be done only before **compilation**. To explain further, *any SW is created from text files with source code*. **Compiling** is the complex and relatively slow process of converting source code into executable SW. As Linux is **Free and Open Source Software** (**FOSS**), any developer has its complete configuration at their disposal before compilation. Thus, they can decide which **built-in** modules should be added to their final compiled kernel version. To give you some idea of the flexibility, the smallest version of Linux with almost no modules is *several MB*. The biggest (with all possible built-in and loadable modules) can reach *2 GB*. The biggest version includes thousands of drivers and modules, never found together on one system; it is used only for theoretical checks. The Manjaro-used kernel version *6.1* is *11 MB*, while version *6.4* is *12.1 MB*.

To conclude, Linux is a **configurable**, **monolithic**, and **modular** OS kernel.

## The CPU, preemption, task scheduler, and CPU operations per second

The supermarket metaphor is excellent for imagining the global OS-level picture, kernel space, and user space. To explain the CPU itself, we will switch to another metaphor.

Imagine a small furniture factory where each **CPU core** is a single worker who can do only one task at a time. We have *hundreds* of **tasks** but only *eight CPU cores* (workers). The workers periodically change their tasks between cutting wood, mounting parts, painting the furniture, packing, processing orders, and many others.

The Linux kernel is like the perfect manager who allocates tasks and trolleys with parts (data and programs) to each worker. The tasks in the kernel can include, for example, calculating the next network packet addresses, preparing the next Bluetooth packet, accepting and processing the next microphone packet, and processing JPEG-compressed data to be displayed.

Each time a high-priority task comes in, the kernel can stop any of the CPU cores (**interrupting** a process), tell them to save their progress on the current task (*saving its* **context**), and work on another, higher-priority task. The period a CPU core works on a task is called a **time slice**. The kernel manages task switching and the length of each **time slice**.

**User tasks** are equivalent to the factory design and customer service bureau orders. The bureau designs chairs and tables, specifies the assembly steps, takes orders, knows the general factory capacity, and so on. However, they aren't explicitly concerned with the workers and how their manager distributes and organizes the tasks. They periodically provide orders and later wait for feedback about completed tasks. To relate the metaphor to a real example, imagine a long shell script, which will need half an hour to finish, and user SW that converts a double-layered DVD into a small **MP4** file we can put on our mobile phone. Neither the script nor the video converter application will know how many **time slices** are necessary and when a CPU core switches to their work or another task. Linux is again so well designed that it can manage task allocation equally well on a weak, single-core CPU and a powerful, 16-core workstation.

In the factory metaphor, *only one thing depends on the user* (bureau) – whether a given order (a user process) is organized into tasks that can be processed in portions in parallel. A SW application designed like this is called a **multithreading application**. However, even in this case, when a CPU core processes a single portion again depends on the kernel.

The interruption of a task for another to be processed is called **preemption**. Linux is, then, a **preemptible** kernel. In this case, the kernel saves the current task **context** (progress, intermediate values, and others), puts it in RAM, loads the next task data, and gives the CPU core a signal to proceed with the new task.

The kernel part that manages task preemption is called a **task scheduler**. Since October 2007, the **Completely Fair Scheduler** (**CFS**) has been in use and is *constantly improving*. At the time of writing, a newer one is in development (the EEVDF scheduler, planned for Linux version *6.6*). Search the internet for more information.

Thanks to a combination of **preemption** and **task scheduling** via **CFS**, Linux is a highly effective **multitasking** kernel.

The **CPU speed** is measured in **basic operations per second**, similar to an engine's **Revolutions Per Minute** (**RPM**). The higher the **RPM**, the faster the engine. For each different engine, we have different lowest and highest values of the RPM. For a CPU, this is measured in frequencies of **Mega** (millions) or **Giga** (billions) of **Hertz**, respectively **MHz** or **GHz**. If a CPU supports dynamic speed management, the Linux kernel changes the speed based on the workload.

Most of the basic OS functions described in this and previous sections also apply to **UNIX**, **BSD**, **macOS**, and **Windows**. They also have kernels and, as computer OSs, have to manage hundreds of tasks and work on multi-core processors. However, *only* **Linux** and several **BSD** derivatives are **open source**. In addition, the **Linux kernel** is the only one completely *configurable* and ported to thousands of devices different from PCs and servers. Finally, as thousands of contributors constantly provide improvements and share them with everyone, **Linux** has been *the most effective* OS kernel for years.

*Figure 16.3* presents the task distribution in a generic and illustrative way. Though not presented, it is essential that the kernel also has multiple internal tasks for the **scheduler**, the **MMU**, for serving **HW** services, and so on. In addition, a typical *2023* CPU can have from 1 to 24 cores. Its cache can be from a single one, a few KB in size, up to three layers and tens of MB. Finally, any process with no active tasks goes to a sleeping or waiting state and takes no CPU time. The scheduler periodically checks whether such a process has something more to do. This is how we can have *hundreds* of running processes yet 0.1% **CPU load** when Manjaro is loaded and operational, but no program has started.

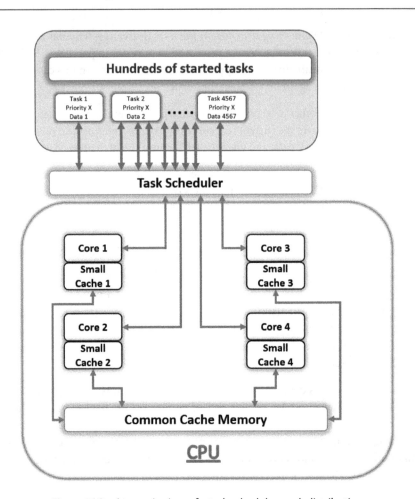

Figure 16.3 – A generic view of a task scheduler work distribution

This ends our metaphoric kernel discussion.

## The kernel from service, driver, and process perspectives

As we learned at the beginning of the chapter, the Linux kernel manages resources. When a task calls an `OpenFile` function, the kernel accepts the request and accesses the HDD via an **ext4** *kernel driver*. If the file permissions allow the user to access it, the kernel opens the file and provides the user task with a RAM address to read the data. If a user task wants to save a file, it again calls a kernel API function, providing the data to be saved from RAM to the HDD. The kernel checks the permissions and, if allowed, takes the RAM-prepared data and saves it in the HDD. While the kernel executes the requests, the given user-space task waits. When the kernel finishes any of the tasks, it returns the control to the user process to continue its execution.

## System Calls

The standard kernel API for operations is named **System Calls Interface (SCI)**. We can list all **System Calls (SysCalls)** with their *unique* **IDs** via $ `ausyscall --dump`. As of September 2023, on Manjaro KDE with kernel *6.1*, I have 363 **SysCalls**. Some of them are `read`, `write`, `open`, `close`, `stat`, `kill`, `mkdir`, `chown`, `socket`, `connect`, and `shutdown`. I have listed these exactly so that you see how many of the basic Terminal commands are translated directly into kernel functions and are, thus, fast and effective. **SCI** is the main API of the few Linux Kernel **APIs** through which applications request resources and interact with the kernel.

## Kernel modules and drivers

We now know that there are *built-in* (requiring re-compilation) and *dynamically loadable* **kernel modules**. To list the currently loaded kernel modules, use the simple `lsmod` command; it has no parameters. To get more information about a module, use a name from the `lsmod` results list and provide it to `modinfo`.

For example, we can use $ `modinfo moduleName` with `ext4`, `usb_storage`, `snd`, and `video`.

To list the built-in modules, we can use the following command:

```
$ cat /usr/lib/modules/$(uname -r)/modules.builtin
```

The $ `(uname -r)` part will be replaced with `your_Kernel_Version`.

Kernel **drivers** are a subcategory of **kernel modules**. Let's see an example. `usb_storage` and `ext4` are kernel **drivers**. The first is used to exchange data with USB storage devices, and the second is used to read/write data to/from any **ext4**-formatted storage partition. Theoretically, a driver is SW that performs low-level communication with a physical HW chip/module, attached via a connector. To make a difference, the `cryptd` and `crypto_user` modules are, thus, not **drivers**. Despite this, many people use the terms **kernel module** and **kernel driver** interchangeably.

There is a **Linux Kernel Driver Database (LKDDB)** web page. While the main page hasn't been updated since *2007*, the web **LKDDB** is *up to date*: `https://cateee.net/lkddb/web-lkddb/`. It includes *29,451* modules as of *2023*! This can explain why, if you include everything, the kernel can reach a final size of *2 GB*.

Kernel modules are saved in `.ko` files and are **compiled** for an *exact kernel* **version**. If you *change the kernel*, all corresponding loadable modules must also be changed. Manjaro takes care of this, and when you add a new kernel version, it downloads the necessary packages. All of them are kept in `/usr/lib/modules/your_Kernel_Version/kernel`, separated into several subcategories. You will find thousands of `.ko.zst` files if you inspect them. The additional `.zst` extension is used, as the files are **compressed** for faster download, to take up less space on servers, and sometimes for faster installation. In *September 2023*, on Manjaro Xfce with kernel *6.4.14-1*, I have *7,065* kernel modules in this directory, totaling *143 MB*. Most of them are not loaded but can be if necessary.

## User space drivers

As the Linux kernel is complex and needs to be *absolutely* **stable**, many people tend to write the less effective but *easier* **user space** modules. The kernel treats them as *regular user applications*. Instead of prioritized processing and direct access to the kernel resources, they access HW resources via **SysCalls**, perform the necessary processing in **user space**, and then communicate with the given HW again via **SysCalls**. *If designed and implemented effectively*, they can offer good performance. An example is the Tuxera user-space **NTFS-3G** driver, which has **issues** only on *weak computers and with high data volumes*. Another example is the **Filesystem in Userspace** (**FUSE**), which NTFS-3G uses. As such applications are like any other user application, the kernel cannot differentiate them, so we rely on the documentation to know whether they are user-space drivers.

## Kernel releases and development

The Linux kernel is developed constantly. Each new release offers fixes, improvements, and often new features. As there are too many intermediate versions, not all are transferred to all distributions. Several stable kernel versions are marked as **Long-Term Support** (**LTS**) releases. This means that whenever updates or critical fixes are necessary, even though newer versions are out, the older LTS ones will get them. Many FOSS projects offer LTS support, and Linux is famous for keeping its LTS support for *four* to *six* years. You can check `https://www.kernel.org/category/releases.html` and `https://www.kernel.org/` for updates. Arch Linux and Manjaro provide a few stable kernel **LTS** versions and a few recent ones. The easiest way to see them is via the GUI. On Manjaro KDE, it is accessed via the **System Settings** application (go to the **Manjaro | Kernel** section). On GNOME and Xfce, it is accessed via the **Manjaro Settings Manager**, in the **Kernel** section. We will also see how to view them from the Terminal. You will see a list of Linux versions in the GUI, which is the same on all flavors. As of *September 2023*, we have the **LTS** versions *4.19, 5.4, 5.10*, and *6.1*. We also have *6.1 RT*, *6.4.6 RT*, and *6.5 RT*. Finally, we have the newer ones, *6.4* and *6.5*. We will explain the **RT** versions later in the chapter; for now, remember that the **LTS** versions are recommended for regular use. The currently running kernel version is marked in the GUI.

It is essential to know that **Manjaro** *will not change the kernel for you*. This is because sometimes SW, drivers, or HW might have an issue with newer kernels. The Manjaro team doesn't want to push you to make changes should you have a custom SW dependent on the kernel version. Each new installation ISO comes with the latest **LTS** version – that is, we can work with the given kernel *for at least four years*. For any regular SW, issues with **LTS** releases are extremely rare, which is why they are recommended for regular use.

## Upgrading/changing the kernel release on Manjaro

As mentioned earlier, sticking to **LTS** versions is the best, and if you have the latest version, you will benefit from newer features. The Manjaro Settings kernel GUI module offers a simple **Install** button – click it, enter your password, and wait for the installation. Then, we will perform a few simple Terminal actions to switch to the new kernel.

I have mentioned the bootloader several times. It is the SW that runs before Linux to start it or any other OS. On most Manjaro versions and many Linux distributions, the GNU **Grand Unified Bootloader** (**GRUB**) is used. It is old (it has been developed since *1995*), versatile, full of features, and it includes dual booting with any Windows version. It has a good Wikipedia page; check it out to learn more: `https://en.wikipedia.org/wiki/GNU_GRUB`.

We will change the GRUB options by editing its configuration file with a text editor:

```
$ sudo nano /etc/default/grub
```

In it, uncomment and change the following lines (which are all among the first 20 lines):

- `GRUB_TIMEOUT=5`: Set the number of seconds the GRUB choice menu is shown before the OS boot. I prefer five seconds; choose a larger value if you wish.

- `GRUB_CMDLINE_LINUX_DEFAULT="udev.log_priority=3"`: If `quiet` or `splash` is present in the quoted string, remove it. `quiet` is used to disable the kernel boot log; `splash` is used to show a splash screen instead of the boot log.

- `GRUB_TIMEOUT_STYLE=menu`: If set to `hidden`, change it to `menu`.

You can read more about the other options in this file at `https://www.gnu.org/software/grub/manual/grub/html_node/Simple-configuration.html`.

When ready, save and exit. The next step is to **obligatorily** run `$ sudo update-grub`. It *updates the actual* **GRUB** *configuration*. This command is a simple call for `$ grub-mkconfig -o /boot/grub/grub.cfg`. Thus, if you need more options, check the `grub-mkconfig` help and `info` pages. Now, reboot, and you will get the GRUB boot menu. Pressing any *arrow* key will turn off its timeout. It usually has at least the following four options:

```
Manjaro Linux
Advanced options for Manjaro Linux
UEFI Firmware Settings
Memory Tester (memtest86+)
```

The *first* option loads the default (last saved) Manjaro configuration. The *second* option is the important one; we will discuss it in the next paragraph. The *third* option is to enter the **BIOS** menu. The *fourth* option, `memtest86+`, is an option that was used extensively years ago when RAM was not that good and failed periodically. If you have more than one OS installation, they will be inserted between options *2* and *3*.

Once you have chosen an option with the arrows, press *Enter*. When we go into `Advanced options…`, we will see two entries for each installed kernel. The first is the kernel itself; the second is its `fallback initramfs` version. Choose the regular kernel version you want. Upon reboot, once you successfully switch to a different kernel, it will be loaded by default the next time. The `fallback` version loads default settings for the given kernel version if you have done a failing custom kernel configuration. With regular usage, you will hardly need it. You can read more here: `https://wiki.archlinux.org/title/mkinitcpio`.

When you boot the new version, you will see the running version changed if you go to the GUI menu. After a few days of testing, you can remove the older version from the GUI.

## Changing the kernel via terminal

We can do this with the following commands:

- `$ uname -r`: Gets the currently running kernel version
- `$ mhwd-kernel -l`: Lists the available kernel version for downloading
- `$ mhwd-kernel -li`: Lists the installed kernel versions and which one is currently running

Unfortunately, these commands doesn't say whether these are LTS versions (unlike the GUI kernel installation module). When called, it will provide you with a list like this:

```
$ mhwd-kernel -l
available kernels:
   ...
     * linux515
     * linux54
     * linux61
     * linux64
     * linux65
     * linux61-rt
   ...
```

Call `$ pamac info linux61` to get more information. However, you can only find out whether it is an **RT** kernel, *not whether it is* an **LTS** release. This may change in the future, but for now, your only option is to read it in the GUI.

The next step is to download and install the kernel of your choice via `$ sudo mhwd-kernel -i linux64` – you can add the `rmc` option *at the end* of this command to remove the current kernel version, but it is not recommended. We should test the new one for some time , for example, for a few days or a week, and after proving it works flawlessly, remove the old one via `$ sudo -r linux54`.

When ready, execute the GRUB configuration steps from the previous subsection.

## Loading and unloading kernel modules

Before going on further, the main question many will ask is *when is loading and unloading kernel modules necessary?*. After all, **mhwd** will detect and load new modules (reviewed briefly in *Chapter 14*). However, **mhwd** works only when invoked. If, for some reason, it doesn't successfully or correctly detect a new device (e.g., a latest-model audio card or USB-attached HW), then we may need to load its drivers (if they exist) manually. In addition, while **mhwd** works for HW, if we want to load a particular cryptographic module, we can only do it via manual loading.

We can use the modprobe command to load and unload kernel modules. As mentioned, we first need the given module to be present,  precompiled for our current Kernel. By default, the Manjaro kernel installation puts the precompiled kernel modules in /usr/lib/modules/your_Kernel_Version/kernel. Then, we can use modprobe directly with a module name. This program comes with an additional one, depmod, used to generate the module dependencies list. modprobe will not load an already loaded module or unload an unloaded one. If an issue occurs inside the kernel after loading (particularly for resolving symbols), the kernel log must be inspected via dmesg or journalctl. Let's see how to use modprobe.

First, you need the module name. For the examples, I will use the asix module, which is necessary for some USB Ethernet adapters. To each command, you can add -n for a dry run – that is, to only see what will be done without actually performing the action. -v is for verbose reporting mode:

- $ sudo modprobe -v asix – Loads the kernel module.
- $ modprobe --show-depends asix – Shows the kernel module dependencies.
- $ sudo modprobe -v -S SPECIFIC_KERNEL_VERSION asix – By default, the uname -r version is used to form the module loading path. With -S, you can substitute the automatic your_Kernel_Version string with a custom one, designated here as SPECIFIC_KERNEL_VERSION.
- $ sudo modprobe -v -r asix – Unloads the kernel module.

For more information, check the help section and the man page.

# What is the Linux RT kernel?

To explain the **RT kernel** version, we must understand more about how the **regular** kernel works. By default, Linux is a *general-purpose kernel*, initially aimed only at personal computers and servers. If we start any complex and extensive task that needs to use the CPU at *99%* load for half an hour, the kernel will do it. Most of the time, our computers stay *idle*, raising the CPU activity when we do something. You can check this with any *system monitor* tool, which shows a graph of the system activity. Due to this characteristic, PCs were often slow and unresponsive years ago when a long and complex task was started.

A perfect example is video encoding. Say you want to encode a full HD movie from Blu-ray disk to MP4. While doing it, starting the browser may be slow. This issue was partially resolved over time, thanks to the **CFS** and its improvements, but it can still happen today. The reason is that a **general-purpose kernel** *tries to do all the work it has*, even if this will slow down some of the other currently running operations.

In contrast, an **RT system** has dedicated **time slices** for each process, ensuring no task will be slowed if another uses its time slices completely. Here are a few examples of **RT** processes:

- The **GSM** communication between your Android mobile and the **GSM** network has strict communication **timing** requirements. The audio quality degrades if data packets are slowed during transfer, encoding, or decoding. In the worst-case scenario, the call drops.

- A **medical scanner** head's movement and **data processing** shall never be obstructed to get the correct image.

- A car **autopilot computer** shall never skip a frame from its cameras; any delay may lead to wrong calculations and an incident.

We cannot perform the described tasks correctly and on time without *guaranteed* **time slices** for each task. **Real time** means a given task has enough guaranteed processor time and OS resources to operate within a predefined period. I have been developing such systems for over a decade. What a team does, in this case, is to add specially developed SW to an **RT** OS. In addition, it preliminary measures and trims the system tasks, sequence, and CPU load. This way, each task's **execution time** and the *whole system's* **performance** are **predictable** and **deterministic**.

The earlier explanation of the main tasks of the OS kernel, the CPU, and the task scheduler was generic. In reality, tasks can be interrupted, and the kernel has modules to deal with resource locks, task preemption, and so on.

Practically, the RT kernel addition influences all kernel systems, along with many modules. Its purposes include the following:

- Guaranteed task scheduling and predictability

- Lower latencies for certain internal operations

- Usage only of high-resolution timers (with a microsecond accuracy)

- Specific interrupt handling

- Specific critical code handling

The result is *potentially higher* **CPU** load and OS behavior predictability (determinism). While this is great for **RT** systems, it will bring no special benefit to a regular desktop installation. In addition, any application running on the RT kernel version needs to consider the RT kernel to achieve determinism. That's why it doesn't bring distinctive advantages to regular desktop users, except for kernel developers working on such applications.

# Summary

In this final chapter, we covered the basics of what an OS does and how the Linux kernel manages all tasks invisibly. We reviewed Linux's primary points by briefly covering SysCalls, kernel modules, and kernel versions. The following practical topic was how to switch the kernel version on Manjaro. We finished with the Linux RT kernel and what it is used for.

I hope this book was as great an adventure for you as it was for me. My sole purpose was to cover all possible essential topics, along with some history and simple explanations.

I'm highly grateful to the Packt Publishing team, who did their best to keep me on track and support me throughout. I am also highly grateful to my technical editor, Kaloyan Krastev, who spent hundreds of hours checking my work and providing feedback on the text's structure, definitions, and quality.

We all owe a great deal to **Philip Müller** (the project director and creator of Manjaro) and the **Manjaro team**, as without Manjaro, this book would never exist.

If you are interested in Manjaro development, here is their **GitLab** site: `https://gitlab.manjaro.org/`.

Manjaro is free SW, but they accept donations at `https://manjaro.org/donate/`.

I also want to express gratitude to the thousands of developers working not only on the Linux kernel but also on all GNU tools, other Linux distributions, and other tools.

An honorable mention are two guys we all owe the biggest thanks to:

- **Richard Stallman**, for coming up with the idea of **Free Software** and the **GNU General Public License**, fighting to popularize it since *1983*
- **Linus Torvalds**, who *continues to manage* the **Linux project** and, apart from this, has given us **Git**, the best version control system, initiated in *2005*

ENJOY LINUX, THE BEST OS KERNEL,

AND MANJARO – ONE OF ITS GREATEST DISTRIBUTIONS –

AS ALWAYS, FOR FREE!

If you find the content so far valuable, I will be eternally grateful to you for a short book review on the platform you purchased it from. With this, you will help people interested in learning more about Manjaro and Linux.

# Index

Packtpub.com

Subscribe to our online digital library for full access to over 7,000 books and videos, as well as industry leading tools to help you plan your personal development and advance your career. For more information, please visit our website.

## Why subscribe?

- Spend less time learning and more time coding with practical eBooks and Videos from over 4,000 industry professionals
- Improve your learning with Skill Plans built especially for you
- Get a free eBook or video every month
- Fully searchable for easy access to vital information
- Copy and paste, print, and bookmark content

Did you know that Packt offers eBook versions of every book published, with PDF and ePub files available? You can upgrade to the eBook version at packtpub.com and as a print book customer, you are entitled to a discount on the eBook copy. Get in touch with us at customercare@packtpub.com for more details.

At www.packtpub.com, you can also read a collection of free technical articles, sign up for a range of free newsletters, and receive exclusive discounts and offers on Packt books and eBooks.

# Other Books You May Enjoy

If you enjoyed this book, you may be interested in these other books by Packt:

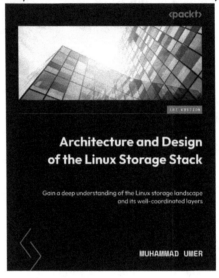

**Architecture and Design of the Linux Storage Stack**

Muhammad Umer

ISBN: 978-1-83763-996-0

- Understand the role of the virtual filesystem
- Explore the different flavors of Linux filesystems and their key concepts
- Manage I/O operations to and from block devices using the block layer
- Deep dive into the Small Computer System Interface (SCSI) subsystem and the layout of physical devices
- Gauge I/O performance at each layer of the storage stack
- Discover the best storage practices

**Fedora Linux System Administration**

Alex Callejas

ISBN: 978-1-80461-840-0

- Configure a Linux environment from scratch
- Reviews the basics of Linux resources and components
- Learn about enhancements and updates to common Linux desktop tools
- Learn how to optimize resources of the Linux operating system
- Understand the best practices in Systems Administration
- Learn how to harden security with SELinux module
- Improve System Administration with the tools provided by Fedora
- Explore how to create open containers with Podman

## Packt is searching for authors like you

If you're interested in becoming an author for Packt, please visit authors.packtpub.com and apply today. We have worked with thousands of developers and tech professionals, just like you, to help them share their insight with the global tech community. You can make a general application, apply for a specific hot topic that we are recruiting an author for, or submit your own idea.

## Share Your Thoughts

Now you've finished *Manjaro Linux User Guide*, we'd love to hear your thoughts! Scan the QR code below to go straight to the Amazon review page for this book and share your feedback or leave a review on the site that you purchased it from.

https://packt.link/r/1803237589

Your review is important to us and the tech community and will help us make sure we're delivering excellent quality content.

# Download a free PDF copy of this book

Thanks for purchasing this book!

Do you like to read on the go but are unable to carry your print books everywhere? Is your eBook purchase not compatible with the device of your choice?

Don't worry, now with every Packt book you get a DRM-free PDF version of that book at no cost.

Read anywhere, any place, on any device. Search, copy, and paste code from your favorite technical books directly into your application.

The perks don't stop there, you can get exclusive access to discounts, newsletters, and great free content in your inbox daily

Follow these simple steps to get the benefits:

1.  Scan the QR code or visit the link below

https://packt.link/free-ebook/9781803237589

2.  Submit your proof of purchase
3.  That's it! We'll send your free PDF and other benefits to your email directly

www.ingramcontent.com/pod-product-compliance
Lightning Source LLC
Chambersburg PA
CBHW060641060326
40690CB00020B/4478